D1117020

SPACE RESOURCES

SPACE RESOURCES

Breaking the Bonds of Earth

John S. Lewis and
Ruth A. Lewis

Columbia University Press
New York 1987

Library of Congress Cataloging-in-Publication Data

Lewis, John S.
Space resources.

Includes index.
1. Astronautics. 2. Outer space—Exploration.
I. Lewis, Ruth A. II. Title.
TL790.L39 1987 333.9'4 86-32677
ISBN 0-231-06498-5

Columbia University Press
New York Guildford, Surrey

Printed in the United States of America

This book is respectfully and affectionately dedicated to the memory of Harold C. Urey, who forsook the science of war to take the lead in the exploration of space.

CONTENTS

PREFACE

Space resources are everything useful that we can find in space. In a sense, they include all the matter and energy of the Universe except what exists on Earth. For practical purposes, this book restricts itself to the use and potential impact of space resources in the portion of the Solar System nearest the Earth, but this restriction is applied only because we are limiting our scope from the present to about 2010.

A vast number of space buffs inhabit this planet, and have inhabited it in the past. Jules Verne, Robert Goddard, Walter Cronkite, hordes of science fiction writers and school children, thousands of astronaut applicants, and the writers of this book all dream or have dreamt of a future in space. Many were caught up in the thrill of President John Kennedy's goal of an American on the Moon by the end of the 1960s. It is frustrating that such heady stuff as manned colonies in lunar orbit or on Mars should be found to be impractical, as they have been. No child wants to be kept from the roller coaster ride because it is too expensive—why else should he have come to the fair?

But our future in space has become economically restricted. Billions of dollars seem necessary to develop even the simplest programs, and politicians and taxpayers cringe. We have the dream, the technology, and the need, both scientifically and socially, to have a permanent manned presence in space, but at the moment the means are too costly.

This book is about how space resources—using space itself and the materials and energy sources present there—can make a meaningful space program affordable. It tells how to make space endeavors pay for themselves. Not only will we suggest ways to reduce costs, using new technologies based on space resources, but we will show that the activities we are most likely to do in space can provide a net economic—not to mention scientific and technological—gain for Earth.

The first chapter introduces the concept of using space resources and shows why this idea is so important in today's decision-making process. Only very recently has the potential role of space resources been realized by planetary scientists, and only now has this knowledge begun to trickle through to administrators and politicians in this country. The potential use of space resources has already led to the beginning of a radical restructuring of both the American and Soviet space programs, and the other foreign agencies that have a space presence, Japan, China, and Europe, may soon follow suit. In chapter 1, we talk about our present stance and our potential goals, and how space resources can influence our choices of future goals and activities in space.

The next two chapters will give a historical summary of the many space races that have occurred to date, attempting to uncover not only what the spacefaring nations are capable of achieving, but also why they choose to do what they do. Chapter 2 focuses on the background history of man's invasion of space from 1860 to 1968, just before the first Moon landing. The details of what actually happened to bring about our space goals in this era is a fascinating study in motives, decision-making, and international politics; this is fully discussed. These details are necessary to understand the decisions that are being made today in regard to mankind's future in space. Technological capabilities and scientific goals, caution and daring, brilliance and dreams, disappointments and tragedy are all part of this story.

Chapter 3 deals with the modern era of the space race, from the actual landing of Americans on the Moon through other less well-known space races to the present era of shuttles and space stations. Again, technical and scientific progress is examined along with the political decision-making processes of both of the countries involved.

Chapter 4 discusses how the space buffs of the late 1950s, from the perspective of events surrounding the birth of the Space Age, foresaw the coming of age of the new space technology. In other words, what did they think we would be doing today? What are we actually capable of doing today? What are we actually doing today? How close were the prophets to being right? What great things have we done that they did not anticipate—and what have we failed to do? Both the great space visionaries and the boys and girls in the classrooms discussing in shocked tones the launch of Sputnik I had their dreams and extrapolations and goals. Are we—who were those boys and girls—on schedule today? If not, why not?

Our visions of a quarter century ago are no longer relevent, either because they were not part of our nation's political needs or because of mistaken conservatism or overenthusiasm concerning our technological or economic capabilities. Chapter 5 looks ahead a quarter century to suggest new visions, and assesses their plausibility. All travel in space requires a way to get going, to keep going, and to stop—gently. Several new technologies are described, including new cheaper—or free—propulsion methods. The principles of space flight and interplanetary navigation are sketched out. Energy is the key: acquiring it, using it, conserving it, dissipating it, being protected from it, replenishing it. What we learn in space, as this chapter will show, has much bearing on how far mankind will ever be able to go in space, and how much Earth will benefit from space pursuits.

Chapter 5 deals with how to do it; the next six chapters deal with where to do it. The answer or answers, already extremely obvious to some, are actually loaded with complexities that will need to be carefully sorted through by decisionmakers. A wrong move now would be costly in dollars and time—dollars because the wrong move would never help to make space pay, and time because any more deadends of the Apollo sort would keep us from realizing a very lucrative potential. The Moon, the near-Earth asteroids, Mars, and the Martian satellites Phobos and Deimos are all discussed for what they can give back and the costs of getting it. It is not a vacation home we are pursuing, but a new industrial site: the discussion is entirely practical, in tone. We will search for sources of metals, water, air, and rocket propellants, the stuff of industrial society. Without them, human expansion into space is doomed.

Chapter 12 discusses in depth the role space resources have to play in getting us into space and making it pay. Specific technologies and products based on space resources are discussed in detail. This discussion concludes with a look back to the previous chapters, asking where we should go next in light of the uses and availability of space resources. We will then be in a position to describe what new space activities would be made economically feasible, or even made possible, by the use of space resources.

Chapter 13 explores the current plans and goals for space development of the space programs of the Soviet Union, the United States, Japan, the European Space Agency, the Peoples Republic of China, Australia, and India. An interesting inside story of the political manipulation of our future in space, the decision to build the Space Station, provides a view of what can happen to a scientific and technological consensus in the hands of the powers-that-be. A critical analysis of costs and benefits is included to pave the way for the final chapter.

The central questions of national and international goals in space will then be before us. By what principles should we design our future in space? Given those principles, what specifically should we choose to do? Should we build a large, permanently manned lunar base? Should we plan a manned expedition to Mars? Should we concentrate on the development of a space industrial park in low Earth orbit? Should we embark on a program of building solar power satellites? And who, precisely, are the "we" in these questions? Space resources have an impact on all of these options in a major way.

The next major goal in space beyond the Space Station may be anything from a $100 billion dead end to a $1 billion investment in a limitless future in space. We must plan our future in full knowledge of the good and bad decisions made in the past, lest today's attractive goal become tomorrow's graveyard of hopes. We fell into that trap with Apollo, and can ill afford to do it again. There is a limit to the amount of folly that Nature can tolerate. We need to be more clever and ambitious than we have been, and make every dollar, ruble, pound, yen, franc, rupee and mark yield the highest possible future dividends. But how can we do so?

Finally, in chapter 14, we will propose a strategy for the cooperative exploration and exploitation of the Solar System that will take into account the distinctive strengths and weaknesses of the various national space programs, as well as tapping the broader scientific and technical strengths of the planet Earth. In this context, we give our proposal for a sane American presence in space. Our plan is balanced in its pursuits; it will do no great economic harm to the nation, but will allow it to develop technologies with the potential to pay well. Our plan requires no unfounded technological leaps, but instead is based on technologies, materials, and energy sources now known, or those extrapolated from present knowledge. Our plan creates a balance between manned missions where these are necessary or desirable, and unmanned, automated missions where these are superior. Finally, our plan provides a meaningful new goal to spark our sense of purpose, built on cooperation with the other spacefaring nations, including the Soviet Union, and leading meaningfully to other more distant goals as a part of the whole of the emergence of human beings into space.

We are excited about what can be done, concerned about what might be done instead, and enthusiastic that the American people should learn the difference. That is the purpose of our final chapter and the main hope of our book.

ACKNOWLEDGMENTS AND CREDITS

We are pleased to acknowledge the generous assistance and advice of James Arnold, David Brin, Tom Jones, and Hans Mark. Useful materials were contributed by Eleanor Helin and Mike Gaffey. We thank NASA for the use of the figures on page 56, 62, 64, 72, 74, 75, 76, 92, 103, 175, 187, 213, 214, 236, 274, 275, 278, 283, 284, 285, 287, 291, 303, and 304, Novosti for those on pages 15, 60, 65, 71, and 182, Japan's NASDA for those on pages 367, 369, Tom Jones for the drawing on page 355, and ESA for the photo on page 370.

Darest thou now O soul,
Walk out with me toward the unknown region,
Where neither ground is for the feet nor any path to follow?

No map there, nor guide,
Nor voice sounding, nor touch of human hand,
Nor face with blooming flesh, nor lips, nor eyes, are in that land.

I know it not O soul,
Nor dost thou, all is a blank before us,
All waits undreamed of in that region, that inaccessible land.

Till when the ties loosen,
All but the ties eternal, Time and Space,
Nor darkness, gravitation, sense, nor any bounds bounding us.

Then we burst forth, we float,
In Time and Space O soul, prepared for them,
Equal, equipt at last, (O joy! O fruit of all!) them to fulfil O soul.

—Walt Whitman

SPACE RESOURCES

1

HOW TO SAVE MONEY WHEN YOU TRAVEL

Good sense avoids all extremes, and requires us to be soberly rational.—Molière

This is the age of the Space Shuttle. It was conceived in an era in which recycling had just been rediscovered (as happens cyclically), and was born amid fanfare and hyperbole centered on how much money it would save. It was to lower the costs of access to space; mankind's dreams, some of them centuries old, would be realized. But our hopes for the Shuttle may well have been born out of season, destined to wither and die.

The reusable Space Shuttle has often been compared favorably to "throwaway" boosters by means of an ingenious analogy to a trip across the continent. Using the old disposable boosters, the analogy goes, was like flying a 747 from New York to San Francisco and bailing out at the end of the journey, letting the plane crash into the ocean. Using the shuttle would, according to this analogy, be more like a normal transcontinental flight, in which the plane is reused.

This is a cleverly posed analogy, worthy of the skill (and salary) of a Madison Avenue executive. After hearing it, no reasonable person could doubt that building the Space Shuttle was clearly the logical thing to do. That's the trouble with a facile analogy: the conclusions are so self-evident that no one would bother working through the details to see if it really is a valid description of reality.

But a student of the space program would note that there is a continuing supply of cheap disposable rockets—old military strategic missiles that have been replaced by newer, sleeker, more deadly models. And space shuttles are not cheap. According to recent experience, they cost about $2 billion each—even if you don't use them. Maybe the analogy should be recast in a fairer light: your real choice is between flying an old, obsolete, but functional small plane and ditching it in the ocean, or flying a 747, landing it at San Francisco airport, keeping it with you for your month-long vacation by the Bay, and then flying it back to New York. If you use the 747, you run up a huge daily cost whether it's flying or sitting idle—and you absolutely have to bring it back! Worse than that, in the shuttle analogy, there's no way you can stay in "California" for more than a few days without the owners of the transportation getting irate about having their expensive hardware tied up.

Even worse, imagine that you had a ready supply of old small airplanes, but the 747 didn't even exist yet! Would you design, develop, and build 747s for this kind of use, at an expense of billions of dollars? We think that, if we were to be soberly rational, we would not choose to do so.

Let us indulge in another analogy concerning a transcontinental trip. Your goal is to take your family from New York to San Francisco by car. Having never driven such a long distance before, you inquire of some of your friends how best to do it. One friend laboriously calculates how much food, fuel, water, and Pampers it would take to get your family to its goal, while also taking into account weather, traffic patterns, speed limits, and the necessity of not exhausting the drivers. He then calculates that you would need a large truck with a double trailer, one a travel trailer devoted to your supplies and sleeping space, and the other filled with sufficient gasoline to propel not only you and your supplies, but also much of your gasoline, across the continent.

He conservatively requires that, since you don't know for a fact where the restaurants and gas stations are all along your route, you had better play it safe by bringing everything you may need. In fact, to play it completely safe, he recommends that you take a few spare tires, an extra engine, a case of oil, and so on. That means you need more gas to move all that weight across the country, which means that you need even more gas to move the gas! At the end, he totals up the bill: you need a very large and very reliable car, $18,000; a new travel trailer, $12,000; prepared nonperishable food; spare parts; bedding; clothing for any weather for the entire family; and about a thousand gallons (four tons) of gas. Oh yes—you also need all the tools to do any conceivable repair work, and complete documentation of the structure and function of every part of your rig so that you will know how to diagnose and repair any problem you may encounter. Total cost: $50,000.

A second friend, who has some experience driving, knows that there are plenty of restaurants, gas stations, service stations, camping grounds, and water fountains along the route. He suggests you bring a limited but appropriate wardrobe, get your eight-year-old car serviced, and set out with $500.

Why the discrepancy? Because one tries to take home with him, while the other makes himself at home wherever he goes. He is not afraid to drink the local water, eat the local food, or buy gas from strangers.

This analogy answers two questions. The first is the time-honored classic, "Why is space travel so expensive?" The answer is that we plan to take home with us wherever we go—and a trip to the planets is a lot more ambitious than driving the Interstate to San Francisco. The second question is rarely asked, and then only by people who know the answer to question number one: "What can we do to make space travel much less expensive?"

There are two kinds of answers to the second question, the technician's answer and the innovator's answer. The former sees us saving money on space travel by, in our analogy, making our cars more efficient and our trailers lighter, carrying dehydrated backpacker's food, using only the lightest down-filled sleeping bags, or switching from gasoline to liquid natural gas. All such changes are indeed beneficial, but they all cost money

Figure 1.1
California or bust

to implement. The central point is that not one of them can make a big difference.

The other way to make travel less expensive is to renounce the idea of bringing everything with us. Instead, we get a tuneup for the old car, buy some travelers' checks, and go. Along the way, we schedule stops to pick up water, fuel, and food. This is the cheap way to travel in space.

But where in space can you possibly find water? It does little good to find it on a planet, since it costs too much to land and take off again. And what about fuel? Are there oil wells on the Moon? What a crazy idea! Surely we cannot expect to find tanks of unleaded gas waiting for us on the Moon or the asteroids, the moons of Mars, or anywhere else nearby. Instead, astonishingly, there is something much better—a limitless supply of water, which provides the very best rocket propellants of all, hydrogen and oxygen. They are not free for the asking, but a few tons of equipment can extract these precious substances from lunar or asteroidal material.

The choice between taking it all with you and making do with local resources has been faced many times in the past. It was faced by the Vikings, by Magellan and Columbus, by the Spanish, Dutch, and English voyagers to the New World. The first intentional voyages were undertaken only with great trepidation and excessive preparation. Later voyagers took much less—and they brought seeds and tools to take advantage of the new lands. The Dutch patroons of New Amsterdam at first prohibited independent agricultural and commercial enterprise by their workers: they wanted to make a profit by importing everything from abroad on their own ships. But human ingenuity prevailed. Workers and soldiers who finished their tour of duty in New Amsterdam refused to take passage back to Amsterdam. Instead, they took up farming or crafts, sold their wares at highly competitive prices, and turned New Amsterdam from a colony into an infant nation, no longer utterly dependent on its founders.

The savings that can be realized by using the resources of space are immense. Every raw material that we could ever want is up there, often in quantities and concentrations beyond anything available on Earth. Using these resources could mean that, with a roughly constant level of spending on space activities, the scale of human activities in space could, as we shall show, be multiplied by a factor of ten to a hundred within a decade or two, and by a factor of a thousand to ten thousand shortly thereafter. The significance of the use of space resources is enormously greater than any mere reduction in launch costs could achieve. It is a qualitative, innovative breakthrough that gives the Solar System to humanity.

This means that learning to use and rely upon space resources will, in the long run, be a much more important step toward making space accessible than the Space Shuttle has been—or ever could be. We need to commit ourselves to basing most space operations in space. We need to learn how to live off the boundless resources available in space, without having to haul our materials and supplies, at staggering cost, out of the deep gravity well of Earth.

This principle has been understood by a handful of pioneers: Konstantine Tsiolkovsky in tsarist Russia, Olaf Stapledon in *First and Last Men* and *Star Maker,* J. D. Bernal in *The World, the Flesh, and the Devil* in 1929, Wernher von Braun and Willy Ley in the 1950s, Dandridge M. Cole and Donald W. Cox in 1964, and modern scientific prophets such as physicists Gerard K. O'Neill and Brian O'Leary. One crucial factor that has been missing until very recently is an understanding of how to find a first step that would be relatively simple and pay for itself. Once such a beginning step has been taken, all the rest will take care of itself.

Astonishingly, without ever really meaning to do it, space scientists and space technologists have already produced virtually all of the technology needed to make the first step in the industrialization of space. Chance discoveries have come together in the last four years to point out a clear path to large-scale space development. These new discoveries neither contradict nor confirm in detail the many earlier speculations and projections. But they do clearly show that the basic idea of the emergence of mankind from its Earthly womb must—and can—be closely linked with severing the symbolic umbilical cord that links every astronaut to Earth. We will, in this book, speak not only of this new philosophy, but also of new science and new technology that permit that philosophy to be put into practice.

The scientific discoveries and the development of techniques of space travel that have emerged from Earth's space programs, especially those of the USSR and the USA, are essential ingredients of this new synthesis. The attitudes, philosophies, institutional prerogatives, and future aspirations of the spacefaring societies are at least equally important elements of the mix, and also require our serious attention. Further, the fundamental question of why nations do certain things in space while failing to do certain other things comes naturally to the fore.

Other crucial issues are the role of military and political confrontation and competition, the role of the military in future space activities, and the practicality of large-scale international endeavors in space. Much has been written on these subjects in the past (some of it good, but most of it utter pap), but it is no longer the past: the new elements of highly advanced Soviet Space Station technology, operational Soviet space weapons, the American Strategic Defense Initiative, the Space Shuttle, impending Soviet development of a shuttlecraft and a superbooster, American plans for a Space Station in the mid-1990s, and a host of other factors make the situation more complex, and more interesting, than ever before.

Planners of scientific space missions have already seen the promise of space resources on the horizon. In fact, the American, Soviet, and Japanese space programs now show the beginnings of a profound reorientation of their basic strategies for the exploration of the Solar System. Instead of addressing purely scientific goals, we see a new, strong, and very distinctive second goal appearing in these programs: prospecting for space resources.

This change has not been widely reported, and seems to have evaded the understanding of many senior bureaucrats, but it is of fundamental importance. It marks the beginning of a transition to a wholly new way of looking at space: as a medium where mankind will live, work, and play. No longer are human beings to be prepackaged, shot into space in little Earthly wombs, and then brought back, still hermetically sealed, to Earth. Instead, we shall be born into life in space. We shall at last break the bonds of Earth.

2

BACKGROUND HISTORY

The Space Races up to 1968

Whenever a distinguished but elderly scientist says that something is possible, he is almost certainly right. Whenever he says that something is impossible, he is almost certainly wrong.—Arthur C. Clarke

Both the United States and the Soviet Union had visionaries who foresaw the importance of space long before the public or responsible officials did. Indeed, thinking about space, both with serious and frivolous intent, began in several nations well before space flight was a reality. The history of space exploration bears the fingerprints of those gifted with vision—and the muddy footprints of those who lacked it! In this chapter we turn our attention to the birth of the space age, the first races in space, the early Soviet domination of space, and the astounding growth of the American space program through the mid-1960s.

1860–1945. ROCKETS: FROM PLEASANT DREAMS TO FEARFUL REALITY

A century before space travel became a reality, some writers wrote "speculative fiction," while striving to remain faithful to the physics known in their time. Jules Verne, in *De La Terre à la Lune* (1865), described the use of a giant gun to shoot a capsule containing three space travelers, their dog, and several chickens from Florida to the Moon. Edward Everett Hale, in 1869, published the story "The Brick Moon" (fig. 2.1) in *Atlantic Monthly,* describing a manned artificial Earth satellite 200 feet in diameter and its

Figure 2.1
An evening stroll on the brick moon

use to observe weather and assist with navigation and communications on Earth.

Others wrote of the physics, chemistry, and engineering of space travel. The earliest Russian pioneer of rocket propulsion, Nikolai Ivanovich Kibalchich, wrote in 1881 of a manned rocket platform fueled by gunpowder cartridges. Kibalchich seems never to have thought of space travel. He was a revolutionary in more ways than one: he wrote in prison, awaiting execution for rebellion against the Tsar. Konstantin E. Tsiolkovsky in Russia had sketched out the main principles and features of the coming Space Age in 1903. By 1909 the American Robert Goddard had already pursued the physics of liquid-fuel rockets to the point that he knew of the superior promise of rockets fueled by liquid hydrogen and liquid oxygen. By 1927 the Verein Fuer Raumschiffahrt (Society for Space Travel) had been founded in Germany, and active rocket research was underway in three nations.

J. D. Bernal wrote in his small book of essays, *The World, the Flesh, and the Devil* (1929), of the opening of the Solar System by rocket propulsion. He emphasized the problem of supplying both reaction mass and propulsion energy for space vehicles, and proposed both microwave transmission of energy and the use of solar radiation pressure to move manned vehicles about the Solar System. He pointed out the immense energy cost of using materials from Earth to develop civilization in space. He proposed the use of solar energy to solve the power problems of Earth, and of the use of space resources and space bases to maintain, fuel, and build space structures. He described huge hollow spherical shells, 10 miles in diameter, made of asteroidal materials, and occupied by tens of thousands of people. And he proposed that these space habitats would be largely occupied with multiplying themselves, using up interplanetary matter and tapping solar energy until the competition between them became so severe that some would embark on interstellar journeys, taking centuries to cross to the nearest stars. We now know that as early as 1918 Goddard had similarly proposed, in a sealed manuscript, that nuclear powered vessels could carry the descendants of man away from the dying sun to refuge among the stars: his note was not opened and published until 1972.

The promise of rocket propulsion became sufficiently obvious in the 1930s to encourage the institutionalization of research programs. The American Interplanetary Society (later known as the American Rocket Society) was founded in 1930, and the British Interplanetary Society in 1933. Soviet research in rocket propulsion became visible for the first time. The competitive aspect of the scene soon became obvious: in 1931, Robert Goddard held the world altitude record for rockets with a mere 1700 feet. A Soviet flight in 1933 of a GIRD-09 rocket, designed and built by M. K. Tikhonravov and Sergei P. Korolev, reached 1300 feet. An ARS flight in 1934 traveled a horizontal distance of 1338 feet.

With success came increased funding and increased militarization. With about 80,000 marks support from Hitler's *Wehrmacht*, a team of rocket experts at Kummersdorf fired an A-2 rocket to a height of 1.5 miles in 1934. In the next year, Goddard fired two liquid-fueled rockets to ranges of 1.8 and 2.6 miles. But by 1935 the *Luftwaffe* and *Wehrmacht* had seen the military potential of rockets: the Kummersdorf operation received a budget jump to 11 million marks. Ironically, German experimentation in rocketry was inspired in part by limitations imposed on them by the treaty that ended World War I: long-range artillery was prohibited, but rockets were not mentioned. By 1937 they were able to open a large, new test range on the Baltic island of Peenemuende.

By this time, with government support, a Soviet rocket had reached a world-record height of 3.5 miles in 1936. But the high German funding level was destined to have even more impressive results. By 1942 the new A-4 rocket had reached an altitude of 53 miles and a range of 118 miles.

During World War II the A-4 project was renamed Vengeance Weapon Number 2 (in German, *Vergeltungswaffe Zwei*, abbreviated V-2—see fig. 2.2). This rocket was produced in a huge underground factory at rates that eventually reached up to 900 rockets per month. A Royal Air Force raid on Peenemuende in August 1943 did severe damage to the facilities there, and killed Dr. Walter Thiel, the designer of the V-2 engine, but was unable to prevent the massive use of rockets against London and other cities in the south of England. As a terror weapon, the V-2 must be counted a success. But the statistics show that the 4300 V-2 rockets fired against England accounted for only 2500 deaths. The guidance system of the V-2 was so primitive that it could not be used against point military targets, but instead had to be launched in large numbers against extended areas such as cities.

Germany had in mind another, much more devastating use for the V-2. The *Luftwaffe* had a secret project under way to develop an atomic bomb, an area-attack weapon well suited for use on a vehicle like the V-2. Again the Royal Air Force staged a raid against the German atomic bomb project at its secret heavy-water plant in Norway, this time with success. The prospect of nuclear missiles in the hands of Hitler was narrowly averted. Yet even the nuclear devastation of England was not enough for the German dictator. Until the demise of the atomic bomb project, the German rocket scientists were working on designs for a huge intercontinental missile called the A-10. It was designed to drop multiton atomic weapons on New York.

Why did German rocket and nuclear technology not come together in

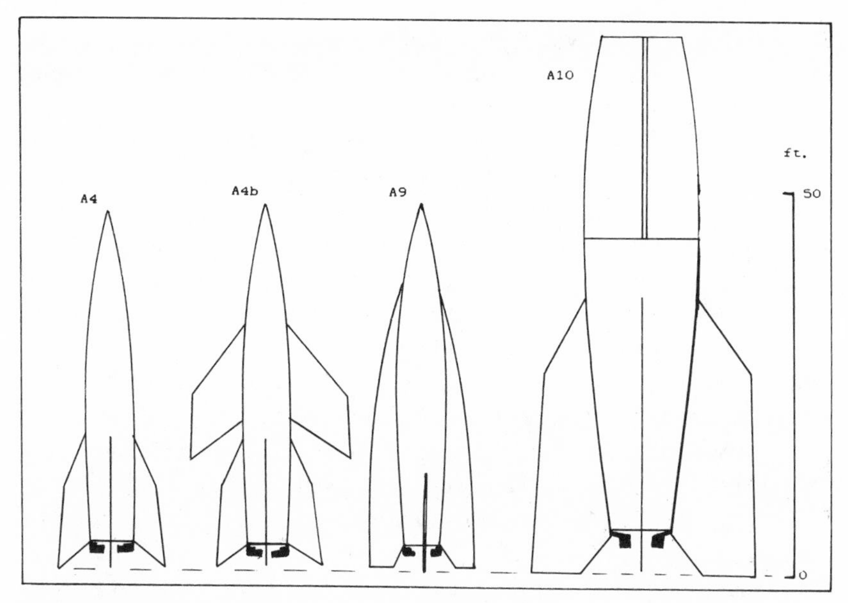

Figure 2.2
German Designs of the 1940s
The basic German military rocket of World War II was the A4 (V2; left). A winged version of the A4, called the A4b, was test flown with limited success. A more advanced modification, the A9, was designed but never built. It was to be used as an upper stage atop the huge A10 booster to give Germany the ability to bombard New York from German soil.

time to win the war? Because of precious lost time: after the first flight he witnessed, in 1939, Hitler, who was highly credulous and superstitious, had a bad dream in which the rocket did not work. On that basis, he withdrew funding from the program. The first full-range test of the V-2, in October of 1942, was conducted on a shoestring. Films of the test were sent to Hitler, who was not interested. Fully a year passed before he could be prevailed upon to look at the film. Once he did, he became a fanatical believer: the V-2 was ordered into production at once. One year later, in September 1944, the first operational V-2s were fired at Paris and London. By then the Thousand-Year Reich was already counting its future in months.

What would the world be like today if Hitler had gone ahead with the V-2 in 1939? He would have had an operational missile by 1941, and with that delivery system, he would have had the strongest reason to push the development of the atomic bomb at top speed. And, of course, there would have been great incentive to cluster several V-2 engines to make a huge first stage, put another V-2 on top, and make the world's first Inter-Continental Ballistic Missile (ICBM), But, fortunately, "reason" appealed little to Hitler.

At the end of the war, Von Braun and most of his team of experts found it expedient to escape from the Soviet occupation of the V-2 test and production facilities in eastern Germany, and defect to the West. This decision was made even though the V-2, although used extensively against England, was never employed on the Eastern front. Evidently, von Braun felt that he could expect better treatment from his intended English-speaking victims than he would receive from the Soviets. The Soviet Union captured what was left of Peenemuende. They were soon given the V-2 underground factory by the Western Allies—after the entire contents of the factory had been shipped off to America! The Soviets captured many production workers, and much equipment, but they did not base their future plans on German hardware and manpower. They studied it, incorporated the best features into their continuing program, and went on from there, using Russian manpower.

In the United States, almost no progress was made on long-range rockets from the time of the test flights of the mid 1930s until the last months of the war. The question of whether rockets could work was answered unequivocally by the V-2.

The United States was by then a nuclear power, and the idea of using rockets as the delivery system for nuclear warheads suddenly became obvious. The Trinity test and the Hiroshima and Nagasaki atomic bombs showed that a package weighing a few thousand pounds could pack the explosive power of tens of thousands of tons of TNT. Such a potent explosive device not only was small enough to be deliverable, but also could wreak havoc on such a scale that even a crude guidance system would be good enough to assure the destruction of its target.

1946–1954: THE FIRST POSTWAR RACE; THE THERMONUCLEAR BREAKTHROUGH

In 1946 the division of the V-2 inheritance between the USSR and the USA had been completed. The United States Army Air Corps let out a contract to Convair to study the possibility of using V-2 and A-10 designs as a basis for construction of an American ICBM. Unfortunately, this program quickly fell victim to the strong opinions of the eminent wartime director of the War Department's Office of Scientific Research and Development, Dr. Vannevar Bush. It was his opinion that intercontinental "high angle" rockets could not hit their targets. Accordingly, President Truman's Chairman of the Joint Chiefs of Staff, General Eisenhower, passed on the directive to end the development program, an act that was to come back to haunt him years later. The Air Corps dutifully scrapped the program in 1947.

At about the same time, in April of 1947, Premier Stalin ordered a start to the development of a Soviet ICBM. The designs of what were to become the first Soviet ICBM (and the launcher of Sputnik) were frozen in 1951–52. The size of the rocket was set by the need to send very heavy atomic bombs over intercontinental distances, and this consideration dictated that the rocket be huge. The guidance system of the first Soviet ICBM design was so crude that warheads with explosive yields of several hundred thousand tons of TNT were needed to assure destruction of the target. Such warheads, using atomic bomb technology, weighed several tons.

Up to this time, all nuclear explosives were atomic bombs, deriving their power from the fission of the heavy elements uranium or plutonium. But the hydrogen fusion bomb promised much greater explosive power, and much lower weight, than giant atomic bombs. The first Soviet test of a fusion device (a prototype hydrogen bomb designed by Andrei Sakharov) came on August 12, 1953. By the time the size and weight of an actual hydrogen warhead was known, late in 1954, the Soviet ICBM project was far enough along so that the hydrogen warhead was sized to fit it, not vice versa. Thus the Soviet ICBM, designed to carry an atomic bomb with a yield of about 500,000 tons of TNT (0.5 megatons), instead became capable of carrying a 20 megaton hydrogen warhead.

On November 1, 1952, the United States exploded a hydrogen device, with the power of 6 million tons of high explosives, on a Pacific atoll. The American ICBM program was revived in 1953, and Convair designed its Atlas ICBM about the American hydrogen bomb technology of 1953 (see fig. 2.3). The advanced state of American hydrogen bomb technology permitted the use of a small ICBM with only 360,000 pounds of first-stage thrust. The six-year hiatus in American ICBM development assured that the USSR would have its ICBM first. Ironically, the advanced state of American nuclear technology, which permitted the construction of ex-

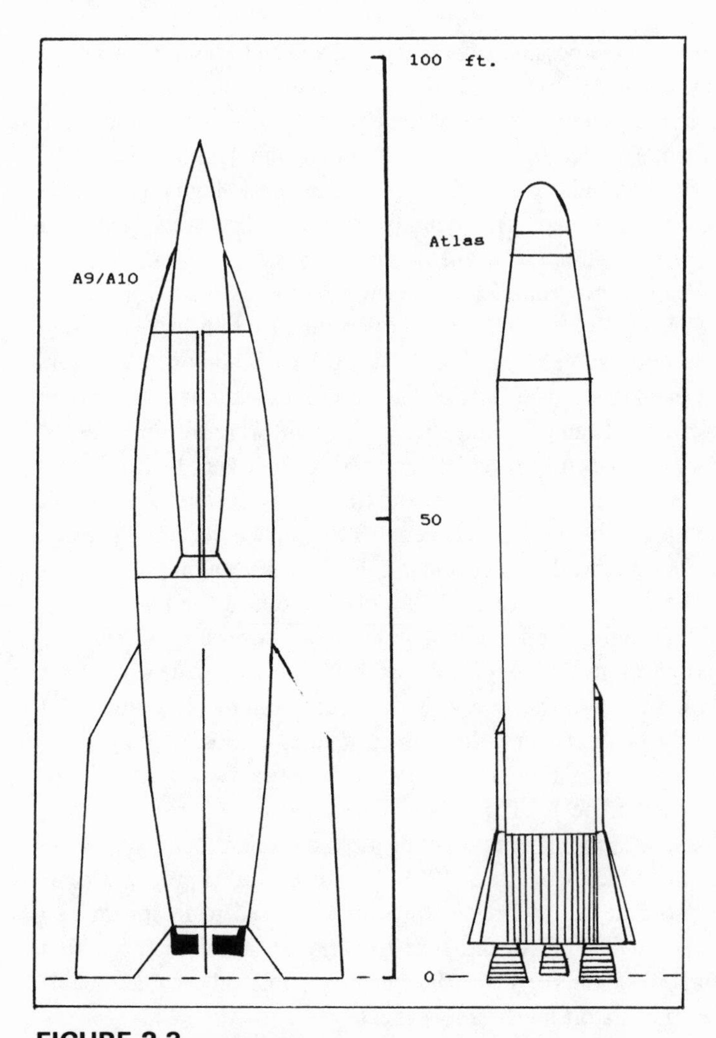

FIGURE 2.3.
Intercontinental Missiles
The German A9/A10 design of 1942 (left) is compared to the Convair Atlas ICBM, designed in 1953. The Atlas had two booster engines attached to a disposable "skirt," bracketing a sustainer engine attached to the main rocket structure. All three engines fired for several minutes after liftoff, after which the skirt and booster engines were jettisoned. The sustainer engine continued to fire until the payload reached the desired velocity. The Atlas is termed a "1½ stage rocket." The A9/A10 was never built. A comparison follows:

Figure 2.4
Engine Clusters on the Soviet A-Class Booster
The A core stage and each of its four liquid-fuel strap-on boosters consists of four main engines, for a total of 20 main engines, each developing a little over 20 metric tons of thrust. The smaller engines are vernier rockets for steering the booster. After about three minutes of flight the four booster rockets are dropped off, leaving the core stage to continue. This particular rocket is an A2 Soyuz launch vehicle.

tremely powerful lightweight hydrogen warheads, dictated that the American ICBM be far smaller than its Soviet counterpart, which had a thrust of about 880,000 pounds at liftoff (see fig 2.4). Thus the payload capacity of the American ICBM, when it finally arrived, would be much smaller than that of its Soviet rival.

Soviet ICBM development progressed in a careful, steady, deliberate manner during the years from 1947 to 1953, while the American effort slept. Those six lost years were of crucial importance, and accounted for

	A9/A10	Atlas
Weight (lb)	300,000	244,000
Liftoff Thrust (lb)	400,000	380,000
Range (miles)	3,000	5,000
Payload (lb)	2,000	4,000
Design Date	1942	1953

most of the terrible frustrations felt by Americans during the first years of the Space Age.

THE VIEW FROM MID CENTURY

In 1947, Lyman Spitzer of Princeton University stated that he expected to see interplanetary travel within his lifetime. In April 1948, Lieutenant General James Dolittle said that he expected manned flights to the Moon within his lifetime.

In February of 1949 a Pentagon briefing on the possibility of launching artificial satellites into Earth orbit was widely reported. The *New York Times*, having evidently missed the work of Goddard and Tsiolkovsky and the introduction of the V-2, editorialized that this was "the first hint that such a fantasy might become a reality." In May, an article in the *Times* also described a scheme to send a manned rocket to the Moon.

In September the wire services carried a report that the Soviets might be building an intercontinental missile at Peenemuende. Such stories were debunked by R. Gibson of Johns Hopkins University in January 1950. He proclaimed that ICBMs could be built, but could not hit their targets "beyond ten feet." In April, aviation pioneer Igor Sikorsky said that he expected to see manned space flight within ten to twenty years.

An article in the *New York Times* of September 4, 1951 predicted that man would reach the Moon, but probably not within the lifetimes of its readers. The first satellite would be launched within ten years, and the earliest possible date for a manned trip to the Moon would be 1976. This aroused the ire of R. A. Smith, who fulminated from London that there would be no space ships for a hundred years. As for military uses of space, his considered opinion was that it was "nonsense." Upon brief reflection, he changed that to "absolute blather."

On October 4, 1951, six years to the day before the launching of *Sputnik 1*, a Soviet official predicted that there would be rockets in space in ten to fifteen years. Alexander de Seversky, the pre-war enthusiast of air power, predicted manned trips to the Moon on atomic-powered spacecraft by the end of the century.

Wernher von Braun, in August 1953, predicted a manned space station within fifteen years. In 1954 Eddie Rickenbacker, the World War I flying ace, said that within twenty years there would be spaceships the size of the largest ocean vessels, carrying 2500 men.

In September of 1954 the USSR announced that it had flown rockets over a distance of 240 miles. V. V. Dobronravov, on October 1, 1955, said in an interview on Radio Moscow that "Soviet scientists expect to be rocketing between the planets in the very near future." He then repeated the claim, phrasing it "within a few years." Two weeks later the existence of the American Atlas ICBM program was announced to the public.

Such sanguine expectations were not unanimous. In October 1952, Milton Rosen, a scientist who later worked on the Earth satellite program at the Naval Research Laboratory, acknowledged reports that a satellite could be launched in ten to fifteen years by calling it a "fantastic scheme" that would be "harmful to US defense." And in April 1955 the Astronomer Royal of England, Sir Herbert Spencer Jones, revealed to the world that there was really no way to get to the planets: "I don't see anyone getting to Mars in this century, the next century, or the century after that." It takes extraordinarily clear vision, or enormous hubris, to predict the future to the year 2200!

In April 1955 a Soviet scientist described plans to send a small automatic tank to the Moon, to be followed by men about two years later. No less an authority than the President of the American Rocket Society, J. Hartford, saved us from this delusion: "Sending a tank to the Moon is so far out of this world from a practical standpoint as to be nothing but science fiction at this time."

In July 1955 President Eisenhower, concerned that the exploration of space would end up dominated by the military, announced the Vanguard project. Vanguard, an open scientific program, would use an evolved version of the Navy's Viking rocket to launch a small artificial satellite in 1957 or 1958 as a part of the International Geophysical Year. The Viking rocket was the renamed Neptune, which was a redesigned and enlarged V-2. V. V. Dobronravov of the Soviet Academy of Science commented in a low-key manner that the Soviet Union would also launch a satellite some time within the next ten years, and that they expected to fly a three-man space ship by 1975 and send men around the Moon in the 1980s. In August 1955, Leonid Sedov, President of the prestigious Soviet Academy of Science, made a more specific prediction: a Soviet satellite would be launched in about two years.

1955-1958. THE FIRST TWO SPACE RACES: ICBMs AND SATELLITES

By early 1956 the rumors were rampant. The New York Communist newspaper *Daily Worker* claimed in January that the Soviets would attempt a satellite launching in 1956. A few days later the Vanguard project announced that it would launch its first satellite on September 30, 1957 from Cape Canaveral, Florida. One cannot help but wonder whether this ridiculously specific date served as the benchmark against which the Soviet Union scheduled its first flight.

But the unbelievers persisted. President Truman, who had sat on the Atlas ICBM program during his entire administration, offered the informative opinion that the business of launching Earth satellites was "a lot of hooey." The Republican view was refreshingly different. Perennial Presidential hopeful Harold Stassen reassured us in February 1956 that "the United States will be the first to perfect long-range guided missiles. Britain will be second, and the USSR third." If you're that good, you don't even have to try!

In September 1956 an Army Jupiter C (a modified Redstone Short Range Ballistic Missile with three small solid-propellant upper stages) was fired 3300 miles downrange from Cape Canaveral. With a slightly lighter payload and different guidance program, it could have been placed in orbit. The Army was, however, explicitly forbidden to put a payload in orbit by order of President Eisenhower: that right would be reserved for the civilian Vanguard program.

By early 1957 the handwriting was on the wall: there was indeed a race to get into space. The main question was when the first satellite would be launched—and by whom. But just as humanity was about to explode into space, the eminent electronics inventor Lee deForest pronounced the death of the space dream: "Man will never reach the Moon, regardless of all future scientific advance." So there.

Many commentators maintained at the time that the launching of *Sputnik 1* was a wholly unexpected bolt from the blue, and that there was no evidence that there had been any space race before *Sputnik.* We choose to answer by the cruel but effective expedient of pointing out a number of articles published in the *New York Times* during the last few months before the launch.

June 2, 1957: A. Nesmayonov of the Soviet Academy of Science: "The satellite and rocket are completed."

June 7: The U. S. Army publicly complains that it has a satellite launching vehicle, the Jupiter C, but has been denied permission to use it.

June 11: Nesmayonov: The first satellite will be launched "in a few months".

June 20: The first Soviet satellite will be superior to *Vanguard.* "Because all the instruments desired do not fit in one sphere, several are planned."

June 23: The first *Sputnik* will be a 20-inch sphere that will be placed in a semi-polar (high-inclination) orbit.

June 27: The first Soviet Moon rocket will be launched in the 1960s, and man will enter space five to ten years later. A satellite will be sent to Mars in the 1965 to 1971 time frame. A Moon shot will then land a miniature tank on the Moon.

July 25: A Soviet rocket will set down on Venus by 1967.

September 18: Moscow radio announces that "the first satellite launching is imminent."

October 1: The first Soviet satellite will carry a flashing light for accurate tracking. It will broadcast on frequencies of 20.005 and 40.002 megacycles.

October 2: There will be no advance warning of launch. The Soviet satellite will orbit 10 to 20 degrees from a north–south line.

October 3: "Several satellites will be used for various experiments," including solar radiation research.

October 4: The Soviet Union asks for American photographic stations to assist in tracking Soviet satellites.

Anyone who was startled by the launching of *Sputnik 1* on October 4, 1957 must have let his subscription to the *New York Times* lapse.

One would have thought that the article of October 28, claiming that a "dog-carrying satellite will be launched in the near future," would attract attention. Strangely, most observers professed surprise when *Sputnik 2* carried the dog Laika into orbit a week later.

Throughout this time, numerous Soviet spokesmen explicitly claimed that the *Sputniks* had been launched by the Soviet ICBM. This was actively questioned by many critics who evidently did not trust the evidence of their own eyes. Millions of people went out at night to "see Sputnik." Every newspaper carried the news that the satellite itself was of 4th or 5th magnitude, about as bright as the faintest star in the Little Dipper. The accompanying carrier rocket was of −2 magnitude, brighter than the brightest star.

Every commentator bemoaned the huge size of *Sputnik 1,* 186 pounds compared to the 20-pound *Vanguard* sphere, claiming that the 9 : 1 weight advantage showed that the Soviets were far ahead in booster power. No one thought to estimate the weight of the huge rocket stage that accompanied *Sputnik* through space: it was about 80 feet long, and probably weighed close to 8000 pounds! Now *there* was something to worry about! This simple observation demonstrated beyond a doubt that Soviet claims about using an ICBM were correct. It also assured us that, even when the much smaller American ICBMs became available, the Soviets would still be able to launch much greater payload weights. Such a large booster could also be adapted, by adding upper stages, to the launching of lunar and planetary probes and of multiton unmanned and manned Earth satellites.

The race to build the first ICBM was won by the Soviet Union with the first flight of their SS-6 vehicle (the launcher of Sputnik 1) in September of 1957. The first successful flight of the American Atlas ICBM over full intercontinental range was on November 28, 1958, and the first use of the Atlas as a satellite launcher was Project SCORE less than a month later. The Soviet lead in ICBM development was 14 months.

President Eisenhower was not greatly impressed with the Soviet lead. He argued, as the nation's leading military authority, that the Soviet satellites had no immediate military significance. As for the Soviet ICBMs, he insisted that there simply were not a significant number of them, certainly nothing remotely approaching the weapons-delivery capacity of the B-47 and B-52 bombers of the Strategic Air Command, and that American ICBMs superior in numbers and quality would soon be entering service. But Eisenhower underestimated the magnitude of the public panic. *Sputnik 1* was frequently called "our Pearl Harbor in space."

Chairman Khrushchev, in his usual energetic and flamboyant style, missed no opportunity to rattle his saber. He spoke ceaselessly of numerous Soviet ICBMs poised to strike America if we caused any trouble. The pose had a powerful effect on America. Khrushchev knew that his ICBMs were great clumsy dinosaurs that required many hours of preparation to launch. They sat exposed on the surface, clearly visible to American photographic reconnaissance aircraft and (by 1959) satellites passing overhead, and vulnerable to destruction by any nuclear explosion within miles of their launchers. Unable to navigate independently, they required free and unimpeded radio links with a long string of ground tracking stations.

But while Khrushchev was preaching instant readiness to the Western press, he was playing the opposite tune to his own Defense Ministry. By his own account, in his memoirs, Khrushchev (as a former Deputy Defense Minister) at once recognized the vulnerability of the Soviet ICBMs and proposed that they be emplaced in hardened underground silos. This was studied and rejected by the Defense Ministry on the grounds that the missile was much too big and cumbersome to make this practical. When intelligence of the development of American silo-based missiles finally reached the Politbureau (actually, Khrushchev's son, an electronics engineer, stumbled across it while reading an American magazine and told his father!) the decision was made to proceed with the development of silos and of more compact ICBMs to fit into them. But the American public was fully convinced of the reality of the Soviet ICBM threat—so much so that John F. Kennedy was elected president mainly on the basis of the fictional "missile gap." A few short months after the election, and long before Kennedy's policies could have had any operational effect, the Department of Defense admitted that there really was no missile gap. Eisenhower's assessment of the Soviet military threat had been correct.

Eisenhower had placed very high priority on monitoring the activity of the Soviet ICBM program. A prototype Ballistic Missile Early Warning system (BMEWS) radar station was built in Turkey to monitor the Soviet launch site near Tyuratam in the deserts of Kazakhstan, and the highest-flying American reconnaissance plane, the U-2, was assigned the task of overflying the Soviet launch center. This in turn triggered an intensive Soviet effort to develop a surface-to-air missile (SAM) capable of reaching the U-2 at its operational altitude. These technologies were to collide with the downing of U-2 pilot Francis Gary Powers near Sverdlovsk on May 1, 1960. This was understandably treated as an outrageous violation of Soviet sovereignty by Khrushchev. Eisenhower first lied about the incursion, not knowing that Powers had been captured alive, and the Soviets responded by canceling plans for a summit conference. But the emphasis by then had already shifted to the use of military satellites for the gathering of reconnaissance data.

As a frantic response to the launching of *Sputnik 1* and *2,* several varieties of smaller American launch vehicles were pressed into service. The Vanguard rocket, rushed to a premature launch, was readied to place a small test sphere in space, under the close scrutiny of the world press. On December 6, before the eyes of a vast television audience, the Vanguard rocket rose three feet off the pad, fell back, and exploded. The tiny test satellite, mocked by Khrushchev as a "grapefruit," was singed but fell clear of the fireball, to lie, beeping pitifully, on the sand. "Why don't they put it out of its misery?" someone quipped. The satellite is now on display at the National Air and Space Museum in Washington.

After this explosion temporarily grounded the Vanguard program, Eisenhower reluctantly gave the Army permission to proceed with a satellite launching using the Jupiter C. The Army promised a satellite by the end of January, and they delivered on their promise with only minutes to spare.

From the launching of *Explorer 1* on January 31, 1958 to the SCORE flight the following December, a blizzard of tiny satellites was launched by small American rockets that were one to three years ahead of the American Atlas ICBM in their development. During that time there were launchings of seven Vanguards, six Jupiter Cs, three Thor Ables, and one Juno II (see fig. 2.5). The Soviets had only a single launch during that period, the 3000 pound *Sputnik 3,* carried by their huge ICBM. It weighed five times as much as all the seventeen American payloads combined.

But American space engineers were not simply trying to put small satellites in low Earth orbit: they had already suffered a humiliating defeat at that game. They tried, instead, to extend the performance of their small boosters to the limit by reaching for a new goal—the Moon. The purpose was clear: to beat the Soviets in a new race.

1958-1959. THE THIRD SPACE RACE: PROBES TO THE MOON

The first American lunar probes were tiny, extremely marginal payloads launched by two different types of boosters. The first, the Thor Able, was a derivative of the Air Force Thor IRBM, a missile with less than half the lift-off thrust of the Atlas ICBM (which had not yet flown), and about one-sixth the payload capacity of the Soviet ICBM. The second type of lunar probe vehicle thrown together by the United States consisted of an Army Jupiter IRBM with the solid-propellant upper stage stack transplanted from the

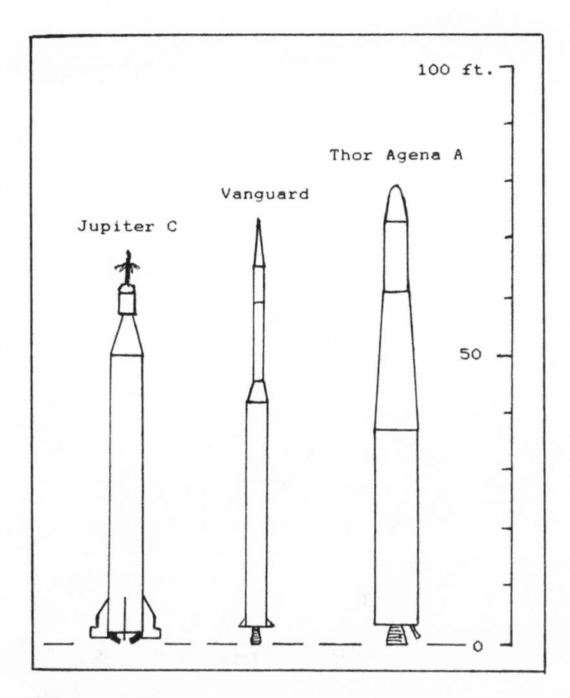

Figure 2.5
Early U. S. Satellite Launching Vehicles
The first U. S. satellite, *Explorer 1,* was launched by the Jupiter C (left), an Army Redstone Short Range Ballistic Missile (SRBM) with an upper stage stack of clustered small solid rockets, for a total of four stages. The *Vanguard,* developed by the Navy for use as a civilian launch vehicle, had three liquid-fueled stages. The first military test satellites, the Discoverer reentry experiments, were launched by the vehicle on the right, the Thor Agena A. The first stage was an Air Force Thor Intermediate Range Ballistic Missile (IRBM), and the second, the Agena A, was evolved from a design for a standoff bomb pod for the B-58 Hustler strategic bomber.

	Jupiter C	Vanguard	Thor Agena A
Height (ft)	71	75	79
Weight (lb)	64,000	22,600	117,000
Thrust (lb)	83,000	28,000	150,000
Payload (lb)	30	30	300
Year Introduced	1958	1958	1959

Jupiter C satellite launching vehicle (which did not, contrary to its name, involve a Jupiter IRBM, but was instead based on the much smaller Redstone SRBM). This was the Juno II.

The first attempt to reach the Moon was the launching of a Pioneer spacecraft by a Thor Able booster in August of 1958. The first-stage Thor rocket, presumably the most reliable component of the vehicle, exploded. The second attempt with the same booster resulted in the payload being thrown about a quarter of the way to the Moon. The third attempt was terminated by a failure in the third stage of the rocket.

Again the Army's Redstone Aresenal leaped to the rescue. As their *Explorer* satellite had saved America's honor from the disgrace of the very public failure of the Navy's Vanguard, so they hoped to recoup our fortunes after the ignominious failure of the Air Force's Thor and the confusion surrounding the startup of the fledgling National Aeronautics and Space Administration. NASA had been founded earlier in the year to provide a home for Eisenhower's civilian space effort.

Because the upper stage stack from the Jupiter C that was transplanted to the much larger Jupiter vehicle was grossly undersized for so ambitious an endeavor as a Moon probe, the payload was an astonishingly small 13 pounds. This payload flew under the NASA logo as *Pioneer 3.* Improbably, it retraced the course of *Pioneer 1* by soaring a quarter of the way to the Moon before falling back to Earth. With only a single Juno II booster and a single Pioneer probe remaining, NASA elected to wait a while and try to figure out a solution to its problems.

On January 1, 1959 a Soviet spokesman, looking ahead to the bounty of the new year, predicted that the first Soviet Moon probe would be launched in 1959. His prophetic credentials were validated a few hours later when the Soviet Union announced the launching of its *Luna 1* probe. This 800-pound payload passed the Moon at a range of about 3700 miles (the radius of the Moon is about 1000 miles) and entered heliocentric (Sun-centered) orbit, the first spacecraft to escape from Earth's gravity into interplanetary space.

In March, *Pioneer 4* was launched on the last Juno II Moon rocket to carry out the same mission. This time the booster and spacecraft worked about as well as could be expected. The tiny 13-pound payload passed 37,000 miles from the Moon and followed Luna 1 into orbit about the Sun. While the targeting of both left something to be desired, the truly important difference was that of size. Small American boosters had done their feeble best, and the era of the ICBMs was now dawning. But even the great size of *Luna 1* did not fully tax the huge Soviet booster. The race to the Moon, even when it became a race of ICBMs, would still be dominated by the USSR until much larger American boosters could be built.

From the Soviet viewpoint, there was much to be complacent about in 1959. They had beaten the United States to the first ICBM, the first Earth satellite, and the first lunar probe. But, to their credit, they did not rest on their laurels. Instead, they continued to exploit their great advantage in payload weight conferred by the superior size of their ICBM.

In September 1959 a second Soviet lunar probe, *Luna 2,* was launched. This spacecraft hit the Moon at high speed on September 14, 1959, to become the first vehicle from the Earth to reach another body in the Solar System. On October 4, on the second anniversary of the launch of *Sputnik 1, Luna 3* was dispatched on a trajectory that carried it out past the backside of the Moon, the portion permanently hidden from view on Earth. The spacecraft coasted out to 300,000 miles from Earth, and then swooped back toward our planet. The photographs, although very crude, did show some identifiable features that survive today on our lunar maps as the Sea of Moscow and the crater Tsiolkovsky.

The first American attempt to launch a lunar probe with its new ICBM was sent aloft in November, but the payload shroud, which is used to protect the payload from aerodynamic forces during the rocket's high-speed ascent through the atmosphere, tore loose. The vehicle, an Atlas topped by the upper-stage stack from the Thor Able lunar probe series, was destroyed. In a further debacle, the next satellite launch attempt of the Atlas, just three months later, failed to achieve orbit because the second stage did not separate. If the launch had been successful, it would have orbited a satellite that taxed the ultimate payload capacity of the Atlas with the conventionally fueled (kerosene/liquid oxygen) upper stages. That payload, a military satellite weighing 4500 pounds, was much larger than the 3000-pound *Sputnik 3,* the heaviest Soviet satellite launched to date (see fig. 2.6).

The failure of the Soviets to launch even heavier payloads with its huge booster was now a matter of comment among Western observers. With their ICBM, 2.5 times the size of the American Atlas ICBM, they ought to have been able to lift about 2.5 times as much payload, about five to seven

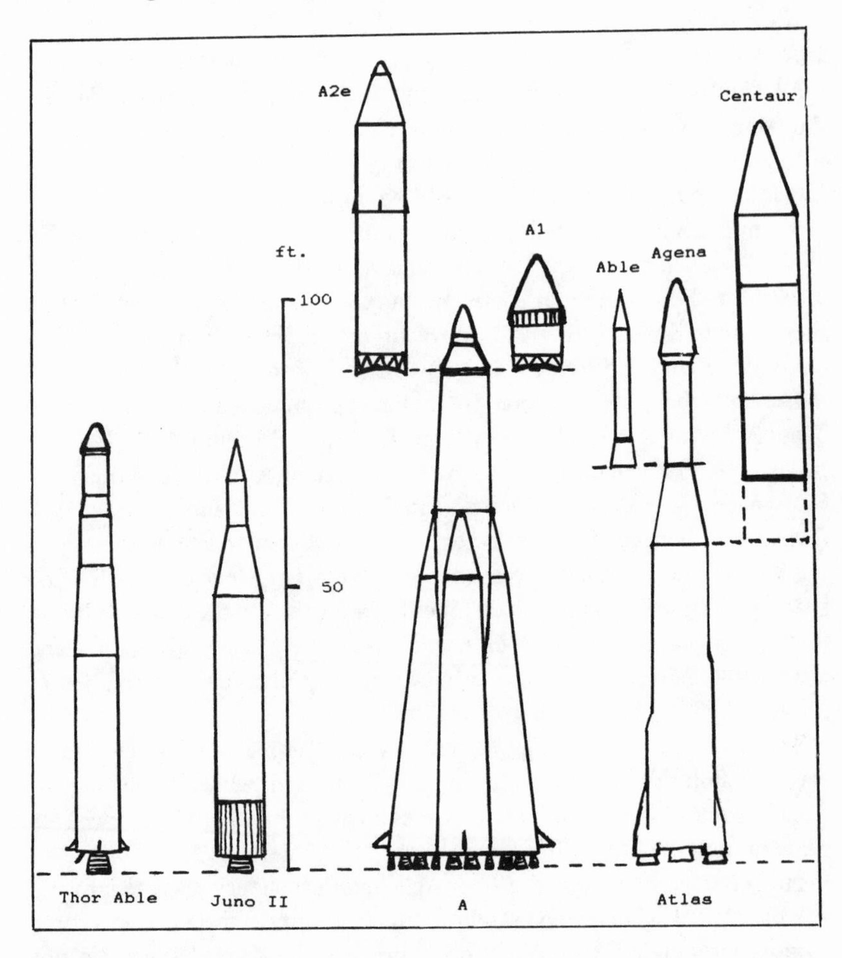

Figure 2.6
Early Lunar Launch Vehicles
The first lunar probe attempts were launched by the Thor Able, an Air Force Thor IRBM with the two upper stages of the Vanguard vehicle. The equally feeble *Juno II* used an Army Jupiter IRBM for its first stage and carried the upper stage stack from the Jupiter C vehicle. The first Soviet lunar probes were launched by the A1, a Soviet A-class 1½ stage ICBM with an added upper stage. When the US Atlas ICBM became available in 1959 it was first used in conjunction with the Vanguard/Thor Able upper stage stack (as the Atlas Able), then with an improved version of the Agena upper stage as the Atlas Agena B. Beginning in 1963, the USSR began launching its second-generation lunar probes with the A2e, using a larger second stage

tons. But they had not yet done so. It was reasonable to suppose that they would soon correct this deficiency, but what would they do with that huge payload capacity when they finally got around to using it?

1960-1968. THE RACE TO PUT MEN IN SPACE

On May 15, 1960, the second anniversary of the launching of *Sputnik 3,* the space race bore a rather strange complexion. The Soviet hare, the first into space, and brought there by huge boosters, had launched a grand total of three satellites into low Earth orbit and three lunar missions. The American tortoise, behind by a score of 2–0 at the outset, had by then launched 19 satellites into Earth orbit and four payloads into deep space, two of which entered orbits around the Sun. The American ICBM had arrived, and bigger boosters were already being planned: the second-generation Titan two-stage ICBM, and an exciting but still remote concept called the Saturn rocket, designed by Wernher Von Braun's team to deliver a huge 1.5 million pounds of first-stage thrust.

Nikita Khrushchev had shrewdly grasped the effect of space spectaculars on Western public opinion—and on the Soviet self-image. He pushed ahead

---

than the A1, supplemented with a smaller "escape stage." By 1966 the high-energy (hydrogen-oxygen) Centaur upper stage was operational for use on the Atlas Centaur. The Atlas ICBM was specially modified for use with the Centaur, strengthened to take the weight of the new upper stage and its payload and widened at the top to accommodate the full width of the Centaur. This vehicle, with only 42 percent of the takeoff thrust of the A2e, could land *three times* as much payload on the Moon.

Booster	Weight (lb)	Thrust (lb)	Lunar Payload Wt. (lb)
Thor Able	114,000	150,000	87
Juno II	122,000	150,000	13
A1	615,000	900,000	860
A2e	674,000	900,000	3600 (220)*
Atlas Able	260,000	389,000	388
Atlas Agena B	275,000	389,000	810
Atlas Centaur	329,000	431,000	2290 (640)*

*Weight that could be soft-landed on the Moon.

with plans to milk every possible advantage from the Soviet lead in booster size and payload capacity. Foremost among his plans was the launching of Soviet cosmonauts into orbit about the Earth.

Speculation about the next Soviet space initiative was fueled by rumors in April 1959 that there might be a Soviet man in space within two years. General plans for Soviet military surveillance satellites and heavy manned spacecraft were also outlined. The Soviets had pointed out the feasibility of launching instrumented probes to Mars and Venus, and carried out tests to improve their ability to launch large payloads. When, on May 15, 1960, the Soviet Union launched the five-ton *Sputnik Korabl 1* (satellite spaceship) vehicle into low Earth orbit, with a dummy cosmonaut, speculation naturally turned toward manned activities. But the attempted recovery of the cabin failed when the retrorockets fired in the wrong direction, lifting the vehicle into a higher orbit.

In August, the race to recover a satellite from orbit ended in a photo finish. The American Discoverer series, launched by Thor Agena boosters, was directed toward development of military space technology, including the ability to recover reconaissance satellite film capsules after return from orbit. The *Discoverer 13* capsule was successfully recovered after it was parachuted into the Pacific Ocean, and just eight days later the *Discoverer 14* capsule and its parachute were snatched in mid-air by a specially equipped recovery aircraft as they fell over the Pacific.

Meanwhile, the second Sputnik Korabl, which was sent aloft on the very day that *Discoverer 14* was recovered, carried two dogs inside the recoverable reentry capsule. The capsule was recovered successfully after one day in space. The third Sputnik Korabl spacecraft was orbited with a dog aboard in December 1960. The recovery attempt failed, and the canine passenger was lost. Nonetheless, it was clear that the USSR was now flying a recoverable satellite with a life-support system and sufficient space for a human passenger.

Reaction to the recovery of the capsule from *Sputnik Korabl 4* in March of 1961 was generally very enthusiastic, despite the widely quoted opinion of the new Astronomer Royal of England that manned interplanetary travel was "utter bilge." In August, both V. V. Blagonravov and Leonid Sedov of the Soviet Academy of Science were quoted as saying that men would soon be in space.

In late March of 1961, just sixteen days later, one more five-ton Sputnik Korabl spacecraft was placed in orbit and was recovered successfully. An unmanned test of the tiny one-ton American Mercury capsule, to be boosted into space by an Atlas ICBM without any upper stages, was scheduled for late April, and manned suborbital flights in Mercury capsules atop Redstone SRBM boosters were to begin in May. Clearly the USSR had to act quickly in order to have the first man in space.

Thus it was that on April 12, 1961, the Soviet Union placed a manned Sputnik Korabl-type spacecraft called *Vostok 1* into orbit, carrying a pilot named Yuri Gagarin. (*Vostok,* meaning "East" in Russian, was a clear reminder of East-West rivalry.) After nearly a full orbit about the Earth, Gagarin was returned safely to a landing in the Soviet Union. Distressingly, the Mercury Atlas unmanned flight two weeks later failed to achieve orbit. The Soviets had won another major race.

Alan Shepard's suborbital flight in May and Gus Grissom's similar flight in July were scorned by the Soviet Union as puny feats, and in August *Vostok 2,* piloted by Gherman Titov, completed 17 orbits of the Earth before landing safely. A month later an unmanned Mercury capsule was successfully orbited and recovered, and in November the Mercury Atlas 5 vehicle carried a chimpanzee named Enos for two circuits of the Earth. In February of 1962 John Glenn became the first American astronaut in orbit, to be followed into space by Scott Carpenter in May and Wally Schirra in October.

On August 11, 1962 the Soviet *Vostok 3* spacecraft entered orbit carrying cosmonaut Andrian Nikolayev. A day later, *Vostok 4,* carrying Pavel Popovich, joined *Vostok 3* in orbit. The two spacecraft passed within several kilometers of each other, but no attempt was made to bring them together to rendezvous and dock. Both spacecraft were recovered after they orbited together for three days.

The last flight of a one-man American spacecraft, Gordon Cooper's *Mercury 9,* was in May 1963. The next planned American spaceflight was to be *Gemini 3* in March 1965. A month after Cooper's flight the Soviets orbited *Vostok 5* and *Vostok 6* two days apart, piloted by Valerii Bykovskii and the first woman to fly in space, Valentina Tereshkova. Again, the two vehicles came no closer than a few kilometers.

A modified six-ton Vostok capsule, with seating for up to three cosmonauts, was under development in competition with the American Gemini spacecraft. By leaving off most of the life-support supplies, the extra seats could be accommodated. Of course, lengthy missions were out of the question without air, water, and food. Khurshchev placed so high a priority on space spectaculars that he felt it essential to fly a multimanned spacecraft before Gemini. This turned out to be possible only if the flight were very short, and if the launch escape rocket (which would lift the cosmonauts and their capsule to safety in the event of a failure of the launch vehicle) were omitted! One might say that this mission, dubbed *Voskhod,* (Russian for "sunrise") was a degraded version of *Vostok.* This newly modified capsule was tested on the unmanned *Kosmos 47* flight, which stayed in space only one day.

Kosmos 47 was just one example of the use of the catchall Kosmos program name to obscure the identity and purpose of spacecraft. Virtually every kind of spacecraft has been launched under cover of the Kosmos name, complete with terse and uninformative launch announcements. Failures and tests in high-profile programs such as lunar and planetary probes and man-capable spacecraft are invariably called Kosmos, as are almost all military missions.

In October 1964, several months before the announced beginning of the Gemini series, and less than a week after the *Kosmos 47* flight, the Soviet Union orbited *Voskhod 1* with a crew of three. The pilot, Vladimir Komarov, and the flight engineer, Boris Yegorov, were accompanied by a physician, Dr. Konstantine Feoktistov. The following March, only five days before the expected launching of the first manned Gemini spacecraft, *Voskhod 2* was launched with a crew of two, Pavel Belyayev and Alexei Leonov, on a mission that featured the first space walk. Each flight (there were only two using the Voskhod spacecraft) lasted only a single day. When Khrushchev was dismissed by the Politbureau, his successors, Leonid Brezhnev and Aleksei Kosygin, discontinued the Voskhod program because of safety considerations. Work was instead focused on the development of a more advanced spacecraft for multimanned flights. (Figure 2.7 depicts some early manned launch vehicles.)

The advent of Brezhnev and Kosygin caused a profound rethinking of the means and goals of the Soviet space program. The emphasis shifted from opportunistic space spectaculars to a much smoother, steadily paced, and systematic long-term program of development of new capabilities. There is convincing evidence that the greatest of Khrushchev's spectaculars was well under way at the time that he was deposed, and that its very existence was disconcerting to the new leadership. We shall return to that program in chapter 3.

During the same time period, the brilliant "chief designer" of the Soviet space program, Sergei P. Korolev, died. At least as serious as the loss of his broad technical mastery was the loss of his ardent and capable political bargaining on behalf of an energetic, exciting Soviet space program.

From March 1965 until April 1967 there was not a single Soviet manned space flight. During that time nine two-man Gemini spacecraft were flown, putting the United States far ahead of the Soviet Union in the number of man-days of flight experience. *Geminis* 6 and 7 flew a joint mission in which they carried out a rendezvous and flew as close as one foot apart. *Gemini 7* stayed in orbit for 14 days, breaking *Gemini 5*'s record of 8 days. Rendezvous and docking techniques were practiced extensively, looking forward to the essential role of orbital rendezvous in the Apollo program.

When the Soviets again appeared in space on April 23, 1967, it was with a one-man flight of a new 6.6-ton spacecraft, greatly improved from Vostok and Voskhod. After successful unmanned two-day tests under the cover of the *Kosmos 133, 140,* and *154* missions, Vladimir Komarov, the veteran commander of *Voskhod 1,* went aloft as the pilot of the new *Soyuz 1* spacecraft. The long gap in Soviet manned flights was taken by Western observers to mean that they were preparing a more capable spacecraft with the ability to carry out rendezvous. Indeed, when Soyuz ("Union" in Russian) appeared, the name seemed a clear confirmation of these ideas.

Quite unexpectedly, Komarov dropped the spacecraft out of orbit after only a single day in space and reentered the atmosphere over the central USSR. Unfortunately, the shroud lines of his parachute became tangled, and the spacecraft smashed into the ground at high speed. Komarov was killed, and the Soyuz program was set back for 18 months as the design team attempted to identify and solve the problems responsible for the tragedy. The loss of that precious time was of crucial importance: by the time Soyuz was again ready to fly, the first manned Apollo had already flown an 11-day mission in Earth orbit.

The eventual revival of Soyuz was less than propitious. *Soyuz 2* and *3* were orbited on consecutive days in October 1968. The two spacecraft unsuccessfully attempted to rendezvous and dock, and both spacecraft returned to Earth after four days in space. But by then an entirely different race for even higher stakes was in full swing, and mere Earth-orbital missions held little interest. That race had two parts; the race to build superboosters, and the race to place men on the Moon.

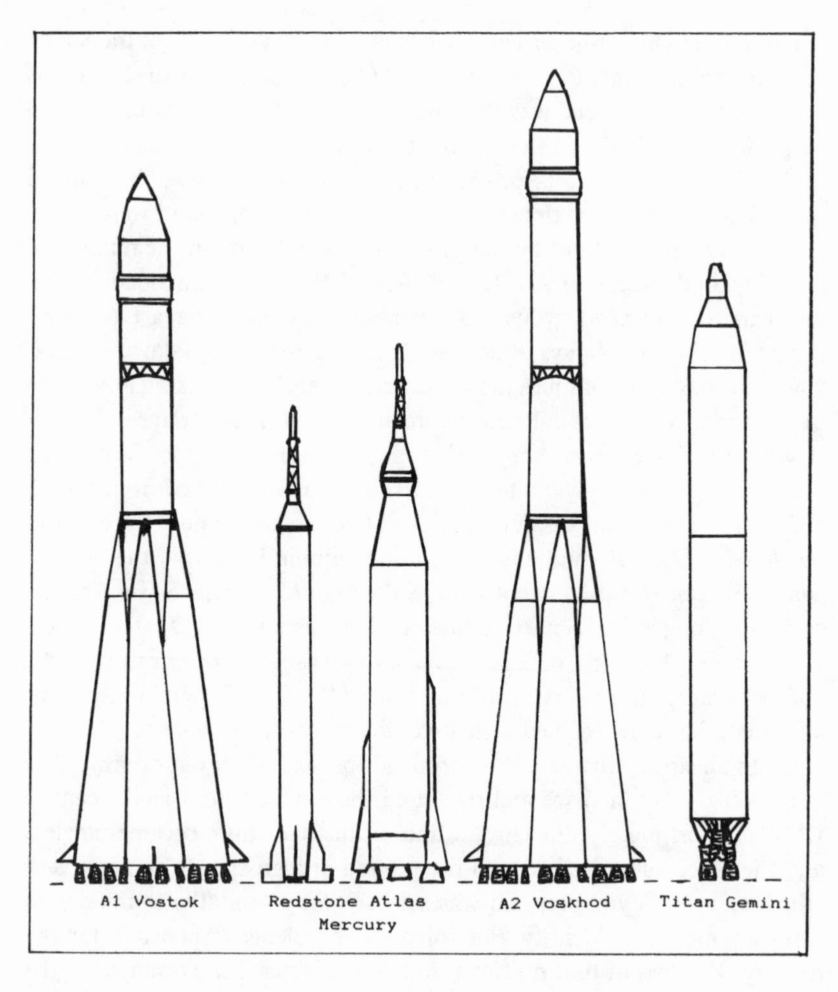

Figure 2.7
Early Manned Launch Vehicles
The first manned spacecraft, Vostok, was lifted by the A1 booster, derived from the Soviet A-class first-generation ICBM by addition of an orbit-insertion stage. The suborbital test flights in the American Mercury program were made on the tiny Mercury-Redstone rocket, and the first American manned orbital flights were made on the Mercury-Atlas, a vehicle far smaller than the Soviet A1. The two Soviet Voskhod multimanned missions were lifted by the improved A2 booster with a much more powerful second stage. The American Gemini missions were lifted by the Titan 2

1962-1968. THE SUPERBOOSTER RACE: PRELUDE TO THE MOON

At the time of President Kennedy's 1961 announcement of the Apollo program it was clear that at least two new boosters, much larger than American ICBMs, would be required to carry out the mission of manned flight to the Moon. The first of these, the Saturn 1, would be used for Apollo flights in Earth orbit. The larger, the Saturn 5 superbooster, would be used for manned flights to the Moon. In order to maximize the payloads of these giant rockets, they would be equipped with upper stages fueled by the ultimate chemical propellant combination, liquid hydrogen and liquid oxygen. Thus it was essential that NASA develop experience with the operational use of hydrogen-oxygen engines well in advance of Apollo so that they would be safe enough for manned use in time for the flights to the Moon.

The test bed for hydrogen-oxygen (H-O) propulsion was the Centaur rocket, and the two giant Saturn rocket boosters were to be developed for the exclusive use of the civilian space program.

The Department of Defense carried out a parallel stepwise development program, beginning with the Titan ICBM, to develop a Saturn 1-class military booster, the Titan 3. Most Titan 3 flights have been launchings of large military payloads, although a few civil missions, including the *Voyager* probes of the outer planets, were also launched by Titan 3 boosters. The Titan 3 family, extremely diverse and complex, is generally based on a two-stage Titan 3 core, a specially strengthened Titan 2 ICBM. Two large solid-

---

ICBM. The launch vehicle and spacecraft data for these programs are as follows:

Name	Booster	Weight (lb)	Thrust (lb)	Payload in Orbit (lb)
Vostok	A1	633,000	900,000	10,430
Mercury	Redstone*	66,000	78,000	*
Mercury	Atlas D	260,000	367,000	3,030
Voskhod**	A2	681,000	900,000	12,500
Gemini**	Titan 2	408,000	430,000	8,370

*Suborbital flights only
**Second-generation spacecraft

propellant strap ons of various sizes, distinguished by the letters C through E, assist the first stage at blastoff. Upper stages such as the Agena or Centaur are frequently carried. The resultant names, though highly descriptive, may sound cumbersome: Titan 3B Agena D, Titan 3E Centaur, etc. The maximum performance versions of the Titan require, like the Atlas, the use of the H-O Centaur stage.

In November 1963 an Atlas Centaur two-stage vehicle sucessfully orbited a test payload of 10,700 pounds (4700 kilograms), a hair heavier than the largest Soviet spacecraft ever launched. The potency of the Centaur stage becomes clear when we realize that the Atlas, which by then had 400,000 pounds (180,000 kg) of liftoff thrust, was competing with a Soviet booster, the A2 space launcher, with about 900,000 pounds (410,000 kg) of thrust. Thus the Centaur gave the Atlas the same lifting power as the 2.5 times larger Soviet launcher! The obvious Soviet response would have been to flight-qualify a hydrogen-oxygen second stage for their standard (A) booster, thus giving it a payload capacity of up to 22,000 pounds (10,000 kg). As of late 1986, this has still not happened.

The rest of the development of the Centaur caused more than the usual number of ulcers and heart attacks. The second Centaur flight test in June 1964 failed to achieve orbit. The next test, in December, was a success, but the fourth again failed to achieve orbit. There was great concern that the H-O engine would not be reliable enough to keep Apollo on schedule.

Centaur's world-record payload of 10,700 pounds (4900 kg) did not stand for long. Two months later came the first launch of the Saturn 1, a giant with eight first-stage engines delivering a total of 1.5 million pounds of thrust. The very first Saturn launch put 37,700 pounds (17,000 kg) into orbit, more than three times the old record. This feat was duplicated the following May and again in September. The first satellites launched by the Saturn 1, the 23,000-pound (14,500-kg) *Pegasus 1, 2,* and *3* micrometeoroid detection satellites launched in February, May, and July of 1965, continued the unbroken string of Saturn successes.

When the Saturn 1's hydrogen-oxygen second stage (the S4) was ready for flight on the unmanned *Apollo 1* mission in February 1966, many observers crossed their fingers. But it flew flawlessly, as did the *Apollo 2* unmanned orbital test in July. That flight injected 58,500 pounds (26,600 kg) into orbit, another record. Apollo and the Saturn 1 were now nearly ready for their first manned mission in Earth orbit.

On January 27, 1967, during a routine ground rehearsal of the impending manned flight, astronauts Gus Grissom, Ed White, and Roger Chafee were caught by a flash fire in their Apollo capsule. All three were killed within seconds. The gutted Apollo was subjected to a most searching investigation to determine the cause of the fire.

Astonishingly, the investigation uncovered so many irregularities and design defects that the precise cause was never identified. However, the two main contributing factors were clearly evident. First, the ground test was conducted in a pure oxygen atmosphere with a pressure of about 16 pounds per square inch (1.06 atmospheres). Such a high oxygen pressure presents an extreme fire hazard: even piano wire would burn in it. Second, a large amount of flammable material had crept into the capsule design; electrical insulation made of organic polymers, plastic netting, etc. The combination was ripe for disaster. Apparently an electrical short shed a few sparks that fell on some flammable material. Within seconds, the spacecraft was a flaming inferno, filled with hot, poisonous gases from partially burned plastics. Finally, quite independent of the fire hazard problem, the escape hatch in the capsule was designed to open inward. In the brief time available, and in the tight quarters of the capsule, the astronauts were unable to open the hatch to escape from the fire. An extensive redesign of the Apollo command module was begun at once, and all manned flights were deferred for more than 20 months.

The next Saturn test, the maiden flight of the giant Saturn 5, was the unmanned *Apollo 4* flight of November 1967. The Saturn 5 must be seen to be believed. Since American designers were reluctant to cluster more engines than the eight in the first stage of the Saturn 1, a huge new single-chamber engine was developed to serve as the building block for the Saturn 5. The new engine, the F-1, delivered a thrust of 1.5 million pounds (675,000 kg), as much as the entire Saturn 1 first stage. The Saturn 5 first stage (the S-1c) used a cluster of five of these giant engines, delivering a first-stage thrust of 7.5 million pounds.

Because of the extremely high reliability exhibited by the Saturn 1, it was decided to go ahead and fly the upper stages along with the new Saturn 5 first stage, the S-1c. Both upper stages were based on a new H-O engine, the J-2, with 200,000 pounds (90,000 kg) of thrust. The second stage (the S-2) carried five of these engines, and the third (S-4b) stage carried one. The first flight was a complete success: it placed a total of 278,700 pounds

(126,500 kg) in low Earth orbit. This was more than seven times the heaviest weight lifted by the Saturn 1. The second Saturn 5 flight was marred by some problems with restarting the engines in the S-4b stage, but the vehicle was considered reliable enough to be man-rated on its third flight. This judgment has since been proved sound: the Saturn 5 performed its mission sucessfully in every subsequent flight.

The Soviet Union had opened the space age with a launcher that, like the Atlas ICBM, was a "one-and-a-half stage rocket." It had a long central core carrying a cluster of four engines, which was surrounded by four large liquid-propellant strap ons. Each of these four modules also carried a four-engine cluster like that on the core stage. Thus there were twenty main first-stage engines with about 44,000 pounds (20 metric tons; 20,000 kg) of thrust each, all burning kerosene and liquid oxygen. The basic engine was about the size of the Vanguard first stage engine. The American Atlas, by comparison, used a central core engine with 60,000 pounds of thrust, and carried a "skirt" with two 150,000-pound thrust engines that was, like the Soviet strap ons, jettisoned not long after launch. The version of the SS-6 ICBM that launched *Sputnik 1, 2,* and *3* was just the 1 1/2 stage version described here, with a payload capacity of about 1.5 tons to low Earth orbit. This is called the A launcher by most Western observers. We should note that this is not the Soviet name for the missile. The A, A1, A2, A2e nomenclature was introduced by an American observer, Charles S. Sheldon II, in the late 1960s. In general, such coined designators must be used because of the deep shroud of secrecy behind which the USSR hides its military launch vehicles. As we found out years later, the Soviet designation of the A booster is the R-7 "Semyorka": this name appears in Khrushchev's memoirs, but never in the Soviet press.

A second stage was then added to the A booster to upgrade its performance. This version, the A1, was used in the early Luna flights and the Sputnik Korabl and Vostok programs. It has a payload capacity of about five tons to low Earth orbit (LEO). With an improved second stage, the A2 version, this same booster was used in the Voskhod and Soyuz programs to orbit payloads of up to 7 tons. Every Soviet manned flight to date has flown on A1 or A2 boosters. A-class boosters are launched from both the Tyuratam and Plesetsk sites (see map), and all manned flights are launched from Tyuratam.

For deep-space missions, those destined for the Moon or for escape from Earth into orbits about the Sun, a third "escape" stage was required. All early Soviet planetary missions used the A2e launcher, with the A core and strap-ons, the advanced second stage, and the added escape stage.

By the end of 1985, over 1100 A-class boosters had been launched. It is, to this date, the workhorse of the Soviet space program. It is good to bear in mind that the heart of this launcher, the A stage, was designed in 1952. Never too much of a good thing!

As manned space activities became more ambitious, and the prospects of manned lunar missions and space stations became more accessible, the Soviets identified a need for a much larger rocket. Also, curiously, they found that their basic A booster was too big for many jobs that needed to be done. Accordingly, they stepped down to two smaller launchers and simultaneously moved to develop a giant booster.

Soviet space launch vehicles are, with only one exception, derived directly from military strategic missiles. We have mentioned that the original *Sputnik 1* "A-class" booster of 1957, which is still in service and being flown at the rate of about 58 per year, is a modified first-generation ICBM.

The Soviet obsession with secrecy is ubiquitous. At the first Strategic Arms Limitation Talks (SALT I) the meeting began with a thorough review of Soviet and American strategic weapons by a member of the American delegation. After the session a Soviet general took the American speaker aside and angrily informed him that he had no business speaking about Soviet weapons systems before representatives of the Soviet Foreign Ministry: they were, after all, only civilians, and the Americans should have the courtesy to avoid leaking such information to them!

Khrushchev also forthrightly stated that Soviet-American cooperation in space was impossible during the early 1960s because they were afraid the Americans would find out about the design of the Semyorka: "We knew that if we let them have a look at our rocket, they'd easily be able to copy it. Then, with their mighty industry and superb technology, they'd be able to start producing replicas of our booster and soon have more than we had. That would have been a threat to our security. In addition to being able to copy our rocket, they would have learned its limitations; and, from a military standpoint, it *did* have serious limitations. In short, by showing the Americans our Semyorka, we would have been both giving away our strength and revealing our weakness."

This attitude, utterly incredible to Americans, is based on the Soviet tradition of "borrowing" useful designs. The Tupolev-4 bomber was a direct copy of the American B-29, which arrived in the Soviet Union through Lend-Lease, too late to be used in World War II. A few years after Powers' U-2 was shot down, a carbon copy of it, affectionately called the "U-ski 2-ski" by American observers, appeared. The history of American "contributions" to the Soviet atomic bomb program is well known. The astonishing similarity of the Tu-144 supersonic transport to the Concorde is also clear: the main difference seems to be that the Tu-144 has never been in regularly scheduled service, and never will be. And now the Soviets are building a Space Shuttle of their own . . . .

The R-7, alias A, was the sole Soviet launch vehicle for several years. In 1962 and 1964, two new but smaller satellite launchers, called the B and C vehicles, were introduced to launch payloads too small for the A booster. Both of these were modified military missiles smaller than the big A-class ICBM, more comparable to the early American boosters such as the Thor and Jupiter.

The largest operational Soviet booster, the D1 or "Proton booster" (after the name of the first satellite it launched) made its debut in 1965. This heavy-lift vehicle is capable of putting 49,000 pounds (22,000 kg) in low Earth orbit. It has also been the mainstay of the Salyut space station and the launcher of the Star expansion modules for Salyut. With added upper stages, as the D1e, it has carried every payload in the lunar and planetary research program since 1972, and is used to place large payloads in geosynchronous orbit. The D1 vehicle has averaged 10 launches per year since 1981. All D-class launches are from the Tyuratam complex.

The D booster is unusual in that it is the only Soviet launch vehicle not derived directly from a strategic weapons system. Nonetheless, photographs and data on the booster are treated as very sensitive information. It is known that the first stage delivers a thrust of about 1.4 million kg (3.1 million pounds), intermediate between the Saturn 1 and the Saturn 5. This stage is structured with a central core and six liquid-propellant strap-ons, analogous to the A vehicle with its core and four strap-ons. The second stage thrust is about 60,000 kg (132,000 pounds), and the third stage, when used, has 15,000 kg (33,000 lb) thurst. Its payload capacity, however, is markedly inferior to both Saturn boosters because of the Soviet failure to develop H-O engine technology. It is also noteworthy that, 20 years after its introdution, the D1 has never been man-rated.

The latest operational Soviet booster, the F1 (based on the SS-9 silo-based ICBM), was introduced into service in 1966. It is smaller than the A-class booster but more efficient, and can lift a payload weight of about 5 tons. Through 1981 the vehicle was used exclusively at Tyuratam, but in 1982 an F1 launch facility was opened at Plesetsk. In recent years a modestly upgraded version, the F2, with a payload capacity of about 6 tons, has been used a number of times. F1 launch rates are highly variable, but the average over the 1980s has been about 10 per year. The F booster has been heavily involved with weapons testing in space. The first two satellite launchings by this booster in 1966 were never publicly admitted by the USSR, although they were tracked and announced by Western sources.

The early use of the F booster in the 1966–1970 time period was devoted to development of the capability to alter the orbit of a satellite by maneuvering it after it was in orbit. The first use of this capability was to orbit apparently innocuous satellites under the Kosmos name and to recall them (bearing simulated ICBM warheads) onto targets on the ground. This "storage" of warheads in space was practiced no less than 18 times.

Kosmos 185 and *198* in *1967,* followed by *209* and *217* the next spring, practiced orbital maneuvering without reentering the atmosphere, as if carrying out interceptions with imaginary targets. *Kosmos 217* maneuvered and then blew up. There then ensued two lengthy series of tests of a satellite interceptor, running from 1968 through 1971 and 1976 through 1982. Since the American announcement in 1983 of impending tests of a small air-launched anti-satellite missile (ASAT), and especially since its successful first test in the fall of 1985, the official Soviet position has been that ASATs are immoral. Nonetheless, no American space-based weapon has ever been developed. Two major Soviet space-based weapons systems have been fully developed to operational status, both using the F-class booster.

James Oberg has claimed in his book *The New Race for Space,* that the A-class booster is being phased out in favor of the F2. There is no clear evidence that this is true. Instead, it appears that the F2 carries mostly highly sensitive military payloads, and has to some degree taken up the slack caused by the retirement of the B1 and the slowing of C1 launch rates. The F2 has carried some payloads of the class previously assigned to the A2, but these appear to be a supplement to continuing A2 launches. The A2 launch rate remains a steady 55 to 60 per year, and only a few F2 vehicles are launched each year.

We can analyze the Soviet Union launch vehicles in very simple terms. The A booster was and is the workhorse of the fleet, with an enormous reservoir of experience. The B booster was a temporary injection of even older technology to fill a need for a small booster, and the C vehicle was intended to fill in the gap between the B and A. The diminishing launch rates of the C booster show that it is playing a lesser role each year. The F booster is a slightly less powerful but more efficient launcher than the A class, but is closely tied to secret military programs. The largest Soviet booster, the D1, with a payload capacity of 22 tons, represents approximately 1960 technology. No larger Soviet booster, and no Soviet hydrogen-oxygen stage, has ever flown successfully. The payload capacities of the large Soviet boosters and their American counterparts are shown in figure 2.8. The heavy-lift boosters are described in figure 2.9.

Nonetheless, a substantial (and often rather speculative) literature exists on a superbooster that was under development and testing in the Soviet Union for a prolonged period. Most of the available information on this G class booster has been released only in small and disconnected morsels by Soviet sources, and in large and often vague chunks by the American intelligence community. Nonetheless, the existence, general properties, and major purpose of the G booster now appear reasonable well understood.

Launch site construction at Tyuratam for the G1 was observed by American reconnaissance satellites commencing in 1964. A plausible conjecture is that in order for construction to have begun at that time, the design studies for the booster could have begun no later than 1962—shortly after the announcement of the Apollo program, and certainly during the Khrushchev era.

By August of 1966 the G1 launch facilities were nearly complete (at least insofar as orbital reconaissance could determine), and ground tests of the booster were under way. All signs of construction activity at the two launch pads ceased in about May 1967, and the pad then sat unoccupied for a year. There is some evidence suggesting that a flight of the G1 was scheduled for April 1968, but was canceled because of problems encountered in static tests of booster components.

Another year passed. In April 1969 the G1 again appeared on the launch pad. On July 4, probably in an attempted launching designed to take the edge off American Independence Day festivities, the first G1 vehicle blew up on launch pad number two, destroying both the rocket and the pad. Later that year there was a reported fueling accident on pad number one involving the second stage of the G1, which, according to some sources, was the first Soviet hydrogen-oxygen rocket. Others assert that all three stages of the booster were fueled with kerosene and liquid oxygen.

During 1970 more static tests were conducted and the number one

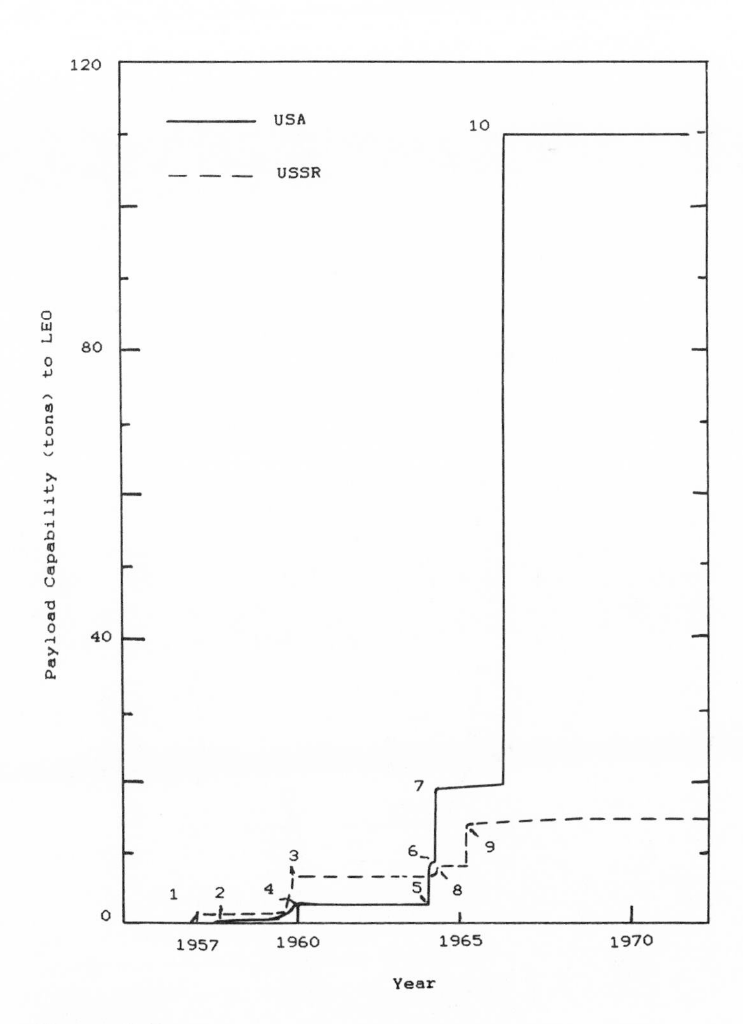

Figure 2.8
Payload Capacity of US and Soviet Boosters 1957–1971
The payload capacities of the largest US and Soviet boosters to Low Earth Orbit (LEO) are shown for the period 1957 to 1971. The launch of the first Sputniks by the A booster is marked with the number 1, 2 is the introduction of the early US Jupiter C, Vanguard, and Thor Agena A vehicles, 3 is the first use of the A1, 4 is the appearance of the Atlas Able and Atlas Agena, and 5 marks the introduction of the Atlas Centaur, which for the first time put the US ahead in payload capability. The first flights of the Saturn 1 and Saturn 5 are labeled 7 and 10, while 8 marks the A2 Voskhhod launcher, and 9 is the introduction of the D1 heavy-lift vehicle. The ill-fated G1 superbooster, which was being developed for manned lunar landing missions in competition with the Saturn 5, never flew successfully.

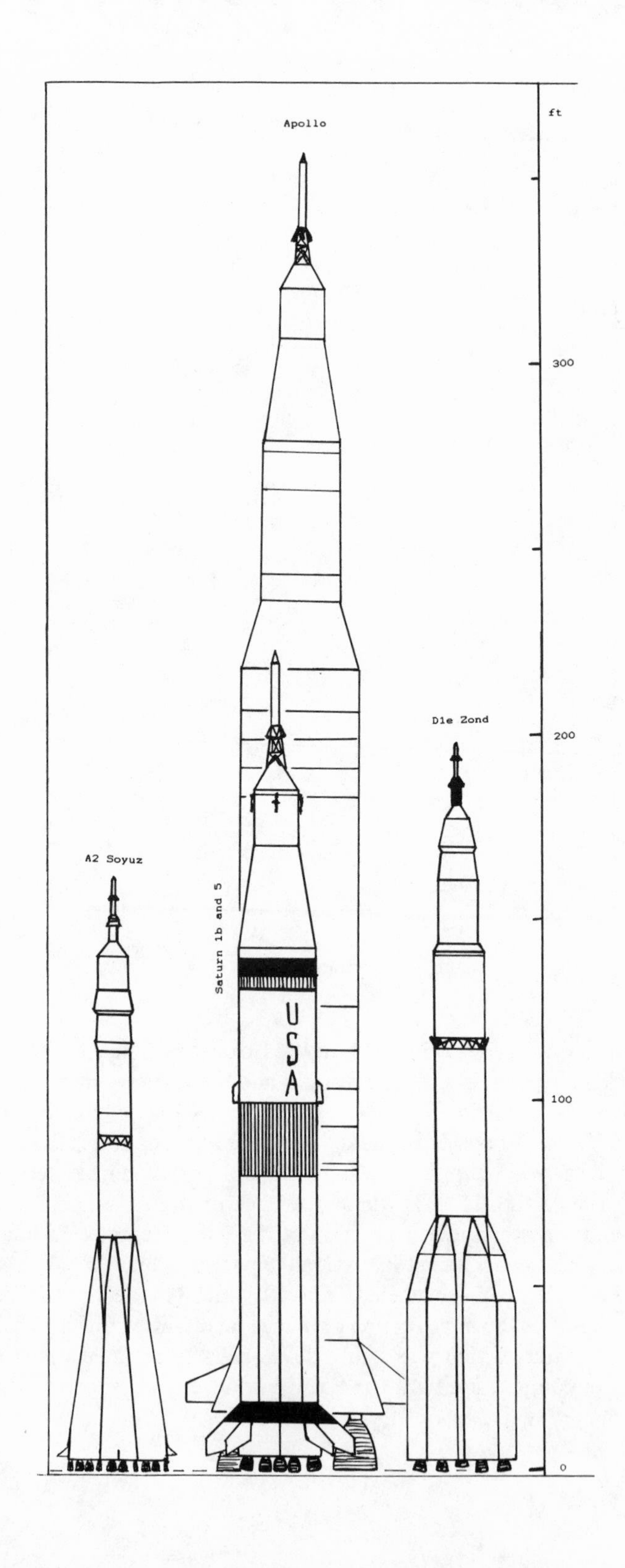

ft
Apollo
300
200
D1e Zond
A2 Soyuz
Saturn 1b and 5
USA
100
0

launch pad was refurbished. In early 1971 the G1 was again seen on its launch pad, and a launch attempt was made on June 24 or 25. The first stage broke up and exploded at an altitude of about 12 kilometers (7.5 miles).

Another year went by before the G1 again appeared on the launch pad. On November 24, 1972 the third launch attempt ended with a second-stage failure at an altitude of about 40 km (25 miles). No further visible activity occured in the G1 program until 1975, when the number two launch pad,

Figure 2.9
Heavy-Lift Boosters and Superboosters
The A2 Soyuz launch vehicle, little changed since the Voskhod launcher of 1964, is still the workhorse of the Soviet space program in 1986. The boosters developed for the Apollo Earth-orbiting program (Saturn 1) and the lunar landing program (Saturn 5) dwarfed the size and performance of the Soviet A2. The booster developed for the Soviet space station and lunar flyby (Zond) program, the D1, is pictured here in its configuration for manned lunar flyby missions, the D1e. The G1 superbooster, under prolonged development for the Soviet manned lunar landing program, was probably similar in overall appearance and design philosophy to the D1, but was larger than the Saturn 5. The G1 never flew successfully, and was scrapped in response to the success of the American Apollo lunar landing program. No pictures of the G1 are available in the West, but various "artist's conceptions" have been published. They do not seem credible enough to reproduce here. Data on these boosters are as follows:

Program	Booster	Weight (000 lb)	Thrust (000 lb)	Weight in LEO (000 lb)
Soyuz	A2	690	900	14
Apollo	Saturn 1b	1100	1500	44
	Saturn 5	5000	7500	220 (108)*
Zond	D1e	2370	3180	42 (14)*
(lander)	G1e	8000	11000	220 (70)*

*Weights launched to the Moon. The figure for the G1e is conjectural, based on the assumption that high-energy upper stages were not used. The weights than can be *landed* on the Moon are much smaller, and depend on the configuration of the vehicle and the mission profile.

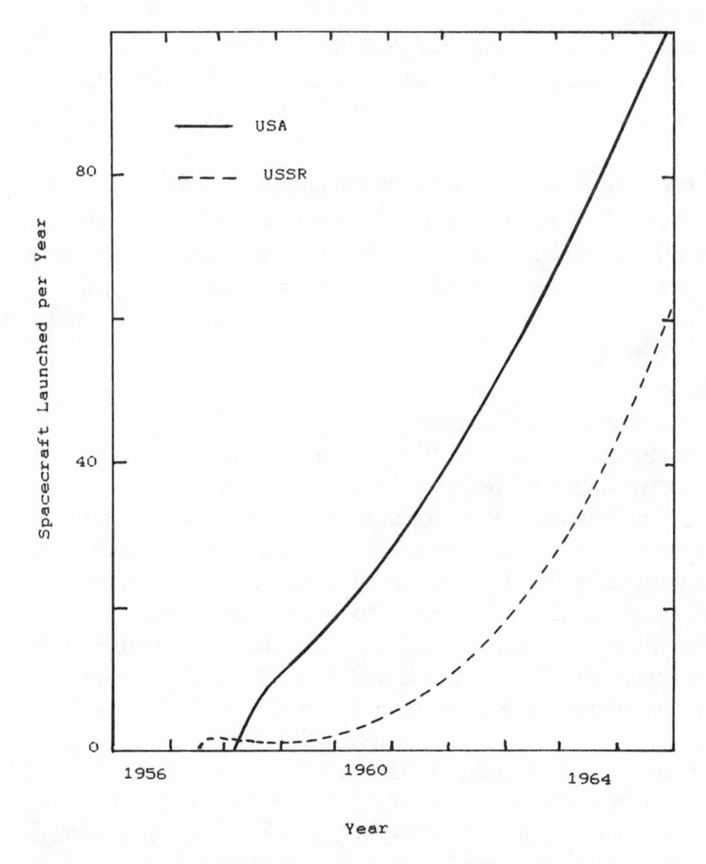

Figure 2.10
Satellite Launch Rates, 1957–1965
The Soviet Union, with its huge A-class booster, launched very few satellites until the introduction of the small B-class booster in 1961 and the midsized C1 in 1964. 1962 also marked the appearance of what was soon to become the largest category of spacecraft in the Soviet space program, recoverable photo-reconnaissance satellites based on Vostok technology. This program was initiated with the launch of *Kosmos 4.* In every year from 1958 through 1964 the USA launched two to four times as many satellites as the USSR.

which had been badly damaged in the 1969 explosion, was rebuilt. However, no G vehicle has ever reappeared on either pad since the 1972 flight failure.

The reasons for the development of the G booster seem intimately connected with Soviet plans for manned travel to the Moon, and only loosely connected with planned activities in low Earth orbit. The huge G1 booster, with an estimated liftoff mass of 10 million pounds (4.5 million kg) and a liftoff thrust of 13 million pounds (6 million kg), had an estimated payload capacity of 100 tons (kerosene-oxygen fuel) to 200 tons (with hydrogen-oxygen upper stages). It was therefore at least comparable with the Saturn 5's useful payload capacity of about 110 tons to low Earth orbit.

The failure of the G1 program spelled the doom of Soviet ambitions to land men on the Moon. But Soviet cosmonauts had other plans in mind for the Moon, and fully expected to beat the American Apollo program. That facinating but little known story is examined in the next chapter.

3

THE SPACE RACES SINCE 1968

For a man to go to the Moon and back is a pinnacle of scientific development. Painful as it is for me to admit, I can't deny that the Americans are now ahead of us in space travel.—Nikita S. Khrushchev

At the time of President Kennedy's announcement of the Apollo program, the Soviet track record in the races that had so far occurred in space was remarkable. The first space race, to build an Intercontinental Ballistic Missile, was won by the Soviets because of a combination of Soviet constancy and American antiwar sentiment. The American missile effort slept while the Soviet program progressed step by step toward its goal. Because of the early American development of the hydrogen bomb, the American first-generation ICBM, the Atlas, was designed to be far smaller than the Soviet ICBM, which had been built to lift very heavy atomic (fission) bombs. The victory of the Soviet Union in the race to launch the first artificial satellite of Earth was won because of their lead in ICBM development.

The third space race, to send probes to the Moon, was won by the Soviet Union because of the availability of their ICBM and because of its great size and payload capacity. They were the first to fly by the Moon at close range, the first to hit the Moon, and the first to photograph its far side.

The Soviets won the fourth space race, to put men in space, by a margin of only a few months. But their five-ton Vostok capsule, far larger than the tiny one-ton American Mercury capsule, promised much greater potential performance. The larger Soviet capsule was used for longer-duration flights than could be managed by Mercury and, stretched to its dangerous limit in the Voskhod version, it carried out the world's first two-man and three-man flights.

But Voskhod was a dead end. When carrying multiman crews, it could stay in space only one day. We have noted that Voskhod pressed its weight limits so severely that the spacecraft did not even carry an escape rocket to save the crew in the event of a launch-phase accident. To the conservative and careful Soviet mind, this was an invitation to disaster. Voskhod was retired after only two short flights, and the greatly superior Soyuz spacecraft was developed.

By a terrible irony, the first flight of this greatly improved and much safer spacecraft ended in the tragic death of Vladimir Komarov in April of 1967, giving a disastrous 18-month setback to the Soviet manned program.

THE RACE TO SEND MEN TO THE MOON

By 1962, faced with the publicly proclaimed American intent to land men on the Moon by 1969, the Soviet Union faced an interesting dilemma. Could it beat the American Apollo astronauts to the surface of the Moon? Even more fundamental, did the Soviets *want* to do so? We believe that there is strong evidence that the Soviet Union did have ambitious plans for manned lunar missions, even including landings. These plans were an integral part of Khurshchev's policy of using space spectaculars for political impact. As part of this effort, the development of the G1 booster and of manned lunar spacecraft were both initiated. At the time of Khrushchev's fall in October 1964, Brezhnev and Kosygin inherited a very ambitious and expensive space program that was already close to its funding peak, and that was already experiencing severe technical problems. The technical demands imposed by lunar landing missions were so severe that there was great doubt in their minds as to whether a slow, orderly Soviet program could beat Apollo. Was there any less demanding task that could be undertaken to "take the edge" off Apollo?

There was indeed a remarkable straight-line path that the Soviets could and did pursue. It would, if successful, have stolen much of the glory of Apollo. Here's what they did.

First, the development of the D1 and G1 boosters was pursued with all possible speed. Second, techniques for rendezvous and docking of satellites were vigorously pursued. Third, a spacecraft, Soyuz, with the life-support capability to carry men to the Moon and back, was to be developed at once.

Table 3.1
Lunar Probe Launchings

Spacecraft	*Launch*	*Mission*	*BR*	*Results*
Pioneer 1	11 Oct 58	flyby	TAb	Reached 70,700 km altitude—missed Moon
Pioneer 2	8 Nov 58	flyby	TAb	Third stage ignition failure
Pioneer 3	6 Dec 58	flyby	J2	Reached 63,600 km altitude—missed Moon
Luna 1	2 Jan 59	flyby?	A1	Flew by Moon at 6000 km distance
Pioneer 4	3 Mar 59	flyby	J2	Missed Moon by 60,000 km
Luna 2	9 Sep 59	impact	A1	Impacted on Moon
Luna 3	4 Oct 59	flyby	A1	Photographed far side of Moon
(Pioneer)	26 Nov 59	flyby	AAb	Payload shroud failed during launch
(Pioneer)	25 Sep 60	flyby	AAb	Second stage malfunction
(Pioneer)	15 Dec 60	flyby	AAb	First stage exploded
Ranger 1	23 Aug 61	test	AAB	Failed to depart from low Earth orbit
Ranger 2	18 Nov 61	test	AAB	Failed to depart from low Earth orbit
Ranger 3	26 Jan 62	impact	AAB	Missed Moon by 40,000 km
Ranger 4	23 Apr 62	impact	AAB	Impacted Moon with experiments dead
Ranger 5	18 Oct 62	impact	AAB	Missed Moon by 750 km
(Luna)	4 Jan 63	lander?	A2e	Failed to depart from low Earth orbit
Luna 4	2 Apr 63	lander?	A2e	Missed Moon by 8500 km
Ranger 6	30 Jan 64	impact	AAB	Impacted Moon with TV inoperative
Ranger 7	28 Jul 64	impact	AAB	Success—returned 4308 photos
Ranger 8	17 Feb 65	impact	AAB	Success—returned 7137 photos
**Kosmos 60*	12 Mar 65	lander	A2e	Failed to depart from low Earth orbit
Ranger 9	21 Mar 65	impact	AAB	Success—returned 5814 photos
Luna 5	9 May 65	lander	A2e	Landing attempt failed—crashed on Moon
Luna 6	8 Jun 65	lander	A2e	Missed Moon by 160,000 km
Zond 3	18 Jul 65	test	A2e	System test—photographed Moon
Luna 7	4 Oct 65	lander	A2e	Retros fired early—crashed on Moon
Luna 8	3 Dec 65	lander	A2e	Retros fired late—crashed on Moon
Luna 9	31 Jan 66	lander	A2e	First successful lunar soft landing

Table 3.1 ***(continued)***

Spacecraft	*Launch*	*Mission*	*BR*	*Results*
**Kosmos 111*	1 Mar 66	lander	A2e	Failed to depart from low Earth orbit
Luna 10	31 Mar 66	orbiter	A2e	First successful lunar orbiter
Surveyor 1	30 May 66	lander	AC	Successful soft landing—sent 11,150 photos
LO 1	10 Aug 66	orbiter	AAD	Lunar photographic mapping
Luna 11	24 Aug 66	orbiter	A2e	Lunar orbit science mission
Surveyor 2	20 Sep 66	lander	AC	Crashed on Moon attempting landing
Luna 12	22 Oct 66	orbiter	A2e	Lunar photographic mapping
LO 2	6 Nov 66	orbiter	AAD	Lunar photographic mapping
Luna 13	21 Dec 66	lander	A2e	Soft lander science mission
LO 3	4 Feb 67	orbiter	AAD	Lunar photographic mapping
Surveyor 3	17 Apr 67	lander	AC	Surface science mission
LO 4	4 May 67	orbiter	AAD	Photographic mapping
Surveyor 4	14 Jul 67	lander	AC	Communications ceased before landing
LO 5	1 Aug 67	orbiter	AAD	Photographic mapping
Surveyor 5	8 Sep 67	lander	AC	Lunar surface science
Surveyor 6	7 Nov 67	lander	AC	Lunar surface science
Surveyor 7	7 Jan 68	lander	AC	Lunar surface science
Zond 4	2 Mar 68	test	D1e	Unmanned test of lunar Soyuz craft
Luna 14	7 Apr 68	orbiter	A2e	Mapped lunar gravity field
Zond 5	15 Sep 68	test	D1e	Circumlunar flyby—spacecraft recovered
Zond 6	10 Nov 68	test	D1e	Lunar flyby—precursor of manned flight
Apollo 8	21 Dec 68	manned	S5	Manned lunar orbital mission
Apollo 10	18 May 69	manned	S5	Lunar orbit test for landing mission
Luna 15	13 Jul 69	return?	D1e	Impacted on Moon after *Apollo 11* landing
Apollo 11	16 Jul 69	manned	S5	First manned lunar landing and return
Zond 7	8 Aug 69	test	D1e	Unmanned circumlunar flight and return
**Kosmos 300*	23 Sep 69	test	D1e	Earth-orbit test of lunar equipment?
**Kosmos 305*	22 Oct 69	test	D1e	Earth-orbit test of lunar modules?
Apollo 12	14 Nov 69	manned	S5	Manned lunar landing
Apollo 13	11 Apr 70	manned	S5	Aborted lunar landing mission
Luna 16	12 Sep 70	return	D1e	Returned lunar surface samples to Earth

(table continued)

Table 3.1 *(continued)*

Spacecraft	*Launch*	*Mission*	*BR*	*Results*
Zond 8	20 Oct 70	test	D1e	Unmanned circumlunar flight and return
Luna 17	10 Nov 70	rover	D1e	Unmanned lunar rover: *Lunokhod*
Apollo 14	31 Jan 71	lander	S5	Manned lunar landing
Apollo 15	26 Jul 71	lander	S5	Manned lunar landing
P&F satellite	4 Aug 71	orbiter	S5	Subsatellite deployed by *Apollo 15*
Luna 18	2 Sep 71	return	D1e	Crashed on Moon
Luna 19	28 Sep 71	orbiter	D1e	Photographic mapping mission
Luna 20	14 Feb 72	return	D1e	Lunar surface sample return
Apollo 16	16 Apr 72	manned	S5	Manned lunar landing
P&F satellite	19 Apr 72	orbiter	S5	Subsatellite deployed by *Apollo 15*
Apollo 17	7 Dec 72	manned	S5	Manned lunar landing
Luna 21	8 Jan 73	rover	D1e	Unmanned lunar rover: *Lunokhod 2*
Explorer 49	10 Jun 73	orbiter	TD	Galactic radio science experiment
Luna 22	29 May 74	orbiter	D1e	Photographic mapper
Luna 23	28 Oct 74	return	D1e	Drill arm damaged; no return attempt
Luna 24	9 Aug 76	return	D1e	Lunar surface sample return

Abbreviations
Spacecraft: K = Kosmos; LO = Lunar Orbiter; P&F = Particles and Fields BR (Booster Rocket): AAb = Atlas Able (Atlas ICBM first stage with upper stage stack from Vanguard booster); AAB = Atlas Agena B (Atlas first stage with Agena B second stage); AC = Atlas Centaur (Atlas first stage with high-energy hydrogen-oxygen Centaur second stage); A1 = *Sputnik 1* carrier rocket (SS6 ICBM with no upper stages); A2 = A1 with large second stage; A2e = A2 with an added kick stage; D1e = "Proton" superbooster with kick stage; J2 = Juno 2 (Army Jupiter IRBM with three solid-propellant upper stages from the Jupiter C satellite launch vehicle; S5 = Saturn 5 superbooster (7.5 million pound thrust liquid propellant S1c first stage, high-energy hydrogen-oxygen second and third stages, and lunar descent and ascent stages); TAB = Thor Agena B (Air Force Thor IRBM with Agena B upper stage derived from the bomb pod engine on the B-58 Hustler); TD = Thor Delta (Thor first stage and Delta second stage)

*Unannounced lunar missions.

Table 3.2
Venus Probe Launchings

Spacecraft	*Launch*	*Mission*	*BR*	*Results*
**Sputnik 7*	4 Feb 61	flyby	A2e	Failed to depart from low Earth orbit
Venera 1	12 Feb 61	flyby	A2e	Communications failure in transit
Mariner 1	22 Jul 62	flyby	AAB	Launch phase failure
**1962 APi1*	25 Aug 62	flyby	A2e	Failed to depart from low Earth orbit
Mariner 2	27 Aug 62	flyby	AAB	Flew by Venus at 36,000 km range
**1962 AT1*	1 Sep 62	flyby	A2e	Failed to depart from low Earth orbit
**1962 APhi1*	12 Sep 62	flyby	A2e	Failed to depart from low Earth orbit
**Kosmos 21*	11 Nov 63	test	A2e	Failed to depart from low Earth orbit
**Kosmos 27*	27 Mar 64	flyby?	A2e	Failed to depart from low Earth orbit
Zond 1	2 Apr 64	flyby?	A2e	Communications failure in transit
Venera 2	12 Nov 65	probe	A2e	Comm. failure just before Venus arrival
Venera 3	16 Nov 65	probe	A2e	Comm. failure before Venus entry
**Kosmos 96*	23 Nov 65	probe	A2e	Failed to depart from low Earth orbit
**Kosmos 159*	17 May 67	probe?	A2e	Possible Venera or Molniya failure: inert in high-eccentricity Earth orbit
Venera 4	12 Jun 67	probe	A2e	Successful atmospheric probe 18 Oct 67
Mariner 5	14 Jun 67	flyby	AAD	Flew by Venus at 4000 km range
**Kosmos 167*	17 Jun 67	probe	A2e	Failed to depart from low Earth orbit
Venera 5	5 Jan 69	probe	A2e	Successful atmospheric probe 16 May 69
Venera 6	10 Jan 69	probe	A2e	Successful atmospheric probe 17 May 69
Venera 7	17 Aug 70	lander	A2e	First successful lander 15 Dec 70
**Kosmos 359*	22 Aug 70	lander	A2e	Failed to depart from low Earth orbit
Venera 8	27 Mar 72	lander	A2e	Landed 22 Jul 72

(table continued)

Table 3.2 *(continued)*

Spacecraft	*Launch*	*Mission*	*BR*	*Results*
**Kosmos 482*	31 Mar 72	lander	A2e	Failed to depart from low Earth orbit
Mariner 10	3 Nov 73	flyby	AC	Flew by 5 Feb 74 en route to Mercury
Venera 9	8 Jun 75	lander	D1e	Landed 22 Oct 75
V 9 orbiter	"	orbiter	"	Orbited about Venus 22 Oct 75
Venera 10	14 Jun 75	lander	D1e	Landed 25 Oct 75
V 10 orbiter	"	orbiter	"	Orbited Venus 25 Oct 75
Pioneer Venus	20 May 78	orbiter	AC	Orbited Venus 8 Dec 78
PV Large Probe	"	lander	"	Landed 9 Dec 78
Pioneer Venus	8 Aug 78	landers	AC	Three small probes landed 9 Dec
PV bus	"	probe	"	Probed upper atmosphere 9 Dec 78
Venera 11	9 Sep 78	lander	D1e	Landed 25 Dec 78
Venera 12	14 Sep 78	lander	D1e	Landed 21 Dec 78
Venera 13	30 Oct 81	lander	D1e	Landed 1 Mar 82
Venera 14	4 Nov 81	lander	D1e	Landed 5 Mar 82
Venera 15	2 Jun 83	orbiter	D1e	Radar mapper
Venera 16	7 Jun 83	orbiter	D1e	Radar mapper
Vega 1	15 Dec 84	lander	D1e	Landed 10 Jun 85
	"	balloon	"	Deployed in Venus atmosphere
	"	flyby	"	Encounter with Comet Halley Mar 86
Vega 2	21 Dec 84	lander	D1e	Landed 15 Jun 85
	"	balloon	"	Deployed in Venus atmosphere
	"	flyby	"	Encounter with Comet Halley Mar 86

Abbreviations

BR (Booster Rocket): AAB = Atlas Agena B (Atlas ICBM with Agena B second stage); AAD = Atlas Agena D (Atlas ICBM with Agena D second stage); AC = Atlas Centaur (Atlas ICBM with high-energy hydrogen-oxygen second stage); A2e = SS6 ICBM with large second stage and third kick stage); D1e = "Proton" heavy-lift booster with added kick stage. Dittoes denote separate payloads launched together on the same booster.

*Unannounced Venus missions.

Fourth, an ambitious program of unmanned lunar missions was undertaken to pave the way for future manned lunar flight.

The American intentions quickly became clear. First, the Saturn 1 and 5 rockets were to be built from scratch as a crash program. Second, techniques for rendezvous and docking of satellites were to be undertaken in a special new manned program, Gemini. Third, life-support systems capable of flying to the Moon and back were to be developed within the Gemini program to prepare for Apollo. Fourth, an ambitious program of unmanned lunar missions, involving impactors, orbiters, and landers, was undertaken to to pave the way for Apollo.

The stakes of the race were enormous, and the cost to both parties was staggering. For those few Westerners who knew what the Soviets were doing, it was an exhilarating race. For the rest of humanity, the race was not nearly so obvious.

By the time of the dead-end Voskhod flights in 1964 and 1965, the race had already begun in earnest. The American series of Ranger probes, a program started well before the Apollo decision, was sent to hit the Moon with its TV cameras running. After six consecutive failures, *Rangers 7, 8,* and *9* struck the Moon in 1964 and 1965, returning over 17,000 superb pictures taken right down to moments before impact.

At the same time, the Soviet Luna series was revived with much more ambitious second-generation spacecraft flown on the improved A2e booster. Launchings in January and April of 1963 failed to reach the Moon, and a flight in March of 1965 was stranded in low Earth orbit when the escape stage of the booster failed to ignite. *Luna 5,* in May of 1965, crashed on the Moon while attempting to carry out a soft landing. A month later, *Luna 6* missed the Moon altogether. *Luna 7* and *Luna 8,* in late 1965, both crashed while attempting soft landings.

Late 1965 also saw the *Gemini 7-6* mission, in which two manned spacecraft rendezvoused and flew together a foot apart, crowned by the 14-day endurance record of *Gemini 7.* The first two satellite launchings (*Proton 1* and *Proton 2*) by the Soviet D1 heavy-lift booster occurred in July and November.

In 1966 came five more Gemini launches, demonstrating spacewalking, numerous rendezvous exercises with unmanned target satellites launched by Atlas Agena D boosters, and repeated successful docking with these satellites.

1966 was also the year that the revived Luna program finally hit paydirt. In January, *Luna 9* became the first vehicle to land safely on the Moon. Another attempt in early March was again left stranded in low Earth orbit, but at the end of March the *Luna 10* satellite was successfully placed in orbit about the Moon. It was the first satellite ever placed in orbit about a body other than Earth.

Two months later, the American *Surveyor* spacecraft, launched by the new Atlas Centaur with its high-performance H-O engines, flew for the first time—and successfully soft-landed on the Moon in its very first attempt. Unfortunately, it was four months behind *Luna 9.*

Later in the year, *Surveyor 2* crashed in a landing attempt. Two Soviet lunar orbiting missions, *Luna 10* and *Luna 11,* and two American photographic mappers, *Lunar Orbiter 1* and *2,* were placed in orbit about the Moon before the end of the year. Their task was to search for safe but interesting future landing sites (see fig. 3.1). In the closing days of the year, *Luna 13* carried out the third (second Soviet) successful lunar landing.

In 1967, there were no Soviet lunar missions. Soviet attention was on other aspects of their program, especially the development of its boosters and the new manned Soyuz spacecraft. The G1 launch facilities were completed, and several unmanned test flights of the Soyuz spacecraft were carried out under the Kosmos cover. Two of these were two-day orbital tests of the Soyuz capsule and its life-support and reentry systems. Both tests were uneventful. *Kosmos 146* and *154,* in March and early April, were man-related flights using the D1 heavy-lift booster. Upper stage stacks with Soyuz reentry capsules were placed in orbit. Data on these flights are incomplete, but it is widely believed that the purpose was to fire the Soyuz capsule back into the atmosphere at the very high speed that it would have if returning to Earth after falling all the way from the Moon.

Soyuz 1 flew in April, with its tragic result. The next flights of this spacecraft were flown under the Kosmos cover in October. Two unmanned Kosmos satellites were automatically rendezvoused and docked, the first time that feat had been achieved by the Soviets, but 15 months behind the Gemini program.

By this time the Saturn 1 and Saturn 5 had flown successfully into orbit. The November test flight of the Saturn 5, carrying the unmanned *Apollo 4,* was a huge success.

The American fleet of unmanned lunar craft continued to lay siege to the

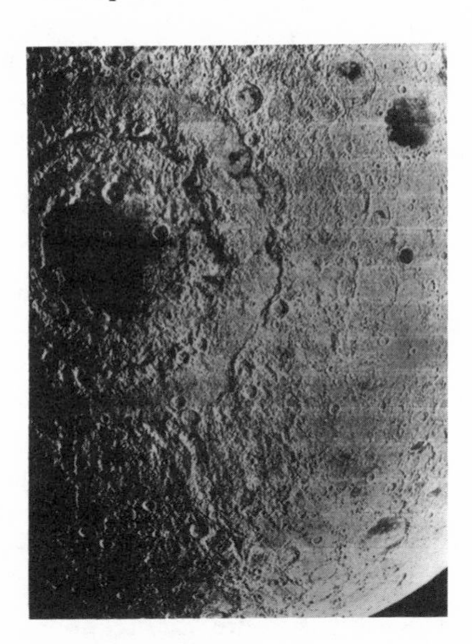

Figure 3.1
Early Spacecraft Pictures of the Moon
The first spacecraft to photograph the far side of the Moon, the Soviet *Luna 3* flyby, returned several low-quality images of features never before seen by mankind, including the dark feature seen in this picture (left), called the Sea of Moscow by its discoverers. The American Lunar Orbiter series produced detailed photographic maps of almost the entire surface of the Moon, including this stunning view of the Mare Orientale basin (right), which is almost completely out of view from Earth.

Moon throughout 1967 and into January 1968. Three more Lunar Orbiter mapping missions were flown, producing a virtually complete photographic map of the Moon and returning detailed views of a large number of candidate Apollo landing sites. *Surveyors* 3 through 7 were sent on lunar landing missions, of which all but *Surveyor 4* worked perfectly. Seven lunar successes out of eight trials: the American space program was finally coming of age.

The year 1968 dawned with the Soviet program in serious trouble. The G1 booster was doing poorly, Soyuz was grounded, not a single lunar scientific mission had flown since 1966, and the future course was a mystery to all. Nonetheless, the Soviet media made numerous allusions to a manned lunar program that might yet beat the Americans to the Moon.

In January, *Apollo 5* flew as an unmanned Earth-orbital test flight of the Apollo lunar lander. *Apollo 6* flew on a Saturn 5 in April for a crucial high-speed test of the Apollo capsule heat shield, simulating a return from the Moon. In this, the American program may have been far behind the comparable *Kosmos 146* and *154* flights a year earlier. Perhaps we shall someday know exactly what happened on those missions.

The puzzle, however, got worse before it got better. On March 2, 1968 the Soviet Union launched a probe named *Zond 4,* again, as with the two mysterious Kosmos flights, using the D1 heavy-lift booster. *Zond 4* was clearly unrelated to the much smaller planetary probe test vehicles launched earlier on the A2e booster under the Zond name. The fate of *Zond 4* is shrouded in mystery. Soviet announcements, commonly lacking in candor, became positively opaque after the launch. Apparently the spacecraft flew out to the orbit of the Moon in a direction opposite to the Moon's position and returned to Earth seven days later. Apparently a massive communications failure occurred, preventing accurate tracking and return of data.

A red herring was interjected in April as the last second-generation Soviet lunar probe, *Luna 14,* was sent to orbit the Moon and measure the lunar gravity field. This flight was not a part of the development of hardware for the Soviet manned lunar program. There is strong evidence that the early Soviet lunar landing attempts were completely preprogrammed, and were able neither to recognize the lumpiness of the lunar gravity field nor to correct for it. This led to errors in the estimated location of the lunar surface, and caused at least two earlier landing missions, *Luna 7* and *Luna 8,* to crash while attempting to land. *Land 14* was the last lunar payload to be carried by the A2e.

Later in April two more Soyuz spacecraft, flying under the Kosmos cover, repeated the rendezvous and docking maneuver. A few days later, according to some Western intelligence sources, another D1 launching of a Zond probe was aborted by a failure of the second stage of the booster.

In August a single Soyuz-type craft was orbited for four days under the Kosmos name. This was taken by many Western observers as evidence that a manned Soyuz mission with a four-day duration would soon be flown. Many observers predicted that there would soon be a dual Soyuz/Soyuz or Soyuz/Kosmos rendezvous mission.

By early fall it was clear that the Apollo program was going so well that a manned flight around the Moon, without landing, could be attempted as early as December. This was earlier than previously planned. The pressure on the Soviet program was now doubly intense.

On September 15, *Zond 5* was launched on a D1e booster. Suddenly, all the questions about the *Kosmos 146* and *152* and *Zond 4* missions were flooded with new light. *Zond 5,* it was announced, was a "cabin" carrying a "phantom cosmonaut," whose tape-recorded voice was played back to test voice communications over lunar distances. The vehicle had the size and weight of a Soyuz reentry capsule, and had an orbital docking port. It flew out to the Moon's orbit, passed behind the Moon on September 18, and headed back to Earth. A mid-course correction on the return to Earth aimed the Zond capsule at a narrow reentry corridor in the atmosphere above the Indian Ocean. The capsule survived reentry, landed in the Indian Ocean, and was retrieved by a Soviet tracking ship. It thus became the first spacecraft returned safely from the distance of the Moon.

The scenario now became clear: a launch of the D1e booster, which had not previously been man-rated, would place the entire Moon-bound stack of hardware, possibly including a manned Zond capsule, in low Earth orbit. (Alternatively, the cosmonaut would arrive in a Soyuz and dock with the Moon-bound hardware.) The cosmonaut would then go off to fly around the Moon. The duration of the mission would be seven days, within the eight days demonstrated by the Kosmos test, and the reentry would be safe, as demonstrated by the *Zond 4* flight with the dummy cosmonaut. Further, *Zond 4* showed that the lunar gravity field was well enough known to allow confident navigation for the delicate return to Earth. All that was needed was faith in the reliability of the Soyuz capsule when used for return from the Moon. By Soviet practice, this would require additional successful unmanned test flights in the Zond series and manned flights in Soyuz.

The first manned Apollo mission, *Apollo 7,* flew an 11-day Earth orbit mission in October. Everything went smoothly. Approval was given to go ahead with a manned lunar orbital mission on *Apollo 8.* The launch date was set for December 21.

Three days after the *Apollo 7* crew returned to Earth, the Soyuz program (fig. 3.2) finally made it back into space. *Soyuz 2* was launched unmanned on October 25, and *Soyuz 3,* following predictions, was launched the next day. The mission was to rendezvous and dock. At stake may have been the entire race to the Moon. Soviet hopes were dashed: the *Soyuz 3* cosmonaut, Georgii Beregovoi, was unable to dock, and the mission was cut short. He returned after four days in space.

The Soviet program was now in a terrible bind. The next lunar launch window was in the second week of November, only ten or twelve days away, and there was no possibility of flying another rendezvous mission before it. The conservative decision was made not to take the risk of sending a man up on the next Zond, but instead to send the Zond out around the Moon empty.

Zond 6, launched on November 10, carried out its seven-day lunar flyby mission and returned safely, via a demanding skip-glide reentry path, to a landing in the Soviet Union. It is likely that a landing in the USSR, although more difficult than a splashdown in the Indian Ocean, was viewed by Soviet officials as essential for a manned flight.

A Soviet lunar launch window was due to open up on December 8, but no launch was attempted. Indeed, it may not have been possible to turn around the D1e pad and Zond hardware in less than two months. In early December a spokesman for the Soviet Academy of Sciences announced, in effect, "Now it's their turn." And so it was.

Apollo 8, commanded by Frank Borman, with a crew consisting of his *Gemini 7* colleague (and *Gemini 12* pilot) James Lovell and rookie astronaut William Anders, blasted off from the Saturn 5 launch complex on December 21 and headed out toward the Moon. This was only the third launch of the giant Saturn 5 booster. They did not simply fly by the Moon at a distance of several thousand kilometers, as the unmanned Zond probes had done, but fired their engines and dropped into a low orbit. From there, on Christmas day, they read to one of the largest live television audiences in history the account of creation in the first chapter of Genesis. Their return to Earth was uneventful.

About five days after the *Apollo 8* mission returned to Earth the Soviet lunar launch window again opened. According to leaked reports from American intelligence sources, another D1e was launched, but suffered a second-stage failure and did not reach Earth orbit. No men were involved (except, of course, for those who may have been planning on flying in a Soyuz capsule to rendezvous with it exactly one day later.)

With that Zond opportunity gone for good, and with at least two months to go before the next expected opportunity, the Soviets concentrated on their elusive goal of a rendezvous of manned spacecraft. On January 14 and 15, *Soyuz 4* and *5* were launched into almost identical orbits. The *Soyuz 4* pilot, Shatalov, rendezvoused with *Soyuz 5*, carrying Volynov, Khrunov, and Yeliseyev. They docked successfully, and Khrunov and Yeliseyev

Figure 3.2
A Soyuz Booster Arrives at the Launch Pad
An unfueled A2 Soyuz booster is transported to its launchpad by train. The large depression in the background beyond the launchpad is the flame pit,

moved over to the *Soyuz 4* capsule. Each spacecraft returned after three days in space. That hurdle had at last been cleared.

But what good did it do to finally see the way clear to a one-man flyby of the Moon, when the Americans had already orbited the Moon with three men aboard? The answer of the Soviet press and their cosmonauts was that the Soviet Union still intended to land men on the Moon in 1969, and they thought that there was still a chance that they would beat Apollo. It is hard for us to imagine that this was a deliberate lie. First, it would appear to history that they were making a promise that they could not keep. Second, although the Soviet space program is conducted largely in secrecy, it is not common for it spokesmen to indulge in a direct lie. When they do, the motive is usually security. No such factor seems evident here.

February and March rolled by without any further Soviet activity. Instead, *Apollo 9* set off in the early March lunar launch window for a thorough Earth-orbiting rehearsal of the lunar landing mission. After a complete runthrough, using the lunar descent and ascent engines, and after ten full days in space, the crew (McDivitt, Scott, and Schweickart) returned elated, full of confidence in their hardware.

Apollo 10 in May was another complete runthrough of the lunar landing mission, this time conducted in lunar orbit (fig. 3.3). The Apollo landing module(the LM—affectionately termed the bug-eyed Lem—after the "bug-eyed monster" of pulp magazine science fiction) was flown down to altitudes so low that it skimmed the tops of the highest lunar mountains. Tom Stafford and Gene Cernan (the former *Gemini 9* crew) and the omnipresent John Young (who had flown in *Gemini 3* and *10,* and was destined to command the *Apollo 16* lunar landing mission and fly the first Space Shuttle

---

which collects the exhaust plume at ignition and directs it away from the booster. The booster is erected by cranes, fueled in the vertical position, and secured until the moment of liftoff by the swing-away towers that can be seen surrounding the launch pad. This picture was taken at the Tyuratam launch site, the only Soviet launch center for manned flights. The Soviets customarily refer to this launch center as the "Baikonur Cosmodrome" after the name of a small desert town several hundred kilometers away. This is an example of Soviet disinformation tactics. It is not obvious whom they mean to fool: Western observers have known of the exact location of the real launch facilities for a quarter of a century, and the facilities can readily be identified on Landsat images.

Figure 3.3
Apollo Above the Moon
An Apollo Command Module in orbit about the Moon. The picture was taken through a window in the Lunar Module.

mission in 1981) felt the ultimate frustration: they went there, did all the right things, and all the equipment worked perfectly—but they weren't allowed to land.

It was now a simple issue. Either the *Apollo 11* flight, announced for July 16, would be a great success, or it would be the best-publicized disaster ever. The late March, April, May, and June Soviet lunar launch windows passed without any activity—although it was rumored that there had been a launch failure of the D1e in June. The next chance was mid-July. But another factor was present: the giant G1 superbooster was on its pad, getting ready for its maiden flight. On July 4, just days before the first Apollo landing attempt, the G1 blew up and demolished its launch pad.

On July 13, just three days before the scheduled liftoff of *Apollo 11,* the Soviets sent a D1e launch to the Moon. But, surprisingly, this was no Zond. It was called *Luna 15.* In fact, it was like nothing we had ever seen before: it was the first flight of the new third-generation family of unmanned lunar spacecraft, much heavier and more sophisticated than the previous round of landers and orbiters launched by the much smaller A2e. On July 17, with *Apollo 11* already en route from Earth, *Luna 15* dropped into orbit about the Moon. It was clear that the stage was set for a dramatic upstaging of the Apollo landing, yet it was also clear that no men were involved. What could they possibly do with this large new spacecraft?

Over the next few days, *Luna 15* twice altered its orbit as if jockeying for the perfect position for a lunar landing. Astonishingly, it was still sitting there in orbit when *Apollo 11* arrived in lunar orbit, and still there when Neil Armstrong and Buzz Aldrin stepped out on the lunar surface. Then *Luna 15* fired its engines again and began a descent to the lunar surface—and crashed into Mare Crisium while still traveling a few hundred kilometers per hour.

The rest is history. *Apollo 11* completed its lunar mission successfully (fig. 3.4). Six more Apollo lunar landing missions were flown, five successfully and one (unlucky *Apollo 13*) forced to return without landing by an explosion in the service module. The Soviet lunar program stumbled along for a while, doing puzzling things. Another unmanned Zond flight, now worse than second best, came in August. Then two D1e missions labeled *Kosmos 300* and *305,* placed in Earth orbit in the September and October launch windows, were stranded in low Earth orbit by engine failures.

A massive manned extravaganza in October of 1969 seemed to epitomize the frustrations the Soviets were experiencing. Three Soyuz spacecraft were orbited on three consecutive days, with the promise of a spectacular linkup in orbit. It never happened. Seven cosmonauts flew around the Earth for five days. The docking attempt failed, and the cosmonauts returned to Earth. But that was not the last of the story. There was one final Zond flight as late as October 1970, but the Zonds by then were just the ghost of Christmas past—the unforgettable Christmas of 1968.

The Soviets continued with nine superbly executed unmanned Luna missions from 1970 to 1976, all involving third-generation heavy spacecraft lifted by D1e boosters. *Luna 16, 20,* and *24* successfully returned to Earth lunar samples collected from the surface. This is certainly what *Luna 15* was trying to do to steal the thunder from Apollo 11.

Luna 17 and *21* landed small lunar tanks (remember them?) called *Lunokhod 1* and *2* that roved about for many kilometers on the lunar surface, driven via television by controllers on Earth (fig. 3.5). *Luna 19* and *22* were advanced photographic mapping missions of the sort that would permit the selection of landing sites for both unmanned and manned missions.

Figure 3.4
Apollo On the Moon
Buzz Aldrin prepares for the "small step" that stands between him and the surface of the Moon. The picture was taken by Neil Armstrong, the first man to set foot on the Moon. The *Apollo 11* Lunar Module rests between flights: the upper half, the ascent stage, will return the astronauts to lunar orbit to rendezvous with Mike Collins in the command module. The lower half, the descent stage, remains on the Moon.

Figure 3.5
Lunokhod I
Two unmanned Soviet Lunokhod lunar roving vehicles were successfully landed on the Moon and driven about on the lunar surface under radio control from Earth. The hat-like thermal shield that protects the tub-shaped body at night can be clearly seen. Because of the time lag for propagation of radio waves between Earth and Moon, the rovers could be driven only at low speeds without risking an accident. Both Lunokhods were extremely successful in fulfilling their exploratory missions.

Luna 18 crashed attempting a sample return mission, and *Luna 23* landed successfully, but damaged its drill arm, making it impossible for it to return any samples.

All told, the third-generation Lunas were a technical masterpiece, remarkably ambitious and remarkably successful. The only trouble with them can be summed up in a single word: Apollo. Planetary scientists recognize the true importance of these missions and the technology developed in them. They perfected techniques for automated landings on the Moon, for the automatic return of lunar samples to earth, for remotely controlled exploration of the Moon by roving vehicles, and for orbital mapping. Their numerous scientific contributions will be summarized in chapter 6. These achievements are greatly respected by Western observers, and for good reason.

But, after all that, what were the Soviets really up to? It is simply beyond question that they did in fact have a manned lunar program. The initial thrust of the program, which almost worked, was to send a one-man lunar flyby around the Moon before Apollo could get there. If Apollo had not been running ahead of schedule, and if *Soyuz 3* had successfuly docked with *Soyuz 2,* then there almost certainly would have been a manned Zond mission attempt in January 1969. If that had been successful, the next milestone in the Soviet lunar program might have been a brief Moon-orbiting mission in a loose, low-energy orbit. Then, if they had been successful in developing lunar landing hardware (and the G1 superbooster to lift it to the Moon), they would have proceeded to a leisurely manned lunar landing after Apollo.

As it was, a Zond flight after *Apollo 8* would never have made sense. An inferior, later mission would not have looked good. That portion of the Soviet program was cancelled on or about January 1, 1969. The later apparently pointless lame-duck Zond flights were simply leftover harware from their manned lunar flyby program.

The hardware needed for a Soviet manned lunar landing was nowhere in evidence. Of course some of the unmanned D1 flights in Earth orbit may have been tests of parts of the system, and the D1-boosted Kosmos tests of late 1970 and early 1971 may have filled a similar role. We may never know.

The forthright belief of the Soviets in their manned landing program was apparently predicated on having a much larger booster than the D1. That means that they fully expected to have their own Saturn 5, the vehicle we call the G1, in use in 1969 or 1970. Instead, the program stumbled along until about 1976, when it presumably was scrapped. The G1 never flew successfully, and no Soviet cosmonaut has ever been on (or near) the Moon.

What would you have done in this situation? Stuck with a superbooster that wouldn't boost, a long-duration spacecraft named Union (Soyuz) that wouldn't fly United, and a superb deep-space capability based on the D1e and the Luna series?

Four answers: use the Soyuz alone to set a spaceflight duration record; practice rendezvous techniques; use the increasingly reliable D1 to launch a space station to be serviced and occupied by Soyuz; and use the mighty D1e (the world champion once the dumb Americans threw away their last Saturns) to do a whopping good job on the planets. Which of these did the Soviets choose to do? All of them!

In June 1970, *Soyuz 9* flew an 18-day mission to recapture the world duration record from *Gemini 7.* On April 19, 1971 the Soviets used a D1 booster to orbit the first space station, *Salyut 1.* Just four days later, *Soyuz 10*, with a three-man crew, was sent up to link up with *Salyut.* Guess what? Yep, they failed to dock successfully. The crew was back on Earth within 24 hours. But the trouble was not with the cosmonauts, it was simply that Soyuz, with three men aboard, was a frightfully marginal vehicle. It had to be flown with its propellant tanks mostly empty in order to carry the weight of the third man into orbit. This meant perilously little fuel for maneuvering and docking. On paper, a three-man flight looks better than a two-man flight, especially after Apollo. But this arrangement carried the seeds of disaster.

And what about the planets? The United States had announced that it intended to land two large Viking probes on Mars on Independence Day 1976, the 200th anniversary of the Nation. What better goal for the revitalized Soviet planetary program than to beat the Americans?

THE RACE TO THE PLANETS

By 1959, as we saw earlier, the Soviets had sent three probes to the Moon. The United States, using much smaller boosters of marginal capacity, had launched five probes—and failed five times.

The exploration of the planets began in 1960, just on the heels of the first round of lunar exploration that had been so convincingly dominated by the Soviets. From 1960 to 1965 the Soviet Union poured huge amounts of energy and money into lunar and planetary exploration, at least twice the level of effort of the NASA space science program. The results were absolutely staggering.

The Soviets launched three Mars probes in 1960, two Venus probes in 1961, three Venus probes and three Mars probes in 1962, at least one known Venus probe in 1963, one Zond probe to Mars, one to Venus, and one other

Table 3.3
Mars Probe Launchings

Spacecraft	*Launch*	*Mission*	*BR*	*Results*
(unnamed)	10 Oct 60	flyby	A2e	Failed to achieve Earth orbit
(unnamed)	14 Oct 60	flyby	A2e	Failed to achieve Earth orbit
*1962BI1	24 Oct 62	flyby	A2e	Failed to depart from low Earth orbit
Mars 1	1 Nov 62	flyby	A2e	Communications failure in transit
*1962BXi1	4 Nov 62	flyby	A2e	Failed to depart from low Earth orbit
Mariner C	5 Nov 64	flyby	AAD	Shroud failed to separate after launch
Mariner 4	28 Nov 64	flyby	AAD	First flyby of Mars: 15 July 65
Zond 2	30 Nov 64	flyby	A2e	Communications failure in transit
Zond 3	18 Jul 65	test	A2e	Flew by Moon as test of Mars spacecraft
Mariner 6	24 Feb 69	flyby	AC	Flew by 31 Jul 69; returned 75 pictures
Mariner 7	27 Mar 69	flyby	AC	Flew by 5 Aug 69; returned 126 pictures
Mariner H	8 May 71	orbiter	AC	Second stage failure at launch
**Kosmos 419*	10 May 71	lander?	D1e	Failed to depart from low Earth orbit
Mars 2	19 May 71	orbiter	D1e	First Mars orbiter: 27 Nov 71
Mars 2 lander	"	lander	"	Crashed in landing attempt
Mars 3	28 May 71	orbiter	D1e	Orbited Mars on 2 Dec 71
Mars 3 lander	"	lander	"	Failed immediately after landing
Mariner 9	30 May 71	orbiter	AC	Orbital photographic mapper
Mars 4	21 Jul 73	orbiter	D1e	Failed to achieve Mars orbit
Mars 5	25 Jul 73	orbiter	D1e	Orbited Mars on 12 Feb 74
Mars 6	5 Aug 73	lander	D1e	Communications lost just before landing
Mars 7	9 Aug 73	lander	D1e	Engine failure; missed Mars
Viking 1	20 Aug 75	orbiter	3EC	Orbited Mars 19 Jun 76
Viking 1 lander	"	lander	"	Landed on Mars 20 Jul 76; surface science
Viking 2	9 Sep 75	orbiter	3EC	Orbited Mars 7 Aug 76
Viking 2 lander	"	lander	"	Landed on Mars 3 Sep 76

Abbreviations
BR (Booster Rocket): AAD = Atlas Agena D (Atlas ICBM with Agena D second stage); AC = Atlas Centaur (Atlas ICBM with high-energy hydrogen-oxygen second stage); A2e = SS6 ICBM with large second stage and third kick stage); D1e = "Proton" heavy-lift booster with added kick stage; 3EC = Titan 3E Centaur (strengthened two-stage Titan 2 ICBM core with two large solid-propellant strap-on boosters and high-energy Centaur upper stage). Dittoes denote separate payloads launched together on the same booster.

*Unannounced missions.

Venus launch in 1964, and one further Zond flight and three Venus probes in 1965. Not one single probe returned any data from any other planet. This was during the same time that Lunas 4-8 and at least three other unacknowledged Luna missions failed. The score for the Soviet lunar and planetary program over the years 1960 to 1965 was no successes and 26 failures. Of course, this counts only the failures we know about!

There seem to have been two major problems with the Soviet deep-space program. First, many of their probes were successfully placed in low Earth orbits but, when the time came for them to fire their escape engines to depart for the Moon, Mars, or Venus, the engines failed to fire. Possibly this problem was associated with the Soviet decision to use only liquid-fueled rocket engines: they may not have properly understood how liquid propellants behave in zero gravity. The second major problem was that all of their planetary probes that were successfully dispatched from low Earth orbit suffered total failure of their electronics while en route to the planets. Time and again, transmissions from their spacecraft mysteriously cut off long before they reached their destinations.

The American run of disastrous luck in its lunar program expired in the last minutes of July 1964 with the success of the *Ranger 7* television impact probe. At that point, the United States had launched 13 lunar probes—with 13 failures.

The triumphs were few, but gratifying. Through 1965, the United States had launched only two Mariner Venus probes, of which one, *Mariner 2,* executed the first successful flyby of Venus. Two American Mariner spacecraft were launched toward Mars on Atlas Agena D boosters in 1964. Of these, *Mariner 4* carried out the first successful flyby of Mars. With the *Ranger 7, 8* and *9* successes, this added up to a promising beginning.

Were the Soviets tempted to give up their planetary program after their string of 26 failures? There is absolutely no evidence of a flagging of their interest or efforts to explore Venus, but the more demanding Mars program, with its longer trip times, was shelved for several years. Of the two principal problems with the A2e-launched planetary spacecraft, their communications reliability problem was ameliorated by the decision to stick with Venus. The other problem, the failure of the A2e kick stage to reignite after insertion into orbit about Earth, was eventually solved.

In the 1967 Venus launch window, two or three Soviet Venera probes and the American Mariner 5 flyby were scheduled for launch. *Venera 4* and *Mariner 5* performed perfectly, with Mariner carrying out a close flyby and

Venera entering the planet's atmosphere. The Venera entry capsule survived the searing heat and crushing deceleration forces of entry, discarded its glowing heat shield, and deployed a parachute. It slowly drifted down through the atmosphere, past the point where its pressure gauges went off-scale (at a pressure of about 6 atmospheres), down to where the temperature was about 550 degrees Kelvin (water boils at 373 Kelvin). There its radio signals abruptly ceased. The Soviet scientists claimed that *Venera 4* had landed on the surface of Venus.

Among the appurtenances of the spacecraft, its designers had foresightedly included an aluminum bust of Lenin and a coat of arms of the Soviet Union. We now know that the vehicle was crushed by the atmospheric pressure far above the hellish surface, at a level where the pressure was about 20 atmospheres. The "air" of Venus, surpassing even Dante's imagination, was a wicked brew of red-hot carbon dioxide, seasoned with traces of carbon monoxide, hydrochloric acid, and a variety of sulfur-bearing gases, decorated with clouds of sulfuric acid droplets. As for Lenin and his coat of arms, the surface heat and corrosive atmosphere would have dealt with them very quickly. Don't go to Venus hoping to pick up any aluminum souvenirs!

In 1969 there were two more Venera probe launchings. Both *Venera 5* and *6* entered successfully, but were crushed long before reaching the surface. The instrumentation on all of these early probes was very crude, and they returned little data; but they worked. Their success gave the Soviets confidence to build more ambitious and competent spacecraft, which they did without hesitation. Two more probes, of more robust design, were launched in 1970, and one of them, *Venera 7,* became the first probe to survive landing on the surface of Venus. It survived less than a half hour before the incredible heat (750 Kelvin, or hotter than a terrestrial self-cleaning oven) killed its electronics.

This brings us to the point, 1969, at which the Soviet manned lunar program was canceled. The Luna series had already switched over from the A2e to the D1e booster, with excellent results. Two 1972 Venus missions, using versions of the earlier Venera design, had alredy been slated for launch on the A2e, which was to be the last use of the A2e in the planetary program. An introduction of the D1e booster to resume and rejuvenate the Mars program was already planned for 1972. The last chance to beat the American Viking probes was the 1973 Mars launch window, so plans were set in motion to launch then as well.

Soviet emphasis on Venus has continued unabated: from 1960 through 1985 only three of the sixteen Venus launch windows were left unused (see Fig. 3.6). By 1985 the Venera series had reached number 18, and still there had been no slacking of effort. After having the first Venus entry probe and the first lander on any planet, they also placed the first satellite in orbit about Venus (*Venera 9*, the first D1e mission to Venus).

The on-again, off-again American Venus program since *Mariner 5* has consisted of a single flyby of the *Mariner 10* spacecraft in 1974, on its way to Mercury, plus a flotilla of entry probes and an orbiter sent in the Pioneer Venus program in 1978 (see Fig. 3.7). The Soviet Union had every reason to be proud of its unstinting efforts to study Venus, and rightly felt that it was competing successfully in that arena. They had built up, slowly and in a stepwise manner, a truly excellent Venus exploration program. But Mars was a bone of contention, with no clear victor as yet evident: it was toward Mars that the race was now directed.

From 1965 to 1970 not a single Soviet mission was sent to Mars. The communications problems encountered with the smaller first-generation planetary spacecraft had forced a five-year gap in the program until the much larger D1e booster was ready to launch much heavier spacecraft with much larger electrical power supplies, transmitters, and antennas. In this period only two American spacecraft were launched to Mars, the 1969 *Mariner 6* and *7* flyby vehicles, both launched by Atlas Centaur boosters. Both spacecraft successfully completed their photographic missions.

In 1971 the skies were thick with traffic heading toward Mars. In May, the United States launched *Mariner 8* toward an orbit about Mars. Unfortunately, the second stage of the Atlas Centaur booster failed during launch, and the payload fell into the Atlantic Ocean. One wag at NASA Headquarters suggested that it be redesignated as an oceanic research satellite. Two days later, a Kosmos satellite was launched on a D1e rocket into a low Earth orbit of the sort used for transfers to planetary trajectories. The escape stage failed to restart, and the payload was left stranded in Earth orbit (shades of the A2e!). Nine days later, *Mars 2* was sent on its way successfully, and another nine days later a third D1e launched *Mars 3*. Two days later, on May 30, Mariner 9 blasted off on its Atlas Centaur booster and joined the convoy.

On November 13, while one of the greatest dust storms ever seen on Mars was covering the entire planet, the one-ton *Mariner 9* arrived. It dropped into a highly eccentric orbit to carry out photographic mapping of

Figure 3.6
Surface of Venus

This panorama of the surface of Venus was taken by the *Venera 10* lander. The horizon appears in the upper right-hand corner. The base of the lander and one of its instruments can be seen at the center bottom. The evenly spaced bars of "noise" in the picture are actually telemetered data from the scientific experiments aboard the lander. The temperature outside is typical, about 470°C (900°F)—hotter than a self-cleaning oven. The atmosphere, made mostly of carbon dioxide with invigorating traces of carbon moxoxide, hydrochloric acid, hydrofluoric acid, sulfur dioxide, and superheated steam, also features clouds of sulfuric acid droplets. The pressure outside is 90 times as high as on the surface of the Earth.

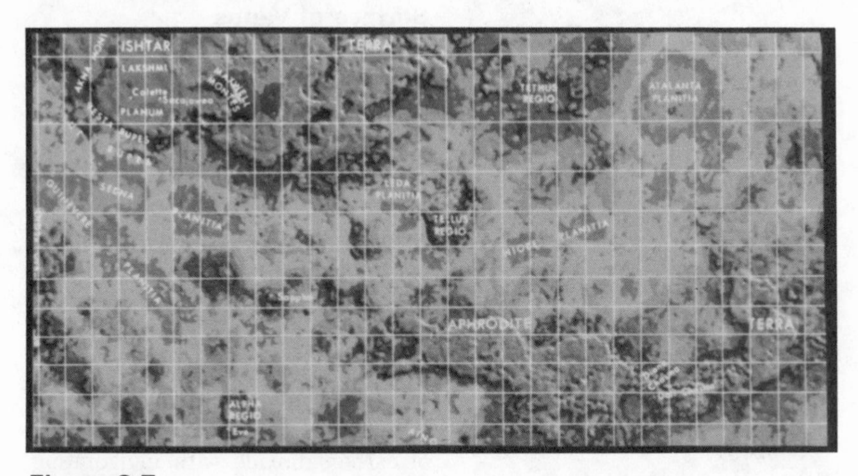

Figure 3.7
Radar Map of Venus
A radar altimeter on the *Pioneer Venus Orbiter* produced this topographic map of Venus.

the planetary surface (and the Martian satellites Phobos and Deimos as well). On November 27 *Mars 2* joined *Mariner 9* in orbit about Mars, and *Mars 3* dropped into orbit on December 2. The two Soviet payloads were very large, about 5 tons. Each spacecraft consisted of two parts, an orbiter and a landing module.

The landing capsules were dropped on Mars-impact trajectories just before the orbital insertion burn of the mother craft. Despite the dust storm, they had no choice but to drop the capsules in at once. The spacecraft were preprogrammed to enter, and could not be reprogrammed. Further, the orbital insertion engines were incapable of carrying the heavy entry capsules into orbit. The *Mars 2* capsule crashed and was destroyed in its landing attempt. The *Mars 3* landing capsule entered and landed successfully. After sitting on the surface about 90 seconds, it opened up to permit the start of scientific measurements, and began a television panorama of the surface. The transmissions terminated abruptly only 20 seconds later, after transmission of a fraction of a single (featureless) picture.

The American opinion was that the lander, having been compelled to descend in high winds in the midst of a major dust storm, was blown over and wrecked by winds traveling up to 300 miles per hour. The *Mars 2* and *3* orbiters continued to function for months. *Mariner 9* waited out the dust storm, whiling away its time by observing the Martian moons, and then, when the dust settled, took and transmitted more than 7000 excellent mapping photographs of Mars showing huge volcanoes, ancient river beds, immense chasms, and a vast range of other striking features.

The net result was, to the Soviets, a very expensive disappointment. But the race was not over: the United States had no plans to launch a Mars mission in 1973, but was planning to launch the Viking spacecraft in 1975. The secret of Soviet success was to do its Mars missions in 1973 and clearly win the race. To this end, they had decided upon another complement of two orbiters and two landers—also the plan for the Viking missions.

However, there was a problem. The lowest-energy trajectory to Mars in 1973 would involve launching from Earth near August 1, with arrival at Mars in February of 1974. At that time, Mars was near aphelion, so that the distance, trip time, and launch energy requirements would all be unusually large. So difficult was that particular mission that it was found necessary to split the payloads up among four D1e boosters, two carrying landers, and two with the orbiters. The spacecraft themselves were extremely well instrumented, and represented a large investment even without considering the costs of the four heavy-lift launchers.

In 1973 the Mars spacecraft issued forth like cans off an assembly line. *Mars 4, 5, 6,* and *7* were launched in sequence between July 21 and August 9. On February 10, 1974 the *Mars 4* orbiter approached its destination, but its orbital insertion engine failed to fire, and the spacecraft flew on by Mars without stopping. Two days later *Mars 5* arrived, and entered orbit successfully. It was needed to serve as a radio relay link between the landers and Earth.

The *Mars 7* entry capsule was released to impact Mars on March 9, but completely missed the planet. The *Mars 6* landing capsule arrived last, with all the hopes of this ambitious mission riding on it. The capsule survived entry, and was tracked while descending under its parachute. Mysteriously, the capsule ceased functioning before it reached the surface. The enormous gamble had failed.

Now it was America's turn. *Viking 1* and *2* were launched in the 1975 launch window by Titan 3E Centaur launch vehicles (see fig. 3.8). Each four-ton spacecraft deposited an orbiter in eccentric orbit about Mars, and each successfully placed a spidery, heavily instrumented lander on the sur-

Figure 3.8
Viking to Mars
The *Viking 1* launch vehicle, a Titan 3E Centaur, blasts off from the Eastern Test Range at Kennedy Space Center. The solid rocket boosters are jettisoned about two minutes after blastoff. Note the "hammerhead" configuration to accommodate the bulky hydrogen tanks of the Centaur stage atop the Titan 3 core vehicle.

face to search for Martian life and study the local climate. The mission (all four missions, actually) were astonishingly successful. The landers were kept alive for years. Only one thing went wrong with the Viking program: the superb cameras on the Viking orbiters, while checking out the preferred landing sites chosen for the two landers, found both sites to be strewn with boulders. The planned first landing on the Bicentennial Independence Day had to be postponed 16 days to pick smoother landing sites and assure a safe landing.

This was truly the end of the race to Mars. Since the summer of 1976, not one spacecraft has been sent from Earth to visit Mars. The Soviet planetary

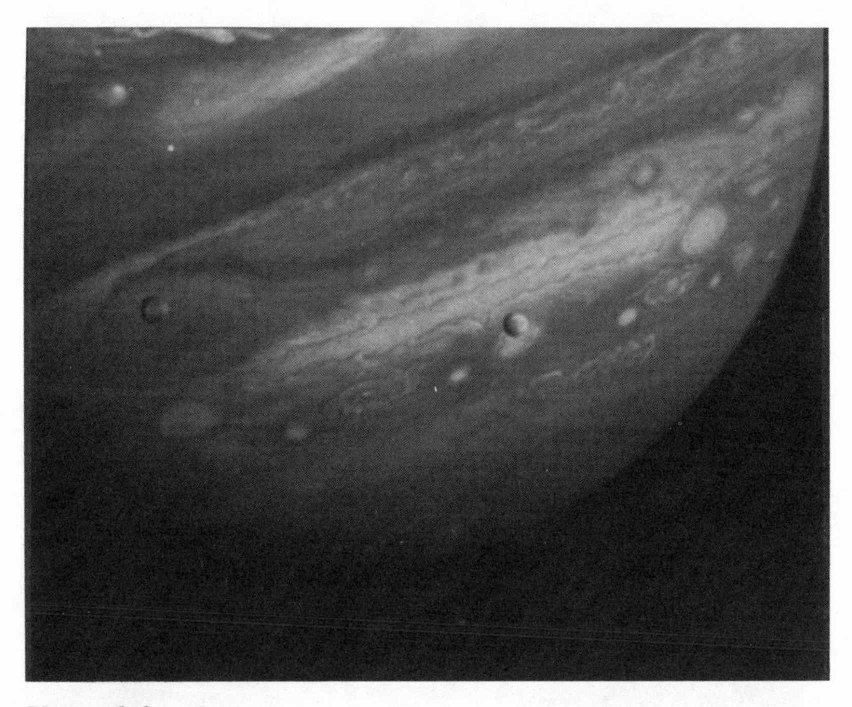

Figure 3.9
Jupiter from *Voyager*
This *Voyager* picture shows the jovian moons Io (left) and Europa silhouetted against the turbulent ammonia clouds of Jupiter. The "eye" behind Io is the famous Great Red Spot, a storm the size of the Earth that has been raging for at least 300 years.

program retreated to Venus, while the American program ranged from Mercury to Jupiter, Saturn, Uranus, and beyond (figs. 3.9–3.10).

THE GREAT SPACE STATION RACE

We have already followed the development of rendezvous technology in the American Gemini and Apollo programs and in the Soviet Soyuz program. The terrible misfortunes of the early days of the Soyuz program, starting with the death of Vladimir Komarov on *Soyuz 1* and the rendezvous and

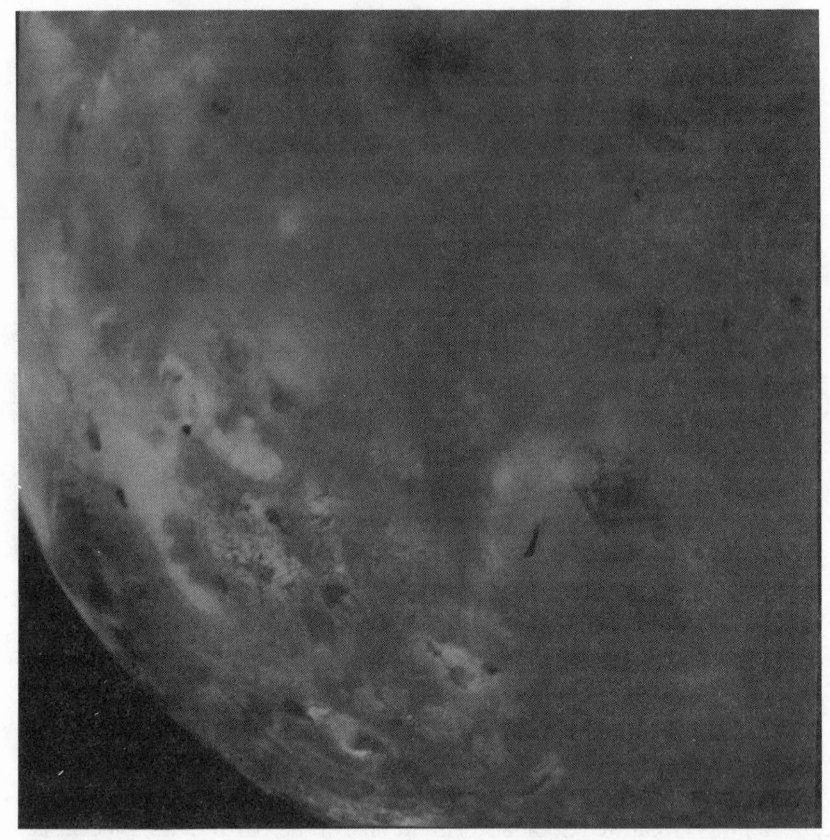

Figure 3.10
Volcanoes on Io
This *Voyager* photograph shows the surface of Jupiter's moon Io. Sulfur lava flows and drifts of white sulfur dioxide "snow" are partially obscured by huge plumes of gases and dust from two major volcanic eruptions. All craters visible on Io are volcanic, not impact, features. Io's volcanoes are powered by the tidal interactions of Jupiter and its four largest (Galilean) satellites, Io, Europa, Ganymede, and Callisto.

docking failures of the *Soyuz 2–3* and *6-7-8* missions, served as a warning that the path to a space station would not be an easy one.

When the *Salyut 1* space station was launched in 1971 it appeared that the Soviets were sure they had solved the problem. The 20.4-ton station was placed in orbit by a D1 booster, which was not man-rated. Four days later, *Soyuz 10* failed to dock with it and returned to Earth, its thruster fuel perilously close to exhaustion.

In June, *Soyuz 11* was sent up, again with a crew of three, for what the Soviet press hinted would be a month in space. They docked successfully with *Salyut 1* and spent 24 days aboard, setting a new space endurance record. The press, both in the Soviet Union and abroad, was full of the exhilarating news of their flight. On the 24th day, without warning, they returned to Earth. When the recovery team reached their capsule they opened the hatch—and found all three men dead.

The loss of these three brave young men was of course a devastating blow. The tragedy was traced to a faulty valve that apparently dumped the air out of the spacecraft cabin just before they entered the atmosphere. Since the cosmonauts were not wearing pressure suits, they must have suffocated before they could ride through reentry. The emotional impact of their deaths can best be brought home to Americans by reminding them of the recent Challenger tragedy. But, at the time of the *Soyuz 11* accident, the comparison that came to mind was with the Apollo lunar landing program, which was then in mid-stride, moving from success to success.

The Soyuz capsule did not appear in space again until an unmanned Kosmos test flight a year later. After six days in space, this Kosmos was successfully recalled to Earth. The Soviets felt a great sense of urgency about getting their manned program back on track. Further delays in the program could be extremely costly, because the American Skylab was due to be launched by the last Saturn 5 booster on May 14, 1973. Something had to be done before then.

The second Soviet attempt to establish a space station began with the launch of the 20-ton *Salyut 2* on April 3. The space station entered an unstable orbit, and started tumbling. It was judged too hazardous to try to send a manned mission to it, and just 11 days later *Salyut 2* fell from orbit and burned up in the atmosphere. There was now just one month left before Skylab.

On May 11 a Salyut-sized payload was launched into a Salyut-type orbit by the Salyut booster, the D1. Strangely, absolutely nothing was said about the functioning of this satellite. Shortly after launch it was named *Kosmos 557*. Eleven days later it fell from orbit and burned up.

While the crippled Salyut was still orbiting silently overhead, the *Skylab* space station was launched. The space station was a converted S4b stage, a cavernous hulk with room for a small army. It weighed 82.2 tons, or roughly four Salyuts. Eleven days later, a Saturn 1b booster lifted up the *Skylab 2* mission, a shipment adding another 22 tons, and containing three astronauts, Pete Conrad (commander of the *Apollo 12* lunar landing mission and *Gemini 11,* and crewman on *Gemini 5*), Joe Kerwin, and Paul Weitz. There they stayed for 28 days, eclipsing the record of the ill-fated *Soyuz 11* crew on *Salyut 1,* and then returned safely home.

While the American Skylab astronauts were still in space, the Soviets flew an unmanned Soyuz on a two-day qualification flight under cover of the Kosmos name. A similar manned mission was expected to follow.

The *Skylab 3* crew occupied the station for two months, setting another world endurance record. They returned to Earth on September 25. Two days later, cosmonauts Lazarev and Makarov flew a two-day shakedown mission on *Soyuz 12*. The new ground rules for Soyuz were noteworthy: no more three-man flights. The hazard of flying an overweight spacecraft with an inadequate fuel reserve was not worth taking.

In November, Skylab was again visited by its third and last team of three astronauts. They stayed for 84 days, again eclipsing the world endurance record. Unfortunately, only one Apollo capsule and one Saturn 1b booster remained. They were to be dedicated to a political spectacular, a rendezvous with a Soviet Soyuz in July of 1975, and were not available to extend the operational life of Skylab.

With the end of Skylab and the Apollo-Soyuz Test Project, American manned space flight went into total eclipse for almost six years. The Saturns were used up, their assembly line dismantled. Standard launch vehicles were being discontinued and production runs canceled to assure that NASA would have no choice but to build the Space Shuttle. Skylab was left, unmanned, with its orbit slowly decaying. With the arrival of the shuttle in 1979, the logic went, we could go back up, save Skylab and resume space station operations in a big way. In the meantime, however, the mere existence of any other means to travel in space would erode the imperative to build the Shuttle. Disposable boosters were therefore banned, and manned spaceflight ceased, leaving the heavens to the Soviets.

What a golden opportunity to work freely in space! No Americans to compete with, no more Apollos, no more Skylabs!

Since it was necessary to get all the kinks out of Soyuz, first came a two-month Kosmos systems test of an unmanned Soyuz. Then cosmonauts Klimuk and Lebedev took *Soyuz 13* up for an uneventful eight-day flight. A ten-day Kosmos test was flown in April 1974, and a Kosmos flew a two-day mission at the end of May. This successful four-flight series apparently restored Soviet faith in Soyuz.

The 20-ton *Salyut 3* space station was launched on June 24, 1974 into a stable orbit. Nine days later *Soyuz 14* followed it into space. The rendezvous and docking, refreshingly, went well, and the crew spent 16 days aboard Salyut carrying out a variety of experiments. That's all we know about the equipment and work done on *Salyut 2* and *3*: the details of these missions have been held in strictest secrecy. Not even external photographs or drawings of these two Salyuts have been released because of the sensitive nature of the military work done aboard them.

Soyuz 15 was launched on August 26 to take over the duties of running *Salyut 3*. The two-man crew tried but failed to dock with Salyut, and had to return to Earth after only two days in space. The unsuccessful cosmonaut team, like the two senior *Soyuz 10* crew and the *Soyuz 3* pilot, who also failed on rendezvous missions, never flew again. Nor was *Salyut 3* ever reactivated, for reasons known only to the Soviet leadership. The next Soyuz mission was a six-day free-flying test, unrelated to Salyut, in preparation for the Apollo-Soyuz mission.

On December 26, 1974 the *Salyut 4* space station was launched. Two weeks later the *Soyuz 17* crew paid a one-month visit to Salyut. *Soyuz 18* arrived in May with two men aboard. They stayed on *Salyut 4* for two months before returning to Earth. The *Kosmos 613* flight had shown that two months was a safe amount of time for a Soyuz capsule to spend in space.

Soyuz 19 met and docked with *Apollo 18*, commanded by Tom Stafford (*Gemini 6* and *9*, *Apollo 10*), Deke Slayton (one of the original Mercury astronauts), and Vance Brand. This mission was flown at the lowest altitude ever used by an American manned flight. This constraint was imposed by the limited fuel margin of Soyuz, which, even with only a two-man crew, had no spare fuel to maneuver with. It had to sit passively while Apollo docked with it. The seven-ton Soyuz returned after six days, and the 16-ton Apollo after nine. Why didn't Apollo get an invitation to dock with *Salyut 4*? There were two basic reasons, and this time military secrecy had nothing to do with it. First, *Salyut 4* had only one docking port, which had to be occupied by the Soyuz craft on which the cosmonauts came up. Second, *Salyut 4* was being "eaten alive" by an incredible invasion of green algae that coated most of its internal surfaces. They were not proud of its appearance!

Nothing much happened for almost another year. There was an unmanned three-day Soyuz test under the name *Kosmos 772,* followed by *Soyuz 20,* an unmanned biosatellite, dedicated to testing the effects of space flight on plants and animals, which was automatically docked with *Salyut 4* for three months.

In June 1976, *Salyut 5* was launched. *Soyuz 21* visited for 49 days, and made an emergency landing without any of the usual prior warning—at night! The *Soyuz 22* mission was flown in an orbit with a 65-degree inclination, a unique feature never fully explained, and it came nowhere near Salyut. *Soyuz 23,* in October, suffered a guidance system failure just before rendezvous, and was unable to dock with *Salyut* under manual control. The crew carried out an emergency landing, and had the bad fortune to land in the midst of a blizzard. Even worse, they landed in a lake!

Kosmos 869 in November ran through an unusual 18-day Soyuz-type mission. Then, in February 1977, *Soyuz 24* flew an 18-day visit to *Salyut 5*. The crew purged the air from the station (perhaps a hint why the *Soyuz 21* crew was in such a hurry to get home) and replaced it, and returned a reentry capsule from the station to Earth before shutting down the station permanently. As with *Salyuts 2* and *3,* a veil of secrecy surrounds the *Salyut 5* mission.

In July, before *Salyut 5* had fallen from orbit, the D1 booster was used to lift the heavy *Kosmos 929* into a Salyut-type orbit, but one somewhat higher than used by the earlier Salyuts. Nothing was done with this pseudo-Salyut, but in September the *Salyut 6* space station was launched into a virtually identical orbit. *Salyut 6* was an improvement on the earlier space stations in that it had two docking ports. It could thus exchange crews without leaving the station unmanned between visits, or it could accommodate unmanned resupply vessels that could dock automatically with the station, a technique demonstrated by the *Soyuz 20* biosatellite flight.

Salyut 6 was destined to be a stunning demonstration of all the things the Soviets had learned how to do in space, an object lesson in how to operate a space station. But its beginning days were anything but encouraging. *Soyuz 25* attempted docking with Salyut, but the docking mechanism failed to work. Although no blame attached to the Soyuz crew, the failure was a bitter disappointment. It seemed that the new, improved Salyut might have a minor flaw that could prevent it from being used. *Soyuz 25* returned to Earth after two days in space instead of the intended three months.

The remaining experience with *Salyut 6* was like a story book. The *Soyuz*

26 crew was able to get aboard Salyut through the "back door" and fixed the "front door" docking mechanism. During the spacewalk, when he checked out the suspect docking port, cosmonaut Grechko was surprised to see his colleague Romanenko coming out of the airlock to watch him. He was astonished to see the end of Romanenko's safety line pull free and start to drift away. He made a grab and caught Romanenko just in time to keep him from drifting off in his own orbit. Grechko (*Soyuz 17*) and Romanenko then spent 96 uneventful days on board Salyut. They tended a little garden, conducted medical experiments, and even entertained guests. Visits were essential because of limits on how long the Soyuz return capsule could be stored in space: fresh capsules had to be brought up on a regular schedule and swapped for the older ones.

Salyut 6 functioned superbly from September 1977 to mid-1981. *Soyuz* flights 26 through 40 docked successfully with Salyut, excepting only *Soyuz 33,* which suffered a failure of its onboard rocket engine shortly before the scheduled docking. The Soviet and Bulgarian cosmonauts aboard (Rukavishnikov and Ivanov) survived a hair-raising emergency landing. Their precipitous return meant that they could not swap their fresh capsule for the aging *Soyuz 32,* then attached to Salyut. It was accordingly replaced by *Soyuz 34,* which was launched and docked unmanned.

Several novel features marked the *Salyut 6* mission. First, the presence of the second docking port permitted the docking of unmanned resupply vehicles at the kitchen door even while the residents' taxi was parked at the front door. The spacecraft used for this purpose, essentially an automated Soyuz freighter, was called *Progress*. No fewer than twelve *Progress* freighters were docked with *Salyut 6* over its lifespan. Further, the improved *Soyuz T* (for Transport) was introduced, apparently qualified by a flight test labeled *Kosmos 1074*. Whereas the earlier Salyuts had accommodated one crew and their Soyuz vehicle at a time, *Salyut 6* was manned on three separate occasions by long-duration crews that stayed in orbit for a time much longer than their Soyuz craft had ever been flown in space.

The need to "recycle" the older Soyuz vehicle docked with Salyut at first required that a replacement be sent up every two months. These were undemanding, routine flights in which a pilot brought up the new Soyuz (complete with mail and gifts from home), visited with the Salyut "resident" crew for a few days, and then returned to Earth in the old Soyuz. Since there was a second seat on the Salyut, the Soviets came up with the idea of bringing along a "guest cosmonaut" from another nation on each flight. So far, guests from eleven other nations, including India and France, have flown in this program.

The *Soyuz 26* "resident" crew, cosmonauts Grechko and Romanenko, spent 96 days in space and were visited by two "guest" crews. The *Soyuz 29* crew, reoccupying the Salyut station left empty after the return of the first residents, stayed 140 days, a new world record, and were visited by two international crews with fresh Soyuz capsules. From November 2, 1978 until February 25, 1979 Salyut was unoccupied. Then it was reoccupied by the *Soyuz 32* crew of cosmonauts Lyakhov and Ryumin, who set yet another record of 175 days in Salyut. From August 19, 1979 to April 9, 1980 Salyut again was unoccupied. It was reopened by the *Soyuz 35* crew of Popov and Ryumin, who stayed for 185 days. This brought Valerii Ryumin's space experience up to almost exactly one year. They entertained four visiting crews, and left the station empty upon their departure on October 11.

Salyut 6 was revisited for only thirteen days by the three-man *Soyuz T3* mission in November, and was again left empty until the following March. *Soyuz T4,* with cosmonauts Kovalyonok and Savinykh aboard, then arrived to set up housekeeping. They also entertained two visiting crews before returning to Earth. After almost four years of use, *Salyut 6* was finally due for replacement.

In 1979, in the midst of the glorious career of *Salyut 6,* the only American space station, *Skylab,* finally ran out of steam. Its orbit had long been slowly decaying. Plans to rescue it by using one of the first Space Shuttle flights to lift it to a higher orbit were deferred, as the Shuttle program slipped and slipped again. Finally *Skylab* fell, two years before the Shuttle made its maiden voyage. The Shuttle's original purpose had been to shuttle astronauts and their supplies back and forth between Earth and space stations. Now it had nowhere to go.

Salyut's replacement, when it came, was the very similar *Salyut 7.* Berezovoy and Lebedev, arriving on *Soyuz T5,* spent 211 days as resident crew, conducting a wide range of experiments on materials science, space medicine, and growing plants in orbit. They were visited by two guest crews (including a French astronaut, Jean-Loup Chretien, and a Soviet woman cosmonaut, Savitskaya, who was insultingly treated like yet another nonfunctional foreign dignitary) and by four unmanned Progress resupply vehicles. The inaugural mission aboard *Salyut 7* was a tremendous success.

Salyut 7 drifted, unmanned and empty, from the time of the return of Berezovoy and Lebedev on December 11, 1982 until the launch of *Kosmos*

1443 the following March. This was no mere Kosmos coverup of something important. *K 1443* was a 20-ton "Star" module related in design to Salyut, and about the same size. It followed the pattern of the much smaller Progress resupply vehicles by carrying out an automatic rendezvous and docking with *Salyut 7,* doubling the size of the space station. This module not only carried supplies, but also added new equipment and two more "rooms" to Salyut. A month later, the *Soyuz T8* crew came up, full of plans for another marathon flight, and full of enthusiasm for the great variety of research projects they anticipated doing during their months in space.

Three-man crews had been banned after the *Soyuz 11* disaster in 1971, and it was not until the improved Soyuz T spacecraft had been flown successfully that a three-man crew was again flown. Soviet spokesmen asserted that the new guidance system of Soyuz was so superior that it could fly docking missions with very small expenditures of fuel, and they accordingly felt that a flight with a three-man crew could safely be flown within their total weight allowance. The fourth such flight, *Soyuz T8,* was commanded by Vladimir Titov (unrelated to Gherman Titov of *Vostok* days), with a crew of Aleksandr Serebrov and Gennadii Strekalov. But on this flight the rendezvous radar dish failed to deploy properly, leaving the navigation computer "blind." Attempts to control the docking manually were ineffective, and the crew ran low on fuel. After two days in space, *Soyuz T8* was back on the ground.

The next mission to reoccupy *Salyut 7* was *Soyuz T9* in June 1983, two months after the *Soyuz T8* debacle. It was apparently a previously scheduled Soyuz-capsule exchange flight. Only two cosmonauts were flown; all went well, and Lyakov and Aleksandrov spent 150 days aboard *Salyut.* The Soviets felt that the *Soyuz T* was good for at least 100 days in space before it would need replacement. A scheduled Soyuz replacement flight in September, however, never reached the station. Shortly before launch, a fire broke out in the first stage of the A2 booster. The booster was fully enveloped in fire and the umbilical cables were burned through before the cosmonauts could receive word of the fire and activate the escape rocket. After a hair-raising ride out of the fireball, cosmonauts Titov and Strekalov landed safely. This incident demonstrates two principles: one, even the most experienced booster sometimes has bad days; and two, planning ahead of time how to deal with foreseeable disasters makes it possible to cope with them when they finally occur. (Unfortunately, the Challenger failure mode seems to have been completely unforeseen).

Lyakov and Aleksandrov talked things over with their ground controllers and decided that their Soyuz was probably good for at least 180 days. They were correct, as they proved when they rode their original Soyuz T9 safely back to Earth after 150 days.

Soyuz 7 was reopened in February of 1984 by the *Soyuz T10* crew of Kizim, Solovyov, and Atkov. They were to stay in space for an incredible eight months. There was a visit and capsule exchange after the first two months with an Indian guest cosmonaut, and another at the five-month mark, with a second visit by cosmonaut Savitskaya.

In June 1985, *Salyut 7* was reopened by the *Soyuz T13* crew of Dzhanibekov and Savinykh, with rumors circulating that the Soviets would soon be using genuine crew rotation so as to keep their space station permanently manned. They were visited by a routine Soyuz-exchange mission in September, which also for the first time rotated part of the crew. Vasyutin and Volkov from *Soyuz T14* stayed aboard Salyut with Savinykh. Grechko from *Soyuz T14* and Dzhanibekov from the *Soyuz T13* crew returned to Earth in *Soyuz T13*. With such rotation, Salyut could be kept permanently manned. But the goal of permanent occupancy was not yet to be realized. Illness struck Vasyutin, and the entire crew returned to Earth in late November. Volkov and Vasyutin had each spent 65 days in space, and Savinykh had spent 168 days. In light of Soviet hints that they intended to continue to build up to even longer-duration flights, it seems likely that a duration of at least eight and probably nine months was intended.

By the end of 1985, eleven Soyuz missions had been devoted to support of *Salyut 7*, of which nine were successful. Progress vehicles 13 to 24 had docked with Salyut and replenished its supplies of fuel and life-support materials. The *Kosmos 1443* and *1669* "Star" expansion modules had also flown lengthy missions attached to Salyut. In all, 26 launchings had been carried out in the program, compared with 34 launches in the *Salyut 6* program. No indication had been made by Soviet sources whether the career of *Salyut 7* was drawing to a close. *Salyut 6* had functioned for 44 months. *Salyut 7* reached this milestone in December of 1985.

Central to Soviet space station are: mission durations that have been built up steadily to about eight months; the use of the Star module, which had made long-duration missions possible with far fewer resupply vehicles; the Star technology, which lays the groundwork for assembling larger space stations carrying more cosmonauts; and the increasing ability to store the Soyuz T spacecraft in space for many months at a time, which makes frequent manned visits much less important. This all adds up to a greatly increased level of autonomy for Salyut. The time is not far away when a Salyut crew will be able to fly a six- to nine-month mission without any visits whatsoever.

The vast body of space experience gathered in the Salyut flights completely overshadowed all that was gained in the American manned space program (fig. 3.11). The pinnacle of the American space station activities was the 84-day mission of *Skylab 4* in 1973. To date, no other American mission has remotely approached that mark, while numerous Soviet missions have greatly surpassed it. From the Apollo-Soyuz mission of 1975 to the first Space Shuttle flight in 1981, six precious years were lost. We have not yet recovered from the folly of sacrificing our Saturn boosters and Apollo spacecraft long before we had their replacements in hand.

The planned demise of Apollo and the Saturn boosters left Skylab stranded in a slowly decaying orbit, and the Space Shuttle was repeatedly delayed until it finally came along far too late to save Skylab. And of course, with the abandonment of the Saturn 5, the United States lost the ability to replace Skylab.

It is depressing to compare what the United States abandoned in 1973 to the capabilities developed by the USSR during use of their first seven space stations (table 3.4). The *Salyut 7* complex, with Salyut, a Star module, a Progress resupply vehicle, and the occupants' Soyuz return capsule attached, masses 54 tons. The Skylab complex, with an Apollo attached, weighed in at 104.5 tons. The *Salyut 7* complex, with its additional solar panels attached, now delivers up to 6 kilowatts of electrical power. *Skylab* had 12 kilowatts. The Salyut-Star complex provides about 5000 cubic feet of living space. Skylab had 10,000 cubic feet.

The bright side, NASA tells us, is that the Space Station is coming. When it finally enters use, around 1994, it will deliver 64 kilowatts—and be almost as big as Skylab. But how big will *Salyut 9* or *Mir 3* be by then? Maybe they won't even call it by such names any more. Something like "Cosmograd" might be more appropriate.

SOVIET STRATEGY AND PRIORITIES IN SPACE

The popular image of the Soviet space program is one of a highly secretive and inscrutable peaceful scientific effort, conducted with callous disregard of human life. The American public is also treated to the view that the

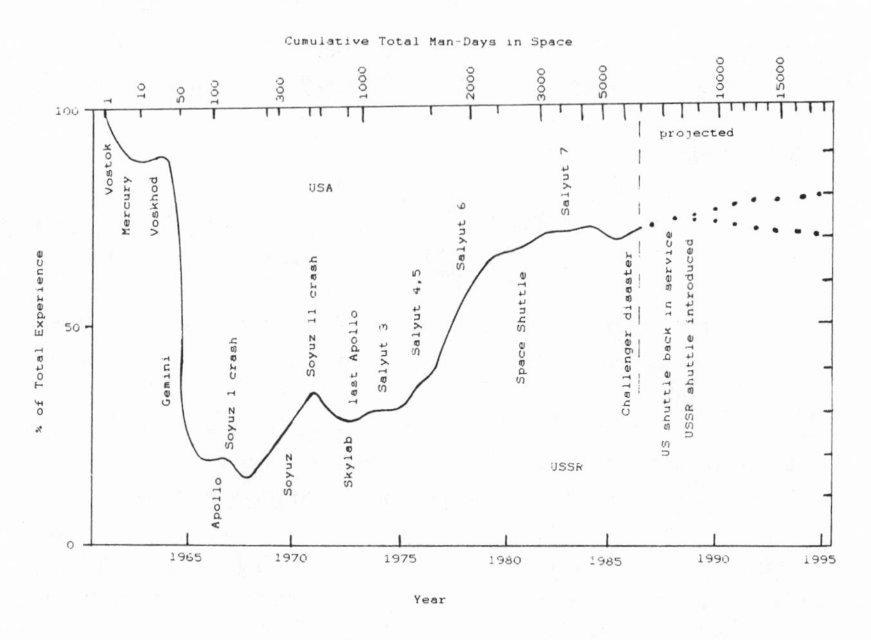

Figure 3.11
Manned Space Flight Experience
This diagram shows both the total amount of experience gathered in space on Soviet and American flight missions and the breakdown of that experience between the two spacefaring nations. The cumulative number of man-days spent in space is marked across the top of the diagram, and the breakdown in terms of the specific flight programs of the two nations is shown within the frame. The early period of Soviet dominance (nearly 100% of the total experience) was brought to an end by the Gemini program. The United States dominated manned activities in space during the Gemini, Apollo, and Skylab eras, but the Salyut space station progrram (especially the highly successful *Salyut 6* and *7*) returned the lead to the Soviet Union. In recent years the Soviet Union has had about 70 to 75 percent of the total manned spaceflight experience; however, the Soviet lead in long-duration space flights is even more striking. If we consider only flights with durations longer than 90 days, the Soviets have all the experience. The Space Shuttle, with its practical limit of about a week in orbit, launches huge numbers of people but gathers very little useful data on the biomedical effects of lengthy space flights.

Table 3.4
Manned Flight Landmarks and Duration Records

Year	*Spacecraft*	*Crew*	*Duration*
1961	*Vostok 1*	Gagarin—first man in space	1.5 hours
1961	*Vostok 2*	Titov	1 day
1962	*Vostok 3*	Nikolayev	4 d
1962	*Vostok 4*	Popovich—first dual flight	3 d
1963	*Vostok 5*	Bykovskii	5 d
1964	*Vostok 6*	Tereshkova- first woman	3 d
1964	*Voskhod 1*	Komarov, Yegorov, Feoktistov—first multi-manned flight	1 d
1965	*Voskhod 2*	Belyayev, Leonov—first EVA	1 d
1965	*Gemini 5*	Cooper, Conrad	8 d
1965	*Gemini 7*	Borman, Lovell	14 d
1967	*Soyuz 1*	Komarov—first fatality	1 d
1968	*Apollo 8*	Borman, Lovell, Anders—first men to orbit Moon	7 d
1969	*Apollo 11*	Armstrong, Aldrin, Collins—first men to land on Moon	8 d
1970	*Soyuz 9*	Nikolayev, Sevastyanov	18 d
1971	*Soyuz 11*	Patseyev, Volkov, Dobrovolskii—endurance record; all die	24 d
1973	*Skylab 2*	Conrad, Kerwin, Weitz	28 d
1973	*Skylab 3*	Bean, Garriott, Lousma	56 d
1973	*Skylab 4*	Carr, Gibson, Pogue	84 d
1977	*Soyuz 26*	Romanenko, Grechko (on Salyut 6)	96 d
1978	*Soyuz 29*	Kovalyonok, Ivanchenkov (Salyut 6)	140 d
1979	*Soyuz 32*	Lyakhov, Ryumin (Salyut 6)	175 d
1980	*Soyuz 35*	Popov, Ryumin (Salyut 6)	185 d
1981	*Soyuz T5*	Berezovoi, Lebedev (Salyut 7)	211 d
1984	*Soyuz T10*	Kizim, Solovyev, Atkov (Salyut 7)	237 d

This table divides naturally into three eras: 1961–1965, in which the Soviets were essentially uncontested; 1965–1973, in which there was very vigorous competition and the United States scored triumph after triumph; 1974–1986, in which the United States abdicated its position of leadership and the USSR registered a long series of impressive records.

peaceful Soviets are rightly alarmed by American plans to militarize space, which they wisely claim should be reserved as a medium for dispassionate scientific research. Almost everyone knows American plans to deploy giant weapons in orbit, thus violating the weapons-free status of space so long espoused by the Soviets. The Strategic Defense Initiative, mocked by the epithet of "Star Wars," is an attempt by a fascist American administration to develop an American ability to carry out a surprise first strike against the

Soviet Union. Ballistic missile defenses are really offensive in purpose, and are immoral.

This view is a hopeless conflation of truth, guesswork, interpretation, and outright lies. Before sorting things out, we need to review a few facts about the Soviet space program.

The detailed history of Soviet development of manned spaceflight and space stations permits the strong conclusion that their program is conducted with great care and conservatism. Detailed reporting of manned activities, including even live television coverage of some launches, has come to the Soviet program. Their long-standing practice of flying unmanned test missions to check out all innovations in their manned spacecraft before committing men to them tells us several things. First, that are being extremely safety-conscious. They deviated from this tradition only briefly, during the Voskhod program, when they crammed so many men into the capsule that they had to omit the escape rocket—but they did that in full knowledge of the excellent reliability of their booster. Second, their manned flights (except on Salyut) do almost nothing that cannot be done automatically. Indeed, their success with automated docking of the unmanned Progress resupply vehicles and their unmanned Soyuz flights is far more impressive than the indifferent record of their manned missions. The case could be made that they would be better off tying the hands of their cosmonauts. Third, the attempts to cover up the test flights suggest that they do not want to project an air of extreme conservatism, but rather of casual competence. Why have failures, including four inflight deaths, occurred in their program? In part, because they have flown a great number of flights, and done a great number of demanding things. But another factor is the historic lag of Soviet miniaturization and electronics technology behind Western standards. They have made great strides in recent years to narrow that gap, but it still exists. They were driving 1938 Packards while we were building Porsches, which confers certain obvious advantages—and some disadvantages, too.

The secrecy issue is not a Soviet or Communist aberration. From the very first, the openness of the American space program, with its live coverage of every launch failure, was the cause of astonishment and anger among our friends. Almost everyone on Earth thinks it simply foolish to air one's dirty linen in public.

The obsession with openness is an American phenomenon. Americans have the tradition of public discussion of every problem, of contempt for secret management of news, and a fierce tradition of independence of the press. The ability to publish information critical of a space program contains within it the ability to criticize the decisions and actions of the government running the program. Not every country can stand such freedom. Soviet secrecy is a spur to Americans (and to Soviet citizens) to read between the lines, to break the code, and figure out what really is going on. Moscow street humor is rich in jokes about official lies and secrecy. Nonetheless, even with such Soviet secrecy as exists, informed foreign observers of the Soviet space program have habitually anticipated and predicted most of their major developments before they were acknowledged by the Soviets.

We also should not forget that the Soviet space program has been conducted almost entirely with boosters that were developed for military purposes. The Soviets have a deep, pathological hatred of spying and a fear of invasion. It is commonplace for Americans to dismiss this as "Soviet paranoia." This is a very fundamental and possible disastrous error. Russia was invaded by Sweden in the seventeenth century, by Napoleonic France in the nineteenth century, and by Kaiser Wilhelm and then by Nazi Germany in the twentieth century. They have not yet forgotten the Mongol hordes, against which they, in their own words, "saved European civilization from extinction." Tens of millions of Soviet citizens died in World War II; indeed, most of the war was fought on Soviet territory. Until very recently, all of the Soviet leadership had lived through those years of horror, and remembered it very well indeed. Invasion is not an irrational fear; it is a memory of personal experience. Their fear of a revived German militaristic state is most genuine. It is for this reason more than any other that they have not, and may never, relinquish their stranglehold on East Germany. Poland, despite the press reports, is not repressed for its own sake: it must be held because it is the highway to Germany. And if you are ever in Moscow, check to see what is playing at the movies. At least a quarter are about World War II. Finally, did you know that Soviets have generally never heard of World War II by that name, and know next to nothing about Pearl Harbor, Mussolini, and the Free French? In their book it was a German invasion of the Soviet motherland—and they call it the Great Patriotic War.

And what about the Soviet obsession with spying? Well, how would an American feel if a Soviet spy plane were shot down on the outskirts of Kansas City? We have included an excellent photograph of the original launch site of *Sputnik 1* and *Vostok 1* (fig. 3.12) to remind you of the facts: the United States carried out routine spyplane overflights of the USSR, using the remarkable U2 high-altitude aircraft, from 1957 until Francis Gary Powers crashed near Sverdlovsk on May Day of 1960. The overflights had been ordered by President Eisenhower out of growing concern over Soviet insistence in 1957 that they had an ICBM. Call it American paranoia—except that here too the fear was fully justified. It was the need to keep an eye on Soviet missile development that led to the requirement for the first American military reconnaissance satellites in 1959, and motivated the accelerated development of orbital reconnaissance after the Powers incident.

In 1960, *Tiros 1,* the first American meteorological satellite, was vigorously denounced by the Soviets as a spy subterfuge. Actually, the low-resolution cameras on *Tiros* were even incapable of detecting New York City (see, by contrast, fig. 3.13).

The issue of weapons in space is another matter altogether. Of course all ballistic missiles pass through space; that is not what we are talking about. We refer here to weapons placed on satellites, in orbit. The Space Shuttle was fervently denounced as a secret American weapon by the Soviets. This may sound ludicrous, even hilarious, to Americans who watch it on television. But the Soviet media picture it as sneaking up on some unsuspecting Soviet satellite and gobbling it up. They have heard of the recovery for reflight of failed communications satellites by shuttle astronauts. They are unaware of, and are constitutionally unable to understand, that this is just a job done for money, a service to various national governments (and insurance companies!). To them, it drips with sinister significance. Part of the reason is that, for several years, the Soviets have had a mini-shuttle program which some observers believe was invented by their military for the purpose of inspecting and destroying American satellites. Could our purpose be any less noble? But the very idea of risking a $2.5 billion Space Shuttle, watched from the start by the press, on a mission to run down and eat a $0.03 billion Soviet reconnaissance satellite that may carry a $10 explosive booby trap is just pricelessly absurd.

Yet the Soviets have carried out large and lengthy programs that have actually developed two completely illegal space-based weapons systems, the Fractional Orbit Bombardment System (FOBS) and their Anti-Satellite (ASAT) system. Every step of both programs has been carefully watched from the outset, yet the Soviets brazenly deny that they have done any such

Figure 3.12
U2 Photograph of the A1 Vostok Launch Complex
Khrushchev's extravagant claims that the Soviet Union had hundreds of giant ICBMs led President Eisenhower, who doubted the veracity of these claims, to authorized photographic reconnaissance of the "Baikonur Cosmodrome" (Tyuratam) and other Soviet strategic bases. This photograph was one of the last returned before Francis Gary Powers was shot down in central Russia, near Sverdlovsk, bringing the U-2 overflights to an end. One crucial benefit of the illegal U-2 flights was the demonstration that the number of operational ICBMs was about five, not several hundred. Khrushchev, unhappy with the vulnerability of the "Semyorka," limited its use as an ICBM while favoring it as a satellite launching vehicle: over 1100 have been launched to date. This particular launch pad was the starting point of the early Sputnik and Vostok flights.

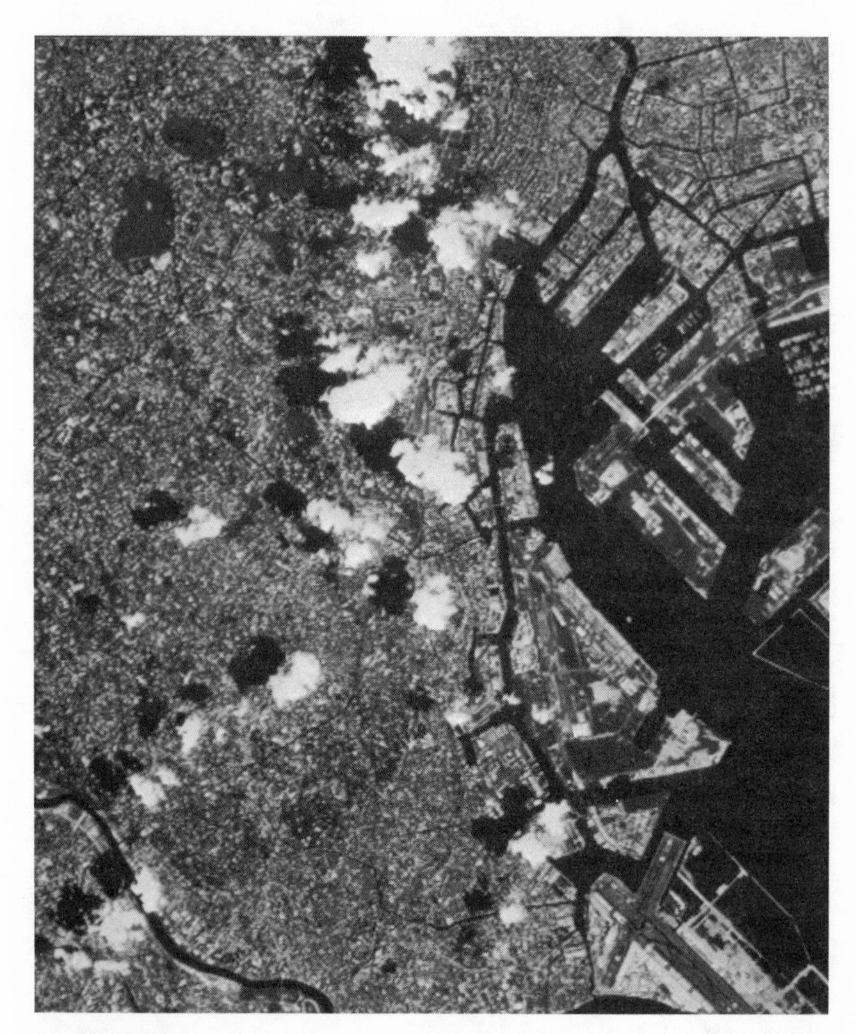

Figure 3.13
Earth from Space
High-resolution photographs taken by military reconnaissance satellites are not available for publication. This photograph was taken by a Landsat Earth-resources satellite. The image, a false-color composite of images made at several wavelengths in the visible and near infrared, has a resolution about 100 feet. This scene shows the center of Tokyo. Ships in the harbor are visible, and airplanes on the runways at Narita airport (bottom) are barely discernable. For comparison, a typical military photographic reconnaissance satellite may have a resolution of 6 inches: a single point in this Landsat image would be seen as 40,000 points in the military image.

thing. The thousands of fragments of debris made by the explosions of their ASATs cannot be discussed with them: they don't officially exist.

And then there is the statistical evidence on the nature of the Soviet space program. The very people who accuse the Americans of militarizing space have developed some interesting abilities themselves. Here are the facts. Judge for yourself. The Soviet Union has recently averaged 92 military satellites per year vs. 7 per year for the United States. Military launches make up 77 percent of the Soviet total vs. 29 percent of the American total. The number of military reconnaissance (spy) satellites launched each year by the USSR has been two to three times as large as the total number of American satellites of all types! You can measure space program priorities quite objectively: by far the largest category of American launches is civil communications satellites. The three leading Soviet categories are all military (see table 3.5).

Ballistic missile defense and antisatellite weapons have long been of interest to both nations. Until President Reagan's famous "Star Wars" speech, the Soviet position on anti-ballistic missile systems (ABMs) was clear: offensive weapons are immoral and defensive weapons are moral. The Soviet ABM effort led to the deployment of a "missile shield" around Moscow in the early 1970s. These missiles, given by NATO the singularly stupid code name of "Galosh," are now being extensively modernized. No American ABM system was ever deployed around population centers. The Safeguard ABM was briefly deployed in the late 1970s at a Minuteman ICBM base in North Dakota, but the missiles have long since been removed.

A point of ABM philosophy: intellectually, it is easy to see that, if you are expecting to carry out a first strike, you don't need to protect your ICBMs (which you will already have launched), only your cities. The public Soviet argument for defending their cities and not their ICBMs was from the gut, not the brain: people are more important than missiles. (Don't forget: defense in any form is moral. Offense is immoral.) It is left as an exercise to the reader to decide what the Soviets' true motives were.

From the earliest days of the space age, there has been a chilling prospect that a nation could launch a large number of nuclear warheads into orbit under the guise of civil or scientific satellites, store them in space for years, and then recall them upon Earth-surface targets at will. As early as 1959 the United States was experimenting with a missile called the Bold Orion, a satellite interceptor that could be launched from a B-47 bomber. Because of

Table 3.5
Types of Satellites Launched by the USA and USSR

	Year				
Satelite Type	*1981*	*1982*	*1983*	*1984*	*1985*
USSR total	*123*	*119*	*116*	*115*	*121*
(military subtotal)	(92)	(94)	(90)	(91)	(100)
reconnaissance	58	62	58	56	65
communications	10	20	19	20	21
navigation	21	10	13	15	14
ASAT	3	2	0	0	0
(man-related subtotal)	(5)	(9)	(7)	(10)	(5)
Soyuz	2	0	0	0	0
Soyuz T	1	3	2	3	2
Salyut	0	1	0	0	0
Progress	1	4	2	5	2
Kosmos/Star	1	0	1	1	1
Shuttle test	0	1	2	1	0
(civil subtotal)	(26)	(16)	(19)	(14)	(16)
communications	18	14	13	10	12
weather	2	2	2	2	3
research	6	0	4	2	1
USA total	*19*	*21*	*28*	*33*	*37*
(military subtotal)	(6)	(6)	(6)	(11)	(10)
reconnaissance	5	6	4	6	2
communications	1	0	1	2	5
navigation	0	0	1	3	3
(manned subtotal)	(2)	(3)	(4)	(5)	(9)
Space Shuttle	2	3	4	5	9
(civil subtotal)	(11)	(12)	(18)	(17)	(18)
communications	5	10	11	11	17
weather	2	1	3	1	0
research	4	1	4	5	1
Japan total	*3*	*1*	*3*	*3*	*2*
PRC total	*3*	*1*	*1*	*3*	*0*
ESA total	*5*	*0*	*3*	*4*	*5*
India total	*1*	*0*	*1*	*0*	*0*

the importance of space surveillance after the Powers incident in 1960, President Eisenhower adopted the "open skies" policy, asserting US rights to watch the USSR from space, while assuring noninterference with Soviet satellites. To demonstrate the sincerity of his position, Eisenhower drastically curtailed Anti-Satellite (ASAT) programs. This principle was castigated by the Soviets until 1962, when they began launching their own military reconnaissance satellites. Since then, it has been adopted as a de facto standard by both nations.

President Kennedy revived and augmented two major ASAT initiatives in the early 1960s. The first was the development of the Nike Zeus ABM by Program 505. The Nike Zeus was a 10-ton missile with an incredible 450,000 pounds of thrust at blastoff. It could hurtle up through the atmosphere with an acceleration of over 20 times that of Earth's gravity to intercept incoming missile warheads or passing satellites out to a range of about 160 miles. It was tested eight times at Kwajalein atoll between 1963 and 1967. This missile was similar to the Galosh ABM now deployed around Moscow.

Kennedy's second ASAT endeavor, Program 437, used a Thor IRBM to lift a nuclear warhead into space. The range of this missile was about 700 miles, giving it a far better ability than the Nike Zeus to travel "cross-range" and intercept satellites that were not passing almost directly overhead. Between 1964 and 1970 some sixteen test flights were flown, using simulated nuclear warheads. This missile was maintained in operational status until 1975. Since the late 1970s the United States has not had an operational ASAT or ABM. The USSR maintains its ABM defenses of Moscow as well as its orbital ASAT as operational programs.

No American space-based weapons system has ever been developed or deployed. The Soviet orbital ASAT and FOBS have no American counterpart.

There simply cannot be any room for doubt: the Soviet program is overwhelmingly military in nature. Yet even the minor portion that goes into nonmilitary manned flights and scientific research represents a large funding commitment.

It is to the advantage of both the Soviets and Americans to shift the balance of both programs more toward the civil side. If there is one service we can do that will be of lasting benefit, it is to lower the level of fear and distrust between the two nations. This requires less secrecy, more freedom of communications and travel, more openness and independence of the media, and more reconnaissance by both sides—enought to know with certainty that the other guy isn't up to anything sneaky. It requires on-site inspection with no advance notice. In short, it violates every Soviet principle. So the real task is to find something that will reduce Soviet "paranoia" enough to permit such an accommodation. This means understanding the rational and historical basis of their fears—and sympathizing with them for their traumatic past. It most certainly does not mean accepting Soviet political dogma or pretending to believe them when they lie, but it does mean not shoving Western dogma down their throats and not lying to them!

In the many races that have transpired in space, the Soviets have learned how they work best and how the Americans work best. The Soviet program thrives on constancy of funding, undeviating political support, and long-range planning. It has great difficulty innovating or responding to rapidly changing conditions. The Americans are impulsive and inconstant, are subject to complete rethinking of their programs at budget time every year, change administrations every four years, and suffer wild and unpredictable boom-or-bust cycles of funding and political support. But there still is the basic American ability, so eloquently attested to by Khrushchev, to respond with explosive energy to an overt challenge, and to accomplish the impossible in a few years. Perhaps Pearl Harbor and the Apollo program should be taught in Soviet schools, as Napoleon and Stalingrad should be taught to Americans. The Soviets learned early that, in a face-to-face confrontation, the United States could start from scratch and move from Volkswagen-sized Mercury capsules to landing men on the Moon in eight years.

The formula for success was simple: the Soviets learned not to talk about their long-range goals. They resolved to advance step by step, and never to give the Americans some specific goal to race for five or ten years hence. Much of the public conservatism exhibited by the Soviet program in the post-Khrushchev years arises from exactly this realization. They still have their long-range goals, but it's up to us to deduce what they are from what we see them doing. Fortunately, by now we have a lot of experience analyzing the Soviet program.

The Soviet manned program and solar system exploration program hold the keys to the understanding of their future capabilities and intentions.

THE "MODERNIZATION" OF THE AMERICAN SPACE PROGRAM

It is equally instructive to consider what has (and has not) happened in the US space program since the days of Apollo. Apollo taught several terrible lessions. The first was that American technology, if unleashed on an ambitious goal and funded reliably for more than four years, could accomplish incredible feats. The second was that no program remotely close to Apollo in size and scope could possibly receive funding unless it was adopted as a national political goal. The third realization was NASA's discovery that it could sell its soul to politicians and get rich doing it. The fourth was that when you have a huge, expensive program that is a national goal, and on a tight timetable, cost overruns are unavoidable. The fifth discovery was that overruns in the big program kill everything else. No other program, no matter how small and inconspicuous, no matter how valuable or promising, was safe when Apollo was hungry. Small and intermediate sized research and applications programs died by the score to feed the giant. The entire Mariner and Pioneer families of planetary probes were sacrificed, and all unmanned missions to the Moon were canceled. At NASA, this era is recalled as the "slaughter of the innocents."

After the fiscal, technological, political, and emotional binge of Apollo, NASA awoke to find that it had spend $26 billion on new hardware and technology that it had just used up and thrown away. NASA was destitute, and at a dead end. It was left with no continuing programs, no present, and no clear future. In an attempt to refocus the lunar and planetary program about a single national goal, the Viking program was conceived. Under the aegis of the search for life on Mars, NASA invested $1 billion in the program. It did so with only modest support from planetary scientists, who felt that it was too narrow scientifically, too expensive, too likely to prove a dead end (it probably would not find life) and too prone to unforeseen technical problems because of the unprecedented nature of the biological experiments it would carry. But Viking had widespread public and Congressional support. It was a scientific and technical triumph, but the financial strains imposed by its huge budget overruns cost the lives of a number of smaller science missions of similar value, but of lesser interest to Congress and the public than the search for Little Green Men. But Viking was too small to breathe life into all of NASA. Something bigger was needed; a genuine national goal. It was in this environment that NASA began to crystallize its thinking about the idea called the Space Shuttle.

The principal virtue of the Shuttle was that it promised to be enormously expensive—so expensive, in fact, that it might qualify as a national goal. But one could hardly expect NASA to try to sell the program to Congress on that basis. The strategy adopted was that of claiming that the shuttle would save money.

JSL recalls vividly a meeting of NASA's Physical Sciences Committee (PSC), back in the pre-Shuttle days, when it had already become clear that NASA was committed to building the shuttle. We were supposed to be responsible for translating the scientific objectives proposed for the nation by the Space Science Board of the prestigious (and nongovernmental) National Academy of Science into a real flight program. We were to be fed data on science objectives, budgetary projections, booster performance, celestial mechanics limitations and opportunities, and so on, and integrate it all into a national space exploration program.

Partway through our deliberations, we were given the very strong hint that, after 1979, we would no longer be able to use the stable of disposable boosters that NASA had developed over the years. The Saturns were already museum pieces, but now, we were told, the Atlas Centaur and the Titan 3E Centaur, the only boosters we had left that were powerful enough to launch planetary missions, would be scrapped by 1979 (see fig. 3.14). The reason seemed to be that, if we retained the capability of using those boosters, someone in Congress might notice that the Shuttle was not after all absolutely indispensable, but could be stretched out or cancelled, which would have left NASA without its National Goal. In the political climate of the Carter administration, this may well have resulted in the demise and dismemberment of NASA. Indeed, several such dismemberment plans were widely discussed.

Finding ourselves without any boosters to launch our future missions, we naturally experienced some trepidation. The Shuttle itself poops out just above a standard low Earth orbit. In order to launch deep-space missions, it would be essential for us to understand how the shuttle would be used to launch planetary probes. We asked NASA to assign some knowledgeable expert from the Shuttle program office to give the PSC a briefing on how the shuttle would launch lunar and planetary probes, what the capabilities of potential upper stages might be, and how much weight could be lifted on a variety of basic planetary mission trajectories. The word finally came back that "NASA doesn't want your advice on the Shuttle." After months of pleading, argumentation, and growing anger, we were told informally, "out of channels," that they had no idea that anyone would want to launch planetary probes with the shuttle, and had only a vague idea of how to do it. They viewed their role as one of selling the Shuttle to Congress and the public, and certainly not one of justifying themselves to some mere scientists. We finally got a short, vague briefing based on the hypothetical performance of nonexistent upper stages.

The issue of why any rational person would want to involve a large complement of astronauts in routine launchings of communications satellites, planetary probes, and so on was left unanswered. For such missions, the presence of astronauts added only three things: mass, complexity, and expense. The one role in which a shuttle craft could be convincingly cast was as a resupply and crew-rotation vehicle for a permanently manned space station. Of course, at that time everyone was thinking of the long-term future of *Skylab.* Little did we know. But the Shuttle, it appeared, was defined as "good." To be sure we got it, we had to need it. To be sure we needed it, we had to pretend there was no other way to launch anything.

A few space scientists were so convinced of the lack of merit of the Shuttle, and so afraid that it would overrun in cost and hence steal the fiscal lifeblood of space science programs, that they launched an energetic but futile attempt to stop the Shuttle. James van Allen was the quarterback of the anti-Shuttle team. He mustered all the economic arguments with verve and eloquence. Almost everything he said was right. But, to our tastes, he went too far in one respect: he argued that people have no cost-effective role in space. We thought most of his concerns about the Shuttle were correct, and history has shown him to have been right. But in his blanket condemnation of the role of human beings in space, we felt sure he was wrong. And, of course, Congress paid no attention to his warnings.

JSL remembers the clear feeling that the action wasn't really in the PSC. It must be that all the real decisions were made by the Space Science Board. Several visits with the Board to give briefings and plead special causes such as Pioneer Venus and Voyager had confirmed this attitude. They had the real clout.

NASA went before Congress and sold the Shuttle on the basis of a mission model that assumed monthly flights of five orbiters. This came to 60 flights per year, each orbiting about 60,000 pounds of payload. Simple arithmetic shows that, somehow, we were committed to developing, build-

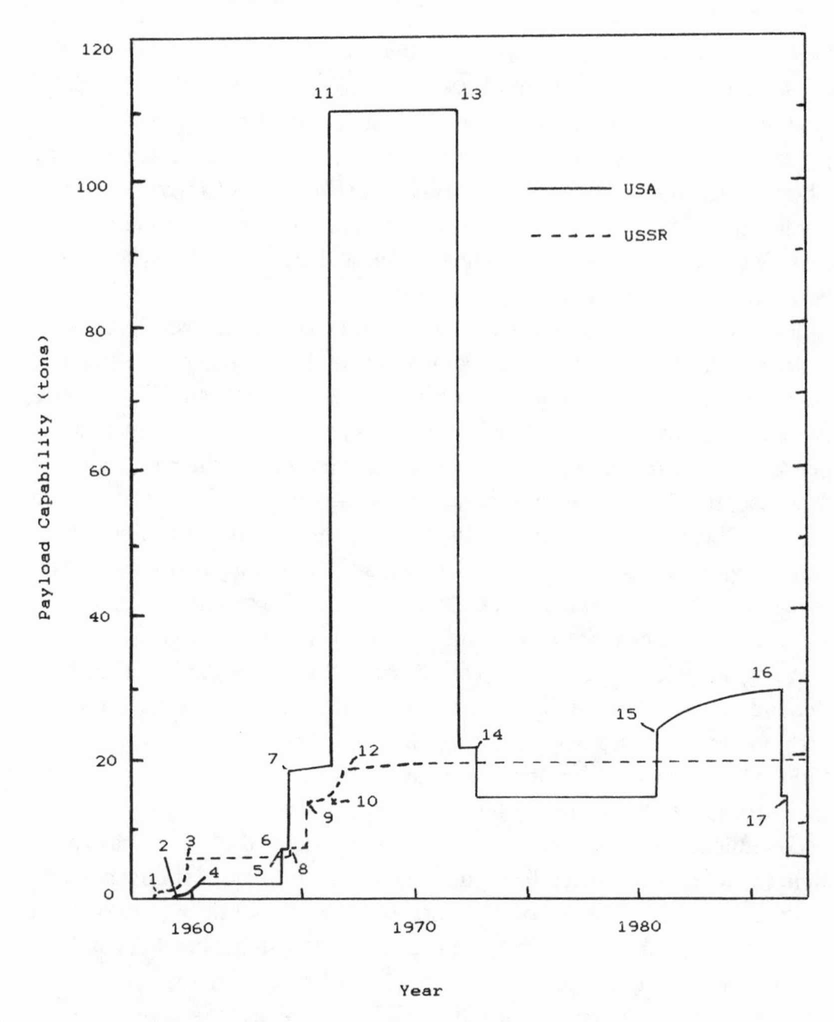

Figure 3.14
US and Soviet Payload Capability, 1957–1987
This diagram shows the decay of American ability to lift heavy payloads after the demise of the Apollo program. The introduction of the Soviet A booster is labeled 1, and the introduction of the early small American boosters is marked with a 2. Step 3 is the introduction of a second stage to the A vehicle to make the A1, 4 is the first flight of the Atlas Agena, and 5 is the debut of the Atlas Centaur, which gave the United States the lead in payload capability for the first time with a flight in 1963 (6). The American Saturn 1 heavy-lift booster (7) and Saturn 5 superbooster (11) gave the Uni-

ing, and launching payloads at the phenomenal rate of 3.6 million pounds per year. Multiplying this by typical payload development, fabrication, and testing costs of $3000 to $10,000 per pound, you can see that we would have to spend between $10 billion and $36 billion per year on the payloads. The dilemma was that this was far more than the already high cost of the Space Shuttle itself. Thus the Shuttle cost for 60 missions per year was not within the $6 billion program cost originally advertised, but entirely outside it—and much larger. NASA argued that the projected flight schedule was so frenetic that the development cost of the shuttle, which ultimately came to roughly $10 billion, would be spread over a vast number of flights, and hence contribute very little to the cost of each flight. But this ignored the payload costs.

But, after a while, it became clear that this was all impossible. The number of orbiters was cut to four, and the number of flights per year was slashed to 12 for the mid-1980s, with the potential to rise to 24 by the end of the decade. It is no longer expected that each Shuttle will fly one mission per month indefinitely. Now, each Shuttle is scheduled for periodic extensive (and expensive) refurbishment. The Columbia orbiter has just returned to service after a two-year rest and major repair and remodeling, and Challenger has been lost. Each orbiter should average about four flights per year over its lifetime.

The Space Shuttle orbiter has an empty weight of about 70 tons. Including a maximum payload weight of about 60,000 pounds, the total weight

---

ted States a commanding lead over contemporary Soviet boosters, the A2 (8) and the D-class "Proton" booster (9). The Titan 3C was introduced at 10. The D booster in its evolved form as the D1 appeared at 12. The last Saturn 5 was launched at 13, and the last use of the Saturn 1 in the Skylab program was at 14. The loss of these boosters gave the Soviet Union a commanding lead for the next eight years. The Space Shuttle was introduced at 15, giving the United States a modest lead in payload capability, but the Space Shuttle was grounded after the *Challenger* tragedy in January of 1986 (16), leaving the Titan 34D as the largest American booster. The Titan 34D was grounded after its second consecutive in-flight explosion in April of 1986 (17), leaving the Atlas Centaur as the largest American booster. This returned America to the 1963 level of lifting power, but without open production lines for new large boosters. A Soviet heavy-lift launch vehicle capable of orbiting payloads of about 100 tons is expected to fly in 1987.

placed in orbit on each flight is 100 tons. If the external tank (ET) were carried along, the total payload weight placed in orbit would actually be raised by about two tons—plus the 35 ton empty weight of the ET and several tons of fuel. The size of the overall booster is set by the ability to orbit 140 tons at a time, more than equal to the payload of the old Saturn 5. However, only 20 percent of that weight is payload that can be left in space. Eighty percent of the mass and cost of each launch is used up lifting the orbiter and the ET, and produces no revenue.

It is common to wax enthusiastic over the high technology exhibited by the Shuttles. These winged, maneuverable, aerodynamic spacecraft are no mere capsules. They are powered by a set of three hydrogen-oxygen engines that are the pinnacle of the art. They are shielded during reentry by thousands of high-performance heat-shield tiles, lifted off the pad by two huge recoverable solid-propellant strap-on rocket boosters (SRBs), and draw their supply of liquid hydrogen and oxygen from an even more immense external tank that rides with them almost all of the way into orbit. The Shuttles always carry out an additonal maneuver to dump the ET into the Indian Ocean rather than carry it (and its residual fuel) into orbit. (See fig. 3.15.)

It is true that the shuttle tiles give constant troubles, chipping or falling off by the dozens. It is true that the large, strong wings have to be shielded with thermal protection tiles to keep them from burning off, and hence add great amounts of weight to the orbiter. It is true that the "recoverable" SRBs always hit the water so hard that they are damaged far beyond original expectations. It is true that NASA steadfastly refuses to carry the ET into orbit and fashion huge space stations out of it, or even cut it up and use its component materials, even though tradeoff studies show that carrying the ET into orbit actually *increases* the payload capacity of the shuttle while also adding in the mass of the ET (35 tons) and its residual propellant (several tons). But none of these are insuperable obstacles to the use of the Shuttle; they are just growing pains.

The obstacle is, quite simply, cost. In the 1960s it was common to project lift costs of $25 per pound for the Shuttle, equivalent to about $100 per pound in 1986 dollars (see fig. 3.16). The Space Shuttle was actually sold on the basis of costs of about $250 per pound. Current charges for the use of the shuttle are $1400 per pound, and current expenses are close to $4000 per pound. This represents no improvement in real launch costs relative to the Saturn 1 and Saturn 5. This does not mean that we are in the same situation we were in when we had the Saturn 5: in sad fact, we are worse off.

Figure 3.15
The Space Shuttle
The Space Shuttle *Columbia* blasts off from the Kennedy Space Center. The luminous exhaust comes from the two Solid Rocket Boosters (SRBs) attached to the immense External Tank (ET). The orbiter is attached to the side of the ET. The SRBs burn for about two minutes and are then dropped off and parachuted back into the Atlantic to be recovered for reuse. The three shuttle main engines on the tail of the orbiter burn hydrogen and oxygen from the ET tanks—the flame is virtually invisible because it contains no carbon compounds or metals that produce luminous particles. The main engines burn from liftoff to the point of jettisoning the ET, just before insertion into orbit. Maneuvering in space is done with the two much smaller Orbital Maneuvering System (OMS) engines in the tail of the orbiter, which are fed by tanks of storable liquid propellant carried in the tail of the orbiter.

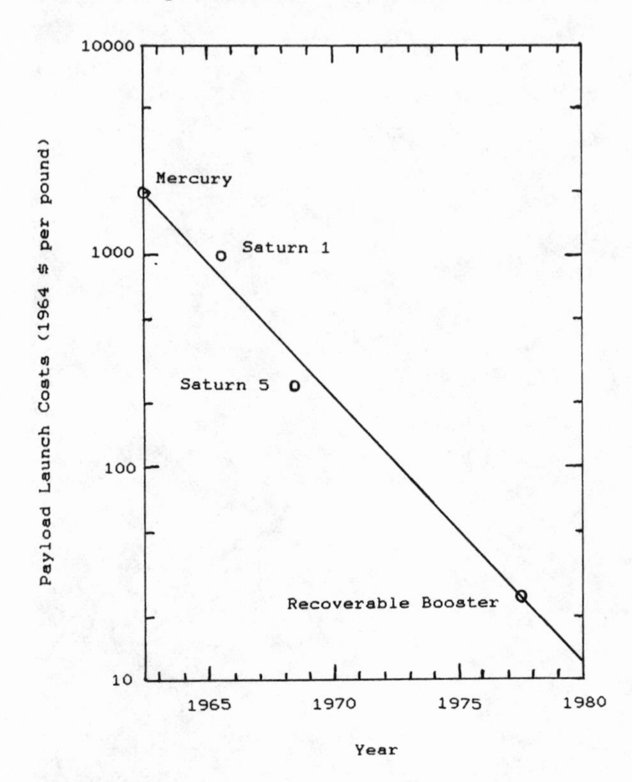

Figure 3.16
Projected Decline of Launch Costs, 1964
This represents general thinking in the early 1960s concerning how development of new boosters would have lowered the costs of launching payloads into Low Earth Orbit (LEO). Note that the cost scale is exponential, ranging from a high of $10,000 per pound at the top down to $10 per pound at the bottom. The recoverable booster, now called the Space Shuttle, was projected to have a launch cost of $25 per pound. It was to have become operational in 1977.

The maximum useful payload we can launch now is 30 tons, vs. 100 tons on the Saturn 5. Also we are out the cost of developing the Shuttle. And finally, on the bright side, we have developed several expensive items of new technology that may have other long-term uses independent of the Shuttle—if only we knew what they were! We have the ability to fly up to seven people at a time for up to seven days at a time. Longer flights could be arranged, but each would unfortunately tie up its precious orbiter for the entire duration. We will not be able to match the Skylab scale of space activities until about 1994—if we are lucky.

In a dispassionate appraisal, the Space Shuttle does not seem to be such a great bargain. It confers on us very few abilities that we would not have had several years earlier if we had evolved from Apollo technology instead of scrapping it, and it has fiscally denied us several years worth of manned operations in space and at least six years of our planetary exploration program. Our best hope may be to salvage some of what we have learned in building the Shuttle to design a more efficient, less costly future launch vehicle.

The ultimate irony is that NASA now revolves around the Space Shuttle, an operational space transportation system. It is the one significant activity that, according to NASA's charter, it should not be doing. NASA was chartered, clearly and unambiguously, as a research and development organization. NASA had to give up their mandated activites for several years in order to pay for the Shuttle, which now hangs about their neck, utterly dominating their attention and monopolizing their budget. There was a preliminary design study of a large Soviet Space Shuttle a few years ago that drew a fair amount of comment in the Western press: with astonishingly penetrating vision, the Soviets chose to name the program "Albatros."

Now we have a Space Shuttle, and we even have a high-energy upper stage, the Centaur G′, built to be carried in the Shuttle payload bay. The first two flights of this stage had been scheduled six days apart in May of 1986, the launchings of the European Ulysses solar polar orbiter and the American Galileo Jupiter orbiter and entry probe. This stage is the latest verison of the original Centaur design of 1961. But, since the Challenger disaster, all such plans are on hold. With the Shuttle temporarily grounded, there is no other way to launch these missions. We put all our eggs in one basket—and the basket broke. In the wake of the Challenger tragedy, the Centaur G′ stage has been cancelled.

In almost every respect, even computer technology, we now find ourselves using hardware and techniques developed in the 1960s. There are more examples of cases in which we have lost ground (solar system exploration, manned lunar operations, superbooster payload capacity) since 1969 than there are examples of increased capability. Let us hope that the Space Shuttle's expensive technical innovations turn out to have broader use. If not, history will record the Shuttle as another dead end.

4

THE PRESENT AS SEEN FROM THE PAST—
And How It Really Is

The only important question is what to do next.—Stanley Kubrick

EARTH SATELLITES AND SPACE STATIONS

Almost all pre-1960 visions of the distant future (and of course to us that means the recent past) were predicated upon an assumed orderly progression of goals and abilities. The consensus of the seers was that small, unmanned satellites were the inevitable first steps that were required in order to develop several new technologies.

The first of the new technologies, and one that goes hand in hand with military development of long-range ballistic missiles, was the recovery of a satellite from orbit. Next came the development of a satellite capable of keeping an astronaut alive for at least one orbit about the Earth. Then came manned orbital flight, initially for a single orbit, but later building up to durations of days and weeks in weightlessness.

The next logical step for unmanned flight would then have been the development of small, high-performance upper stages to enable the launching of lunar and planetary probes. In parallel, the manned program would logically have evolved in the direction of building a large, reusable shuttle craft to permit routine delivery of massive cargo to Low Earth Orbit. This winged shuttle would deliver to LEO a number of modules that would be assembled into a huge, permanently manned space station, capable of housing dozens of astronauts. Then the logical role of the shuttle would have been to service the space station. The most famous visualizations of this concept are the paintings by Chesley Bonestell that illustrate the fifties-era popular space books by Wernher von Braun and Willy Ley.

Since it was generally supposed that delivery of payload to LEO would continue to be very expensive, there seemed good reasons to establish a habitat in space that would be safe and congenial for its residents during visits of many months, and to avoid short, expensive sojurns in space. Weightlessness was one of the two great concerns of the authors of the time (the other was meteoroid impacts), so the entire station was shaped like a giant wheel and spun on its axis to produce artificial gravity. This version of the space station is often referred to as a "von Braun wheel."

It was widely accepted at the time that manned flights deeper into space could not possibly be attempted by direct launch of a rocket from the surface of the Earth because the size of the rocket would have to be prohibitively large. This implied that lunar and planetary missions would have to be assembled in Earth orbit out of modules lifted by a number of small launch vehicles. Thus one major function of the occupants of a space station would be the assembly and refueling of outbound spacecraft.

Once the orbital facilities were complete, the next logical step would have been the assembly and launching of a giant Moon rocket to follow the path blazed by earlier unmanned probes. The huge rocket would have been assembled, fueled, and dispatched on an eccentric orbit reaching from LEO out to the orbit of the Moon. A manned lunar flyby could have been attempted before a landing expedition were dispatched because a flyby would require far less propellant, and would therefore be much lighter than the vehicle needed to carry out a landing. It was common for early writers to assume that all velocity changes, such as return of a lunar flyby to LEO, or slowing down a returning spacecraft for a safe landing, must be carried out by burning huge masses of propellant. When it was realized that aerobraking (braking the return vehicle to LEO orbital speed by passing through the upper atmosphere and letting frictional drag slow it down) was feasible, the masses required in LEO to launch such missions were dramatically reduced.

R. B. Beard and A. C. Rotherham, writing in the "optimism" of 1958, saw the development of unmanned satellites as occupying the 1960s and 1970s. They believed that a space station could be built by the 1980s, and "taking the optimistic view, we can hope that the first men will land on the Moon about the year 2000." They went on to predict a manned visit to Mars by 2020.

DAS MARSPROJEKT

One seminal early idea was the 1952 suggestion by von Braun of a truly massive effort to send an expedition of 70 men to land on Mars and return to Earth. Just assembling the expedition vehicles in LEO would require 950 launchings of superbooster carrier rockets, each with a takeoff weight of 7000 tons and a payload capacity of 36 tons. (By way of comparison, the Saturn 5 booster of the late 1960s weighed about 3000 tons and delivered up to 100 tons to LEO.)

The passage from LEO to Mars would be made by a fleet of ten ships, of which three would carry 200-ton winged landing vehicles. After 16 months on the Martian surface, two of the landers would take off, return to Mars orbit, and consolidate with the ships left in orbit so that seven ships could return to Earth. The entire mission, from LEO departure to Earth return, would last 32 months. All consumables, including propellants, food, water, and air, would be carried along from the start.

The timing of the Mars project was dependent upon a whole chain of "next logical steps," culminating in the availability of a booster that could orbit space station (and Mars expedition) modules. There was hope that a space station could be built by 1975, which would have given us the Mars mission by 1980.

THE NUCLEAR ROCKET

Many writers in the 1950s and 1960s regarded the von Braun scenarios for expeditions to the Moon and Mars as demonstrations of just how ridiculous and unwieldy chemical rockets would be if applied to deep-space travel. The usual solution to this dilemma was to propose tapping nuclear energy.

In the 1960s the United States maintained an active program of experimentation with nuclear rockets. The Kiwi-A nuclear rocket, which used a nuclear reactor to heat a "working fluid" up to very high temperatures and ejected the heated gases through a rocket nozzle, was tested extensively in the 1958-1960 period in Nevada. These tests were continued under the Nerva program, which was to culminate with a flight-ready stage, Rover, suitable for use atop the Saturn 5 or Nova booster. The Nova itself was a direct extension of Saturn technology to a first stage with eight huge F1 engines, totaling 12 million pounds of thrust (the Saturn 5 had five F1s, delivering 7.5 million pounds of thrust).

So convinced were many observers of the merits of nuclear propulsion that they saw it as the means to send men to the Moon by the late 1960s. Despite the great weight of the reactor system and its high level of radioactivity, the conventional prediction from 1958 to 1968 was that a flight-rated nuclear stage could be available in ten years. It would be fair to say that this statement is still true: the nuclear rocket is still at least ten years in the future.

The performance advantages of the nuclear rocket were most exciting. Using liquid hydrogen as the working fluid, the exhaust would be the very lightest gas of all. Temperatures could be achieved in the reactor core that were at least as high as the best achievable through chemical combustion. Because the speed of a molecule increases with the temperature, and decreases with increasing molecular weight, a very hot hydrogen exhaust promised the best possible exhaust velocities for high-thrust rockets. We shall look at the performance and efficiency of this and several other types of rocket propulsion in chapter 5.

SPS: THE GOLDEN APPLES OF THE SUN

An especially intriguing idea for space development was widely discussed during the late 1970s, largely due to the efforts of Dr. Peter Glaser of Arthur D. Little Associates in Cambridge, Massachusetts. He proposed that one of the greatest resources of space, virtually unlimited solar power, be exploited commercially by building a Solar Power Satellite (SPS) system in geosynchronous orbit (GEO).

The satellites in the SPS network would have hundreds of square kilometers of solar cells to capture sunlight and convert it into electricity. The electric power would be broadcast to Earth as microwave energy, where it would be received by huge antenna arrays on the ground and converted back to electricity for distribution over conventional power grids. Each SPS would be in almost continuous sunlight, and hence would be very efficient. Once installed, each SPS would require no fuel or other source of energy, and would need only very infrequent repair.

The principal objection to the SPS idea was that the cost of launching the necessary components from Earth was very high. Initial cost studies showed that it was questionable whether the program would return benefits in excess of costs. In the context of 1977 ideas about the depletion of terrestrial energy reserves, however, the idea of a system that would not consume any fuel once installed was very attractive.

An interesting response to this problem was offered by some space scientists, who noted that the Moon was potentially a less expensive source of construction materials than the Earth. Given the investment of a large amount of capital in an industrial infrastructure on the Moon, the costs of the SPS program could be brought down far enough to be competitive.

However, the feasibility of SPS hung on the cost of launching materials from the Moon. If chemical propulsion were used, than the program was completely infeasible unless a source of chemical propellant could be found there. Alternatively, some method of launching payloads from the Moon without using rocket propulsion would be required.

COLONIES IN SPACE

Gerard K. O'Neill, a physicist at Princeton University, has crystallized a particular view of the future of space around a scenario that concentrates on building huge space colonies in high Earth orbit. The chosen places for the initial colonies are the points on the Moon's orbit located 60 degrees ahead of and behind it. Objects placed in orbit at those positions, called the fourth and fifth Lagrange points (after the French mathematician who discovered them) will stay there forever. In mathematical jargon, the L4 and L5 points are "stable libration points." A body orbiting there will, even if modestly disturbed in its motion, remain near the point where it started but gently "librate" about it. The word "librate" comes from the Latin word *libra,* meaning a balance, or set of scales. When slightly disturbed, scales make this motion (libration) back and forth past the point of perfect balance.

The colonies themselves are hollow watermelon-shaped bodies with thick "skins" of lunar or asteroidal rock. They are spun about their long axis to produce centrifugal force, which provides artificial gravity in much the same way that the von Braun wheel was to work. Typical colonies would be several kilometers in length, and have populations of tens of thousands of people, complete with farms and farm animals.

O'Neill's space colonies are the culmination of a lengthy and complex program of space development. The material of which his multikilometer hollow habitats are built is mined from the surface of the Moon. After being processed into useful materials (or even into finished products), the lunar-derived mass is shot into space by huge electromagnetic accelerators that stretch across the face of the Moon. These accelerators, called *mass drivers*, are powered by electrical energy captured by solar cell "farms" on the lunar surface. Mass drivers are an interesting idea with considerable future promise, and we shall return to them in chapter 5.

It should now be clear that we are discussing a truly huge program, with multi-billion-dollar investments in each of a large number of different technologies. Why should anyone pay this kind of money in order to build a space colony?

The answer comes in two parts: first, the initial cost of getting this civilization established in space could be an incidental expense of a program to build a network of Solar Power Satellites in high Earth orbit, as described above. Second, after the completion of the SPS system, the space colony would be made self-supporting. Those high-technology imports that are necessary to the colony, but which the colony would not be large and complex enough to make for itself, would be paid for out of exports back to Earth. In one scenario, the main export is electrical power.

With a prompt start on the SPS system, the 1980s would have seen the development of a lunar base and ore-processing facilities, the construction and installation of a mass driver on the lunar surface, and the beginning of SPS construction.

ISLANDS IN SPACE

In 1964, Dandridge M. Cole and Donald W. Cox, in the astonishingly far-seeing book *Islands in Space*, wrote about the potential resource value and programmatic importance of nearby asteroids.

Their premise was that there must be some asteroids in orbits that make them extraordinarily accessible to Earth. Landing on the best of them would require significantly less propulsion energy than landing on the Moon. Further, taking off from one of them and returning to Earth could, potentially, be incomparably less demanding than climbing out of the Moon's gravity well and returning to Earth or to LEO.

Cole and Cox cite six basic reasons for the importance of the asteroids to future space activities: basic knowledge of the origin of the Solar System, their collision hazard to Earth, their value as "filling stations" in space, their potential to supply resources to Earth, their promise as sites for extraterrestrial colonies, and their use as vehicles for interstellar travel.

With amazing prescience, they express doubt about the direction of the space program after Apollo, and question whether larger boosters and nuclear upper stages will be built. They point out that it would be a technologically easy task to step from the Apollo program to near-Earth asteroid exploration in the early 1970s, and then propose that this be made the main goal of the future space program.

The remarkable truth is that asteroids with highly favorable orbits, the ones essential to initiation of their scheme, had not yet been discovered at the time they wrote. The first of this class, Anteros, was discovered in 1973, and the first that realized the full benefits of favorable orbital properties was not found until 1982!

Cox and Cole also point out that the two satellites of Mars, Phobos and Deimos, are of exceptional interest as refueling stations and as mineral resources. They emphasize the value of propellants derived from Phobos because of the assistance they can render to any missions going to or beyond Mars.

If this scheme had been adopted by NASA, the first manned mission to an "asteroidal" body would probably have been to Phobos. Anteros may also have been visited by 1980.

FORESIGHT IN RETROSPECT

It is certainly clear that not all of these projects could have been carried out at once. Indeed, several of the options diverge very widely in their technological assumptions. It is interesting that, in every case, their vision of the 1980s was more ambitious than what we have actually accomplished. This is because the authors of these plans did not anticipate a virtual cessation of innovation in 1968.

Another striking observation is that, in general, these programs *seem to be more logical, sequential, and efficient than what we actually did.* The logical flow from unmanned satellites to manned capsule to reusable space trucks to manned space stations to orbital workshops to lunar and planetary expeditions was cut short at an early date by President Kennedy, who elected to leapfrog the delivery system and space station elements to go directly to the Moon. This was done in a fashion designed to achieve a single well-defined goal, a manned lunar landing mission, and then stop. John Logsden has noted the reasons that the Apollo program was regarded so highly by Kennedy's advisors: First, it was something we could do; second, it was the first major landmark that we could hope to beat the Soviet Union to; third, *the program was a dead end that did not commit any funding to any goal beyond the achievement of the lunar landing.*

It is ironic that one of the most attractive virtues of Apollo from the political perspective has turned out to be its greatest liability: it was designed from the start as a dead end—and it was well-designed indeed. This is the reason why there is so much use of, and so little agreement about, the phrase "next logical step" in the post-Apollo world. Apollo itself was such a logical anomaly that there *is* no next logical step. As soon as Neil Armstrong and Buzz Aldrin set foot on the Moon (or, more properly, as soon as they set foot back on Earth) we found ourselves with the need to rethink our future completely. *We had formulated only one goal and we had achieved it.* Now what? Roll over and play dead? Begin again and develop a transportation system, a space station, and the technical ability to visit the Moon and the planets in an affordable way? Pretend that Apollo really was part of some long-range plan, and go on to build a lunar base with Apollo technology? Which is the best "next logical step"?

. . . AND HOW IT REALLY IS

Here we are, at a time when much was expected, and much could have been done. Where do we stand in our pursuit of these exciting goals of the recent past?

First, consider manned space stations. Skylab, the largest, heaviest space station ever built, and the last payload launched by the giant Saturn 5, had an empty weight of 82.5 tons. It was occupied on three separate occasions in 1973 and early 1974, for periods of one month, two months, and three months, respectively. With the 22-ton manned Apollo spacecraft attached, the total mass in orbit was over 104 tons. The Skylab program used three of the four remaining Saturn 1b boosters. The last was used in the 1975 Apollo-Soyuz rendezvous mission. From mid 1975 until the operational debut of the Space Shuttle in 1981 the United States had no manned spaceflight program and no ability to fly men in space. The arena of manned space flight was abdicated to the Soviet Union for six years. And the Space Shuttle, far from being a dedicated servant of manned space stations, has bloated into a monster that lays claim to all payloads destined for space.

A triumph of the Soviet space station program was the September 1985 assembly of a complex composed of *Salyut 7*, the *Kosmos 1686* module, and the *Soyuz T-14* manned return vehicle: the total mass in orbit was about 53 tons, about half the capacity abandoned by the United States a decade earlier.

Why did we not preserve and use Skylab? Because, in order to make certain that we would be committed to development of the Space Shuttle, NASA threw away, *well in advance of the arrival of the shuttle,* all of its capability to launch heavy payloads with other rockets. Thus, when the orbit of Skylab slowly decayed and the Space Shuttle development program slipped badly, Skylab was doomed. It eventually reentered the atmosphere over Australia in 1979, where it burned up because we had intentionally thrown away the means to rescue it.

If all goes well with the Space Station program, we may once again have a 100-ton American manned space platform by 1995, twenty years later than Skylab, and at a cost of roughly ten billion dollars (enough in real terms to have bought at least two Skylabs).

The real triumph of space station use has been the massive, determined effort by the Soviet Union to demonstrate manned activity in space for periods of more than eight months on their *Salyut 6* and *7* space stations. Their biomedical research program at first indicated that the effectiveness of their crews declines steadily after about four months in orbit. The main culprit appeared to be weightlessness, which causes a wide assortment of physiological changes that are mostly adaptive to the space environment, but which greatly degrade their ability to return to normal activities in full gravity. But now the evidence seems to suggest that a plateau is reached after a few months in space. The Soviets are again planning the extension of manned flights to durations of at least nine months. Still, prolonged exposure to zero gravity is a problem for people who must eventually return to Earth.

The prospect of spinning manned stations to provide the artificial gravity believed necessary by the space seers of the 1940s and 1950s seems very remote. The American Space Station will be a completely zero-gravity structure. No von Braun wheel is planned by any nation.

In the realm of probes of the planets, we are in the midst of a truly de-

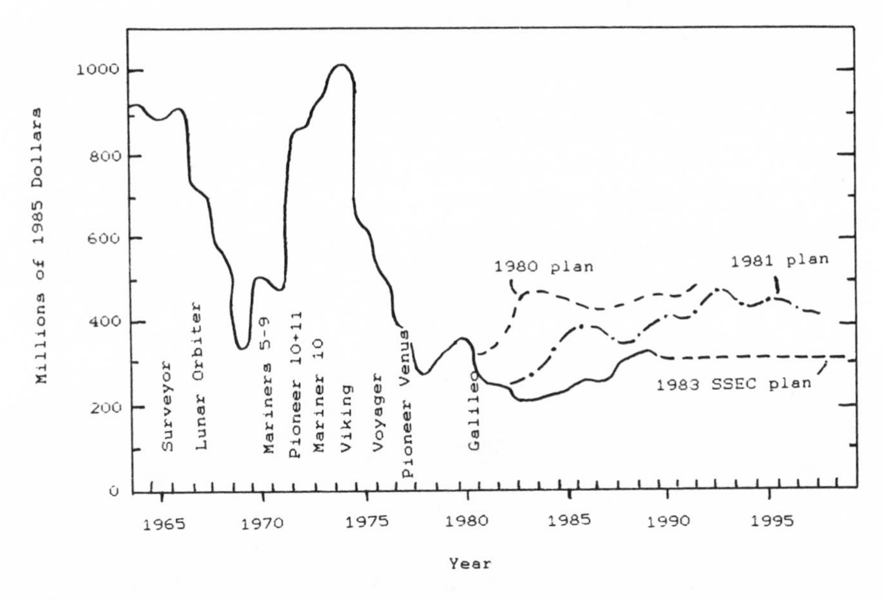

Figure 4.1
US Expenditures on Planetary Exploration
The history of NASA's expenditures on lunar and planetary exploration is presented in 1985 dollars. The early high expenditures (prior to 1968) were to pay for the Ranger, Surveyor, and Lunar Orbiter programs, the unmanned precursors of the Apollo manned lunar landing program. The lull in funding corresponds to the period of greatest flight activity in Apollo, and the second big funding peak was caused mainly by the Viking Mars missions and the beginning of the Voyager program. The only mission started in the 1978–1984 time period was the Gallileo Jupiter orbiter and entry probe. The launch date of Galileo has been repeatedly delayed by budgetary stringencies, and most recently by the grounding of the Space Shuttle after the *Challenger* disaster. It is now tentatively set for 1989. Note the progressively diminishing expectations of NASA's long-term plans as they have evolved in recent years. The United States has recently spent an average of less than $1 per year per person on the exploration and exploitation of the Solar System.

pressing morass of inactivity and fiscal starvation (see fig. 4.1). The Moon was abandoned by the United States at the end of the Apollo lunar landing program in 1972, and by the Soviet Union in 1976. Mars was abandoned by the United States after Viking in 1976, and no Soviet Mars missions have been launched since 1973. No spacecraft has been sent to Mercury since *Mariner 10* was launched in 1973. The last mission to the outer planets, *Voyager,* was launched in 1977, and the last American Venus mission, *Pioneer Venus,* was launched in mid-1978. The only planetary activity in the following seven years was the Soviet Venera program. The flurry of activity directed toward Halley's comet in 1985 was created by the USSR, the European Space Agency, and Japan, without a single American launch. We have made no visible progress on the goal of sending a manned expedition to Mars.

Any new capabilities that the space-faring nations will have depend in large part on the progress of propulsion research. The nuclear rocket, like nuclear aircraft propulsion, is stone-cold dead. The few million dollars that have been spent by NASA for advanced propulsion research in recent years have been devoted almost entirely to squeezing another few percent improvement out of hydrogen-oxygen chemical propulsion systems. Scarcely a cent has been spent on genuine new technology. If you spend all your resources on 1 percent improvements, and ignore anything that might give a factor-of-ten improvement, then of course you are immortalizing a level of technology that otherwise would have inevitably given way to new, superior systems.

So the nuclear rocket never flew. If any single factor killed the nuclear rocket, it was the prospect of a catastrophic core failure or crash of the booster vehicle, showering high-level radioactivity over a wide area. This is a valid fear. There may be technical fixes that could have worked, but with the level of antinuclear sentiment running today, it is utterly unthinkable that anyone would have dared to try to solve the problem. Indeed, anyone who thought he had a solution would have found it impossible for him to acquire funding to develop it.

The mentality at work here is of the sort that argues against nuclear generation of electric power on the grounds of astronomically improbable modes of failure that might release radioactivity into the atmosphere, while favoring coal-burning generating plants that are hideous polluters and that release more radioactivity into the atmosphere than a nuclear plant with the same generating capacity. The assessment of the relative risks of very rare, very intense, and very local contamination versus very wide, unavoidable low-level pollution requires time and work. It's easier to ban nuclear plants than to solve the energy problem.

Space colonization has progressed as far as having eleven people in orbit at once, in two different vehicles. The largest crew ever flown was eight on a 1985 Space Shuttle flight that lasted a week. Space colonization is still science fiction.

Solar power satellites (SPS) were studied intensively for a brief time in the late 1970s. The economics of Earth-derived SPSs looked so unexciting to a panel of the National Academy of Sciences that they issued a very negative report. How negative? Did they recommend more basic research before any decision was made on the desirability of the system? No: they recommended that *no research effort or funding be devoted to SPS studies for at least ten years.* What better way to guarantee that no progress will be made than to deny funding for basic research?

And what of the uses of asteroidal and lunar resources? To date, not a single mission to assess space resources has been flown by any nation. No experiments on processing of asteroidal or lunar minerals have been done in space. Laboratory work on processing lunar materials has been ingeniously eked out on a very small scale as incidental aspects of other lunar research programs. NASA has so little funding available for space resources work that almost none of the competent groups are supported.

Space activity in the Soviet Union has been stable since about 1972 (see fig. 4.2). The United States reached a peak of launch activity in 1966, and went through a prolonged and disastrous decline into the early 1980s under the weight of massive expenditures on the Apollo program, the Space Shuttle, and the Vietnam War, which squeezed out many of NASA's most promising smaller programs (fig. 4.3).

The overall picture is simple to describe: the decision to leapfrog the shuttle and the space station and do the Apollo lunar mission first has left the American space program without either a goal or a sequence of "next logical steps" to pursue. The failure of the Soviet space program to beat the United States in any significant "space race" since the mid-1960s has caused them to adopt a low profile, speak more of means than ends, and develop their skills methodically without overt reference to competition with the United States. This should not be interpreted as a lack of goals, but as a desire to avoid very costly and probably unsuccessful races with the American hare. Moderation of real growth of the Soviet economy has made remote the prospects of new, large infusions of money into their space program. The desire to improve the living conditions of the Soviet people rightly occupies a prominent place, and of course Soviet paranoia (richly deserved, as students of their history can attest) dictates a very high priority

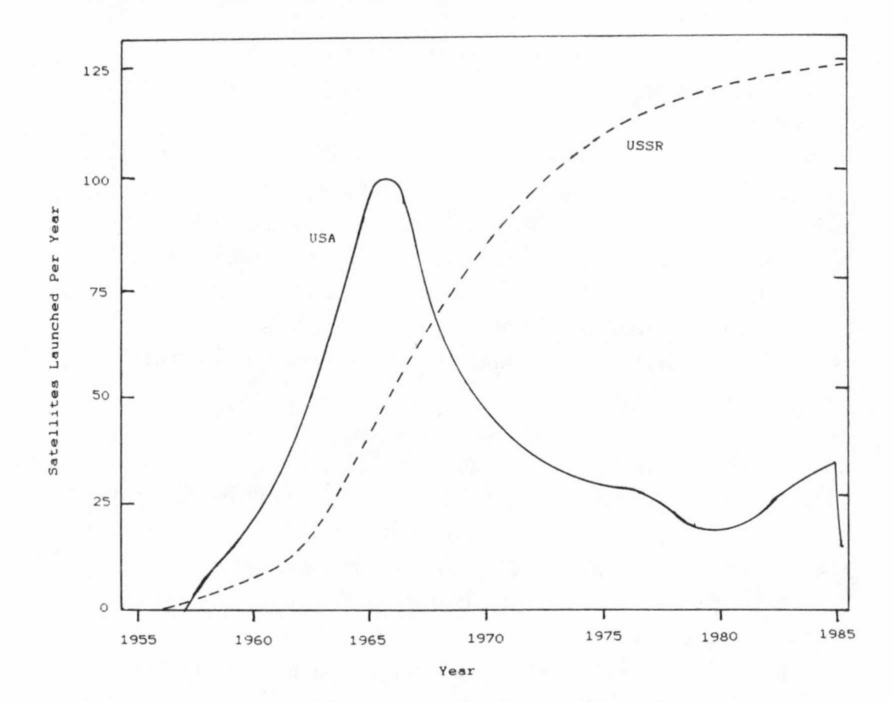

Figure 4.2
Satellite Launch Rates, 1957–1986
The number of satellites launched per year by the United States and the Soviet Union are compared. Since the Apollo era the American launch rate has fallen catastrophically, to be temporarily and partially arrested by the introduction of the Space Shuttle in 1981. Since the grounding of the Space Shuttle and Titan 34D in 1986 the American space program has come to a virtual halt. The Soviet program, on the other hand, has experienced steady growth to a very high level. Our projection for 1986 is 120 Soviet and 5 American spacecraft.

for military expenditures. In this setting, a large, expensive race to land men on Mars would be a disaster to the Soviet economy.

It would have taken real guts for the United States to have eschewed the expensive self-gratification of Apollo in favor of a sounder, long-range, stepwise program of development. But history has shown that Soviet space technology foundered on the rock of the lunar race. The manned missions, except for simple flybys, seem to have been beyond their competence.

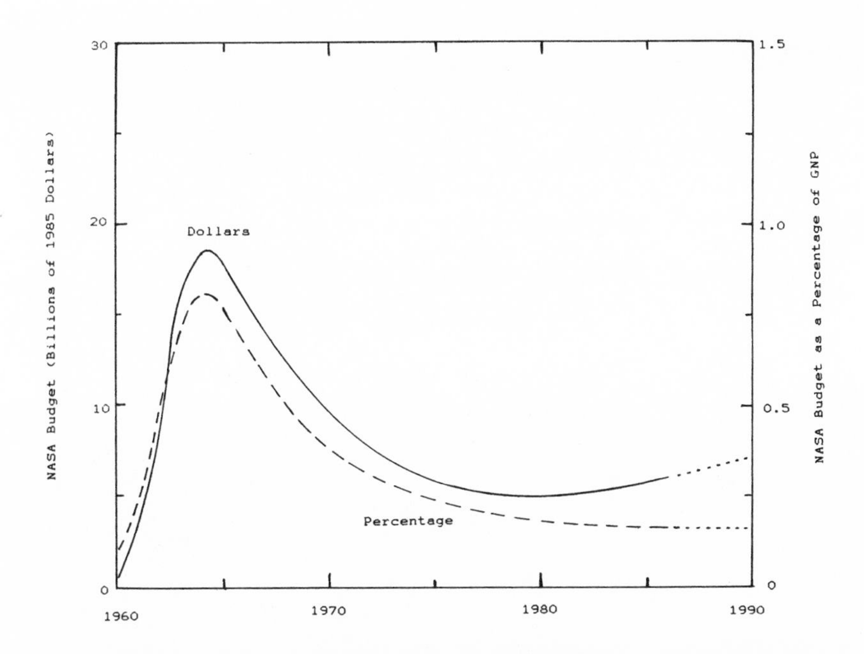

Figure 4.3
The Rise and Fall of the NASA Budget
The solid curve (left axis) shows the size of the NASA budget, in billions of 1985 dollars, for the time period from 1960 to 1990. The dashed curve (right axis) gives the NASA budget as a percentage of the Gross National Product.

Without Apollo's competition, the Soviets would surely have sent the first man to the vicinity of the Moon. Still, it is far from obvious that they would have even then succeeded in being the first to land men on the Moon and return them to Earth. Even today they cannot do it.

In recent years, another factor has entered the competition between the USSR and the United States: the emergence of the European Space Agency, Japan, and China as active participants, competitors, and potential collaborators. The role of the resources, energies, and ideas of these players in the future of space is still rapidly evolving and difficult to predict, but there is here a promise of something new under the Sun.

5

NEW VISIONS:
Barriers, Opportunities, and Costs

. . . a form of space sailing might be developed which used the repulsive effects of the Sun's rays instead of wind. A space vessel spreading its large, metallic wings, acres in extent, to the full, might be blown to the limit of Neptune's orbit.—J. D. Bernal, 1929

Many possible goals confront the space program of the 1990s and the early part of the twenty-first century, including manned lunar bases, Mars expeditions and colonies, space settlements, space industrialization, and solar power satellites. The choice among these possibilities is not easy. What *should* we do in space in the next quarter century? Let us split this formidable question into four manageable parts:

1. What technological possibilities lie within our reach?
2. What flight opportunities are propulsively feasible as a result of astronomical considerations?
3. What attractions do particular destinations in the Solar System offer us?
4. Which of the technically and propulsively possible opportunities contribute to our general goals in space?

In this chapter we shall deal with the first two of these questions. The following six chapters will then introduce our present knowledge of the nearby bodies that might be visited within the technical and astronomical contraints we describe here, and the remainder of the book screens the entire range of possibilities and attractions with a view to reassessing our goals for space exploration and exploitation.

ORBITS

If there were no forces acting on a body, nothing pushing or pulling it about, the body would remain in uniform motion forever: it would travel with constant speed in a straight line. The only reason that orbits of any other kind exist is that forces exist. In astronomy, the only force of consequence influencing the motions of stars, planets, and large satellites is gravity.

Gravity is the spontaneous attraction of all the matter in the universe for all other matter. Any two bodies experience a force that is proportional to the mass of each body separately. This force drops off rapidly with distance: as the distance between the centers of two bodies is doubled, the force drops off by a factor of four. A distance increase of a factor of ten causes a force decrease of a factor of 100, and so on: the force drops off with the square of the distance. The gravitational force between two bodies always acts to pull each body directly toward the center of the other mass. Gravitation is said by physicists to be an attractive inverse-square force.

If an initially stationary body is subjected to an absolutely constant attractive force, it will slowly begin to move. It will not only acquire speed; it will also add more speed at a constant rate. If, after the first second, the speed has increased by one centimeter per second, then it will also gain an additional speed of one centimeter per second during the next second. This speed will be exactly in the direction that the force is impelling the body. There is a special word for a speed that is associated with a particular direction: *velocity.* After 100 seconds it will have a velocity of 100 centimeters per second. Its velocity is changing at a constant rate, one centimeter per second per second. That is not a misprint. The rate of change of velocity is called *acceleration.* In this case, where the force is constant, the resulting acceleration is also constant.

The force experienced by small bodies near the Earth's surface due to the gravitational pull of the Earth's mass is called *gravity* (see fig. 5.1). The normal acceleration of gravity on the Earth's surface is about 981 centimeters per second per second. This is called one standard gravity or one Earth gravity. The exact value varies slightly from place to place because the Earth is not perfectly spherical and because of differences in altitude.

Imagine two bodies started at rest some distance apart. Each will exert a small force on the other body, and they will "fall" toward each other. As they get closer and closer, the strength of the force attracting them increases dramatically. Because the force gets constantly stronger as they approach,

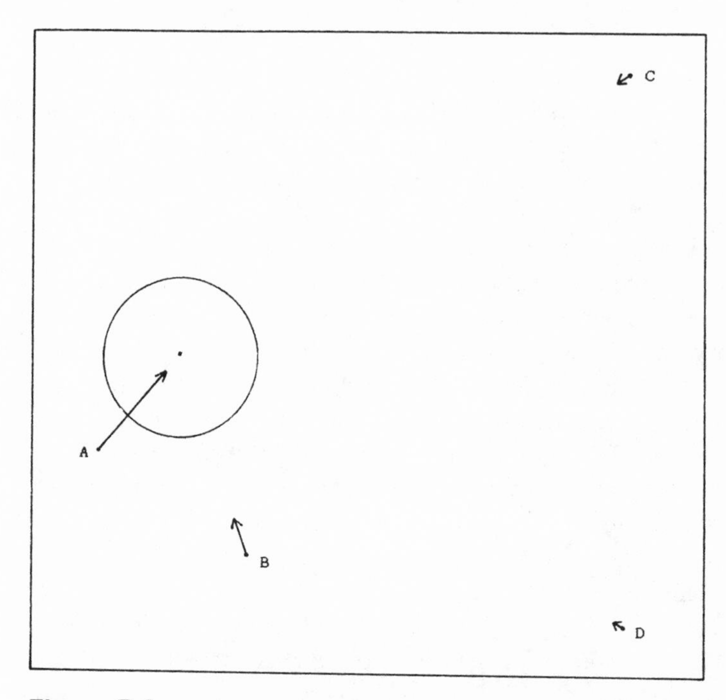

Figure 5.1
Direction and Strength of Gravitational Force
The arrows attached to the equal-mass bodies A, B, C and D show the direction of the gravitational forces exerted on them by the planet P. These forces are all directed exactly toward the center of mass of the planet. The *lengths* of the arrows show the *size* of the gravitational force felt by each small body. B is twice as far from the center of the planet as A, and B feels a force only ¼ as strong as that felt by A. C is five times as far from the center of the planet as A, and feels only 1/25 as much force. D is as far from the planet as C: note that the forces on the two are equal in strength but different in direction—both forces point to the center of P. Generally, the gravitational force drops off as the *square* of the distance from the center of the planet. The gravitational force does not depend on the speeds of the small bodies.

the bodies will approach each other not just with a constantly increasing velocity (which would be true even if the force were constant), but with a constantly increasing acceleration. Eventually they will collide at very high velocity, with catastrophic results.

Why doesn't the Moon fall to Earth? A simple calculation tells us that, at the Moon's distance from Earth, the acceleration due to Earth's gravity is large enough to cause a stationary Moon to accelerate toward the Earth and collide with us in less than three days.

The answer is not that there is something wrong with our idea of force, but that the bodies are never stationary with respect to each other. They always have a sideways component of their velocity. If there were no force between them, they would travel off in straight lines and never see each other again. The force between the bodies curves their paths: because the force of gravitation is attractive, the bodies will follow curved paths that are concave toward each other.

It is not difficult to prove mathematically that the curves followed by gravitating bodies are always ellipses, circles, parabolas, or hyperbolas (figs. 5.2 and 5.3). Parabolas and hyperbolas are "open" curves that do not close on themselves to allow repetitious motions: bodies following such curves are traveling too fast to stay in each other's vicinity. In the language of physics, we say that the bodies have enough energy of motion (kinetic energy) to overcome the attractive (potential) energy of their gravitational interaction. Two bodies that have exactly enough energy to get infinitely far apart before stopping (an infinitely improbable case) follow parabolic orbits. If they have any excess kinetic energy, they will follow hyperbolic orbits.

If the bodies do not have sufficient kinetic energy to get away from each other, then they will follow bound (closed) orbits, in general, ellipses. A circle is just a special case of an ellipse in which the distance is always constant.

Note that we have repeatedly said "their orbits." Both the larger and smaller bodies are in orbits about their common center of mass. The Moon orbits not about the exact center of the Earth, but about the center of mass of the Earth-Moon pair (fig. 5.4). Earth does the same thing; however, because the Moon is much smaller than the Earth (Earth has 81.3 times as much mass) the point about which the Earth and Moon orbit is not very far from the center of the Earth: it is actually inside the Earth.

The point of a satellite's closest approach to Earth is called its *perigee,* and the point of greatest distance is called its *apogee.* For bodies orbiting the Sun, the point of closest approach to the Sun is called *perihelion,* and the point of greatest distance is called *aphelion.*

Most Earth satellites do not orbit exactly in the plane of the Earth's

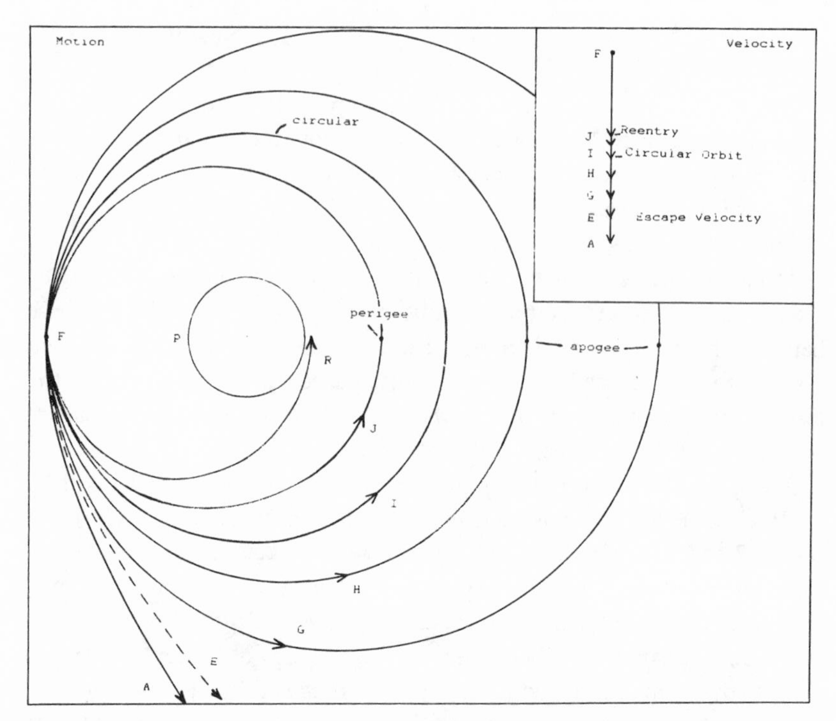

Figure 5.2
Elliptical Orbits

A spacecraft, initially following a circular orbit I about the planet P, briefly fires its engines at F to accelerate it in its direction of motion to a final speed V. If V is slightly greater than the original circular orbital velocity, then the spacecraft will enter an elliptical orbit, tangential to I at the point where the engines were fired, but rising to a maximum altitude (apogee) that lies well above the circular orbit height. The apogee of the new orbit (H) will lie at the point on the orbit that is directly opposite F. A slightly higher velocity increase at F will produce an even more eccentric orbit (G) with the same perigee height but an even higher apogee height. There is a particular value of V at which the spacecraft will depart from the planet altogether on an open (nonrepeating, nonperiodic) parabolic orbit. This value of V is called the *escape velocity*. For any circular orbit with orbital velocity V(C), the escape velocity is 1.414 times V(C). The velocity *change* needed to accelerate a spacecraft from circular velocity to escape velocity is the *difference* between its final and initial speeds, 1.414V(C) − 1.000V(C), or 0.414V(C).

If the spacecraft engine is fired exactly *against* the direction of motion at F, then the spacecraft speed will *decrease*. The new orbit (J) will have its apogee at F and its perigee at the opposite point on its orbit. A slightly longer retarding engine burn would drop the perigee into the atmosphere. The spacecraft could then use atmospheric drag to lower the apogee height from point F, or even reenter the atmosphere (R) to burn up or land. A roughly circular orbit just outside the atmosphere is called a Low Earth Orbit (LEO).

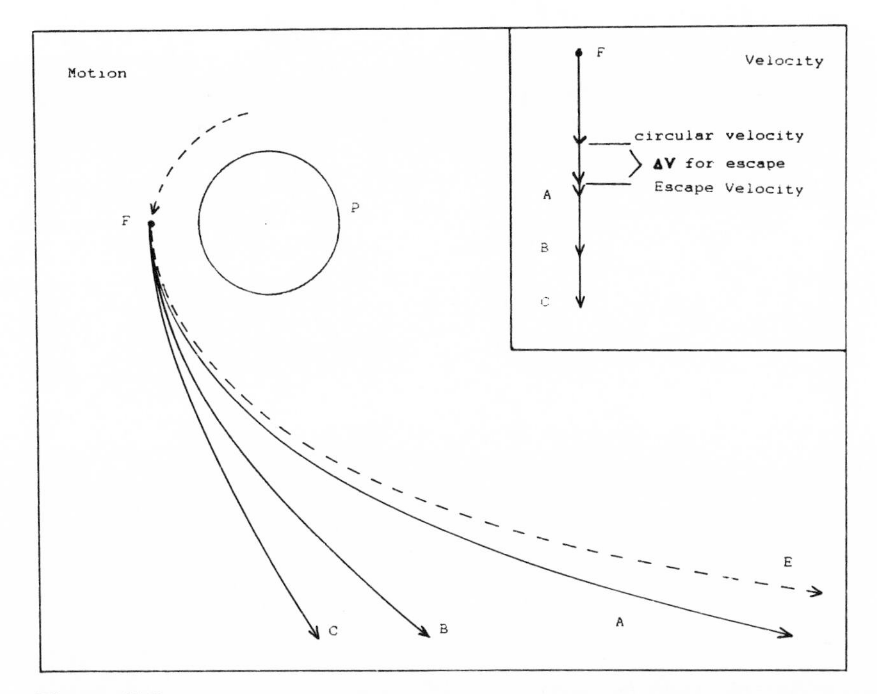

Figure 5.3
Hyperbolic Orbits
A spacecraft initially following a bound orbit about the planet P briefly fires its engines at F to accelerate it in its direction of motion to a final speed V. If V is greater than the escape velocity from the point F, then the spacecraft will depart from the planet on an open (unbound) orbit that is hyperbolic in shape. The spacecraft will slow down as it climbs against the gravitational attraction of the planet, but will never stop and return: its velocity will still be directed outward even when it is at nearly infinite distance from the planet. The inset shows the velocities A, B, and C at E. The fastest spacecraft, C, has its path least affected by the planet, and travels in a most nearly straight line. Spacecraft B, traveling more slowly, is more affected. Spacecraft A is traveling just above escape velocity: it would be traveling outward extremely slowly when it is at immense distance from the planet. Spacecraft E departs at exactly escape velocity and coasts along the dashed line to come to a stop at infinite distance from the planet, after an infinite amount of time. Its path is a parabola, not a hyperbola. Any speed at F that is less than the escape velocity would leave the spacecraft in a bound (periodic, repeating) orbit about the planet. The eccentricity of a parabolic orbit is 1.0000. All hyperbolic orbits have eccentricities greater than 1.

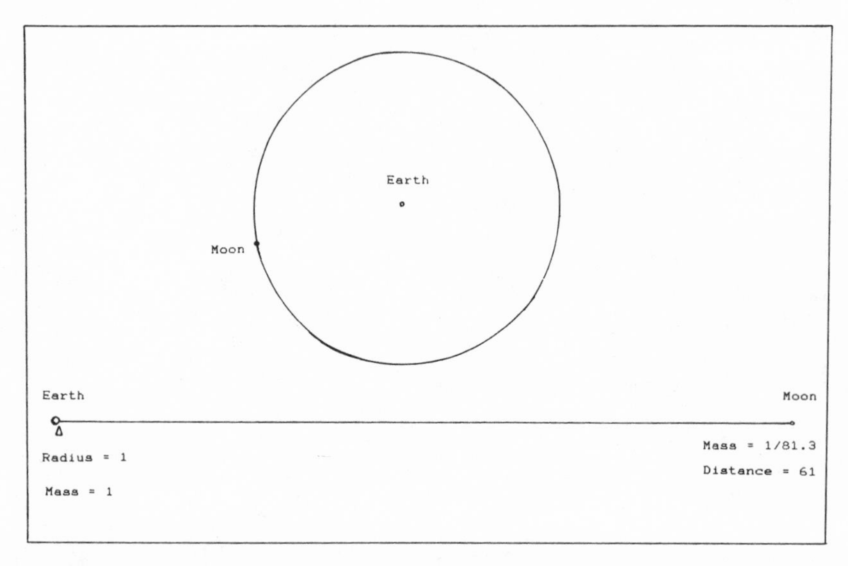

Figure 5.4
Center of Mass of the Earth-Moon System
The Moon pursues a roughly circular orbit about the Earth. Because of the significant mass of the Moon (1/81.3 of the mass of the Earth), both bodies actually orbit about a point that is the center of mass of the Earth-Moon system. That point lies inside the Earth, but closer to the surface than to the center.

equator but rather in planes inclined to the Equator. The angle between the orbital plane and the equatorial plane is called the *inclination* of the orbit (fig. 5.5). If the inclination of the orbit is close to 90 degrees, then the orbit will pass over both poles. This is what is meant by the phrase "polar orbit."

Another handy description of the shape of an orbit is its *eccentricity,* the degree to which it deviates from a circle. A circle has an eccentricity of 0, an ellipse may have any eccentricity from 0 to 1, a parabola has an eccentricity of 1, and a hyperbola has eccentricities between 0 and 1.

If satellites were acted upon by only a single force, then, according to the laws of motion discovered by Sir Isaac Newton over three centuries ago, they would remain forever in uniform motion. A satellite orbiting the Earth under the sole influence of Earth's gravity, and free of all other forces,

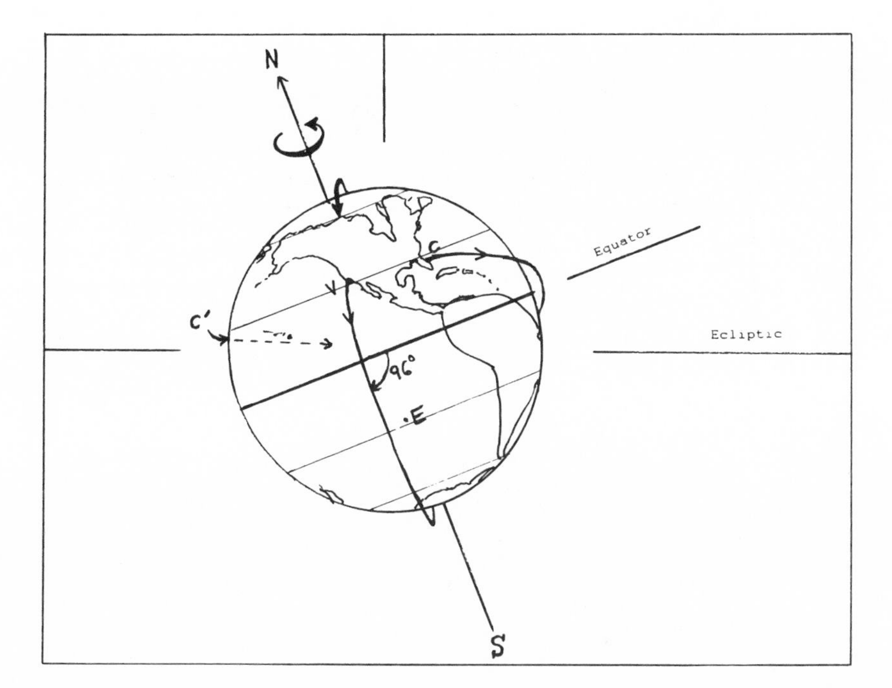

Figure 5.5
Inclined Orbits
A satellite from Cape Canaveral (C) gets the most help from Earth's rotation speed when it is launched due eastward into an orbit with the lowest possible inclination. Launch directions from Canaveral are limited only by the safety requirement that launches not overfly populated areas in the Antilles and on the eastern coast of North America.

Launches from Vandenberg Air Force Base (V) in California can achieve polar orbits (90 degree inclination) or Sun-synchronous orbit (about 96 degree inclination) without the booster overflying inhabited territories, but at two costs: first, launches cannot take advantage of the rotation of the Earth, thus reducing the payload, and second, Space Shuttle launches have almost no ability to reach landing sites in the event of a failure to achieve orbit. Easter Island (E) is the only possible emergency landing site between California and India.

Note that, at the time of day when Canaveral is at C′, a spacecraft can be launched directly into the Ecliptic plane with a due-east launch. Soviet launch sites are at much higher latitudes, which forces them to pay a significant payload penalty on lunar and planetary missions.

Satellites in Low Earth Orbit (LEO) take about 1.5 hours to complete each orbit. In that time, Earth rotates 22.5 degrees to the east, which shifts the ground track of the satellite 22.5 degrees farther to the west on each orbit.

would indeed orbit forever. There are, however, several other forces that act upon satellites. First, satellites in orbits below about 1000 kilometers altitude pass through the most tenuous upper reaches of Earth's atmosphere. The friction (drag) of passage of the satellite through the atmosphere slows it down, and the satellite loses energy (fig. 5.6). Accordingly, the

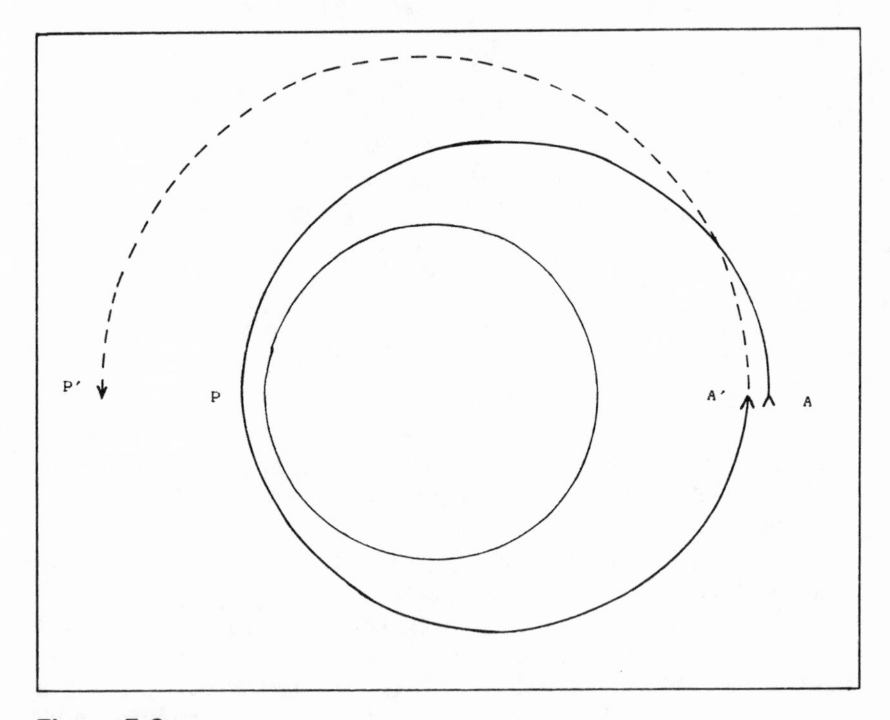

Figure 5.6
Atmospheric Drag
A satellite that dips into the upper atmosphere at perigee (P) will lose a small amount of velocity because of frictional drag. This will lower the height of apogee (A) slightly on every orbit. Since there is no drag near A, the perigee height is almost unaffected. If P stays fixed and A descends, then the orbit must decrease in eccentricity (become more nearly circular) as its orbital period decreases. The orbit will be nearly circular (eccentricity near zero) and very low (P and A both near 60 miles, or about 100 km) just before the "final plunge" into the atmosphere. A satellite with a small engine can greatly extend its lifetime by executing an apogee burn at A′ to lift the perigee out of the atmosphere to P′, where drag is negligible.

orbit slowly "decays," closer and closer to the Earth, until the satellite eventually burns up in the lower atmosphere.

The second complicating, or perturbing, force on an Earth satellite is the gravity of the Sun. This force is very small compared with Earth's gravity for satellites in LEO, but gets increasingly important for satellites in higher orbits. For satellites in low Earth orbits with orbital inclinations of about 96 degrees (polar orbits that actually move slightly backwards relative to the Earth's direction of spin) the effect of the Sun's gravity is to twist the plane of the orbit of the satellite slowly so that the plane makes one complete revolution in one year. A satellite in such an orbit may be placed in a plane that passes over a certain band of latitudes (such as that occupied by a particular nation) at the same time of day every day. A satellite in such an orbit can be made to stay in perpetual sunlight. Because of the special nature of these orbits, they are called "Sun-synchronous."

A third perturbing force is the gravity of the Moon. This may be used to advantage by using the Moon as a "gravitational slingshot" to help accelerate and aim outgoing or returning spacecraft: the Moon provides a velocity change without the expenditure of precious fuel.

A fourth, very faint, perturbing force is the pressure of sunlight. This has nothing to do with the solar wind—the stream of particles constantly emitted by the Sun—which are too few in number to be important. Rather, this is the momentum carried by sunlight itself. A reflective balloon satellite in, for example, a Sun-synchronous orbit would be pushed away from the Sun: the entire orbit would be displaced outward away from the Sun by several kilometers. In effect, the satellite would no longer be orbiting about the center of mass of the Earth, but about a point several kilometers farther from the Sun than Earth's center.

Imagine Venus and Earth orbiting about the Sun in perfectly circular orbits in the same plane (this is very nearly the case). Any orbit lying in that plane which has an aphelion distance at or beyond Earth's distance from the Sun (called one Astronomical Unit: one AU is just under 150 million kilometers) and a perihelion distance inside Venus' distance (0.7 AU) could be used to get from one orbit to the other (fig. 5.7). Picture a spacecraft on Earth's orbit about the Sun, traveling at the exact speed needed to keep it in a circular orbit. If we add to its energy by briefly firing its rocket motor so as to increase its orbital speed, then it will be moving too fast to stay on Earth's orbit. Instead, it will follow an elliptical orbit that always has its point of closest approach to the Sun at exactly the spot that the engine was fired (the

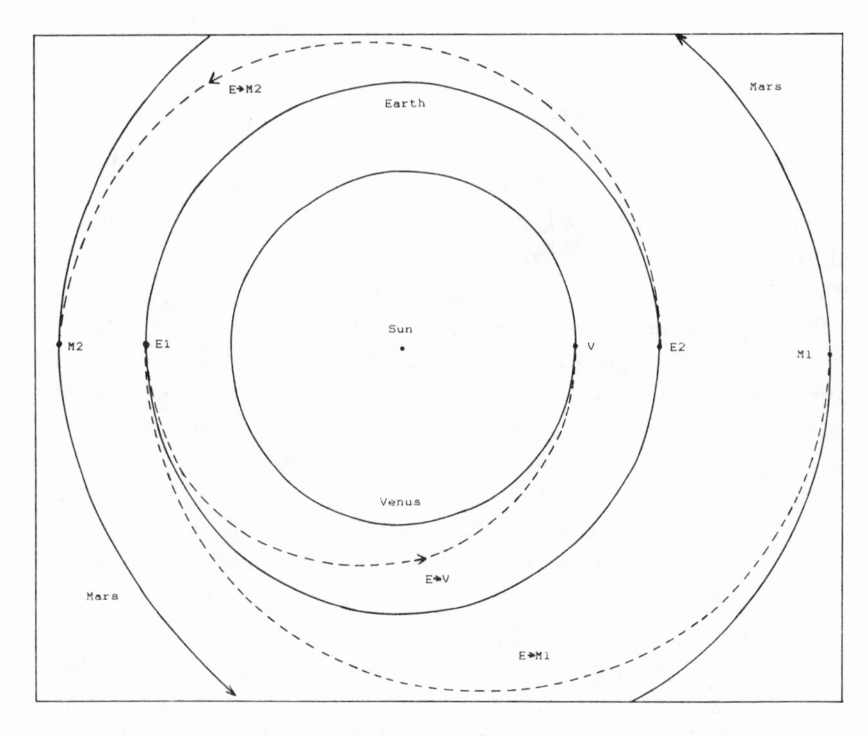

Figure 5.7
Trajectories for Earth to Mars and Venus
Launches of planetary probes from Earth at E1 into minimum-energy transfer orbits to Venus (E→V) and to Mars (E→M1) must be timed to arrive at the orbit of the target planet at the same time that the planet is at the intersection point (V or M1). Note that a "minimum energy" transfer orbit that arrives at Mars at aphelion (M1) requires much more energy than a transfer departing Earth at E1 and intercepting Mars at M2. About a quarter of all Earth launch windows to Mars provide arrivals at Mars when Mars is near perihelion, and these launch windows are therefore unusually desirable. Because of the very small eccentricities of the orbits of Venus and Earth, all launch opportunities for flights from Earth to Venus are nearly equally favorable.

LEO to GEO, and a second engine burn at the apogee of the GTO (at F2) inserts the satellite into GEO by lifting the perigee from LEO to GEO. The transfer time from LEO to GEO is about 6 hours (half a GTO orbital period).

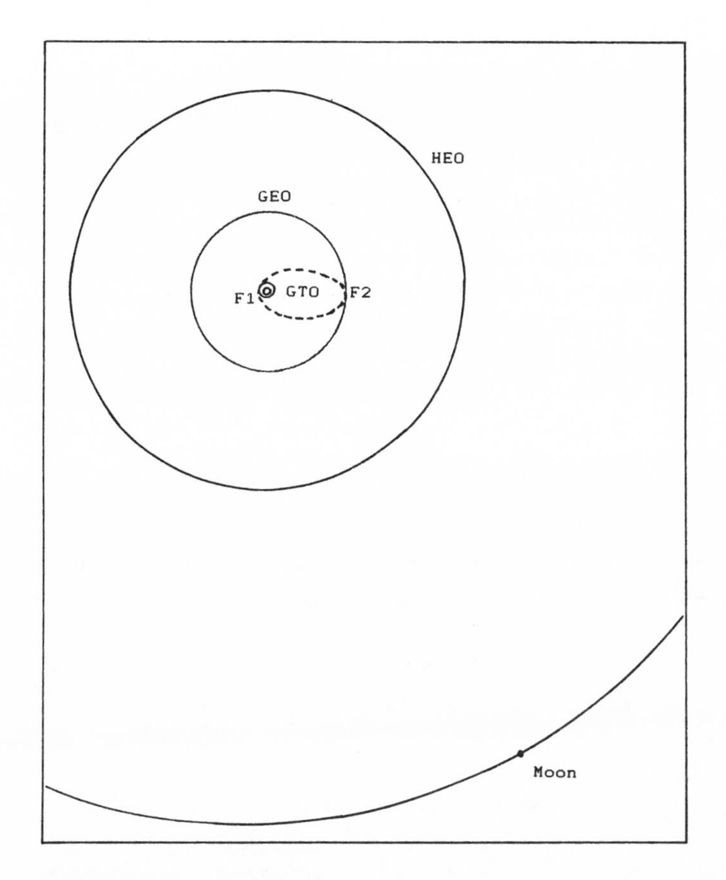

Figure 5.8
Circular Orbit Periods

Orbit	Height (mi)	Period
Low Earth Orbit—LEO	200	1 hr 30 min
Geosynchronous Orbit—GEO	22,400	24 hr (1 day)
High Earth Orbit—HEO	60,000	3.6 days
Moon	238,000	28.3 days
(LEO to GEO transfer—GTO	200–22,400	12 hrs)

Orbital periods of Earth satellites can range from just under 90 minutes for very low circular orbits to several months for satellites far beyond the orbit of the Moon. A satellite with an orbital period of one day circles the Earth in the same time it takes the Earth to rotate upon its axis. Such a "geo-synchronous" satellite can be made to hover forever over a fixed location on the Equator.

We include here a sketch of a typical transfer orbit from LEO to GEO, labeled GTO. A single engine burn at F1 lifts the apogee of the orbit from orbit is bound, and must repeat). Its aphelion will be exactly on the opposite side of the Sun from the place where the engine was fired, which we now see has become the perihelion of the new elliptical orbit.

If, rather than using the rocket motor to add to the orbital speed of the satellite, we instead turn the rocket around and fire the motor to slow the satellite's orbital velocity around the Sun, then the satellite will no longer be able to stay that far away from the Sun: it will have less kinetic energy relative to the gravitational energy of the Sun, and hence will fall inward along an elliptical path. The point of motor firing will then become the aphelion of the new elliptical orbit, and the perihelion will lie exactly on the opposite side of the Sun. A short burn of the rocket means a slightly altered orbit. The more energy that is expended in the burn, the larger the change in the orbit. There is a particular velocity change that will let the satellite just graze the orbit of Venus at perihelion. This minimum-energy transfer orbit is called a Hohmann transfer ellipse, after the German scientist who first discovered the principle behind it.

The basic principle that the new orbit must, if it repeats at all, pass through the point where the velocity change occurred, is a fundamental one. It tells us, among other things, that moving from one circular orbit to another circular orbit requires two distinct rocket burns, the first to get into the Hohmann transfer orbit and the second at the point where the Hohmann orbit intersects the target orbit. The total velocity change (delta V) for such a mission is the sum of the delta Vs for these two engine burns (fig. 5.9). The same principle also tells us that a satellite in an elliptical orbit of Earth, dipping into the upper atmosphere at perigee, would slowly alter its orbit so as to *lower the apogee while leaving the perigee altitude almost unchanged.* This makes the orbit more circular. Another way to express it is to say that atmospheric drag reduces the eccentricity of the orbit.

Each minimum-energy (minimum-fuel) flyby, round trip, or expedition is composed of one or more segments that are Hohmann transfers. The planetary mission designer must conjure up a system of bodies that generally are not in circular orbits and rarely exactly in the same plane (coplanar), and find the minimum energy required to carry out the mission and the best time to launch it.

Consider again the transfer orbit needed to send a flyby from Earth's orbit past Venus. In the real world, we must launch from Earth. Thus the transfer flight must start from Earth's orbit at the time that Earth is at the point of tangency with the transfer orbit. Further, the satellite must arrive

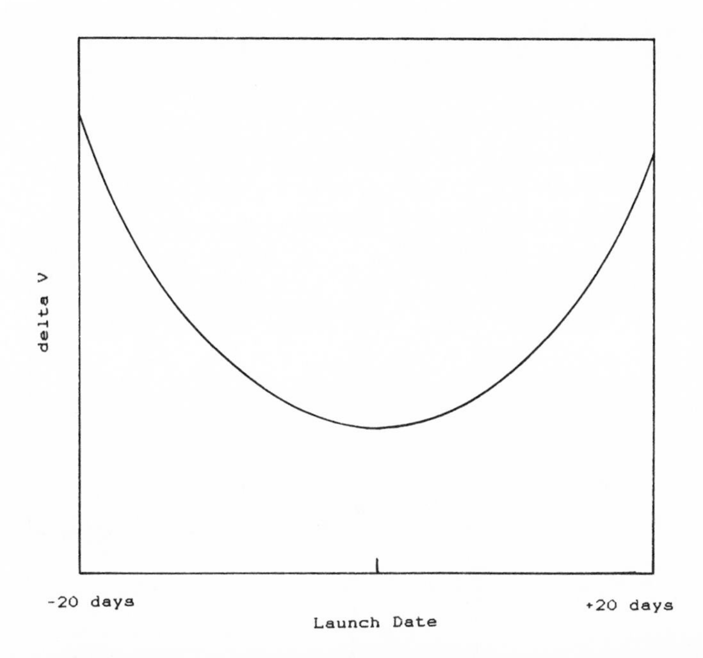

Figure 5.9
Launch Windows
The total velocity change (delta V; Δ V) required for a particular interplanetary transfer is near its minimum value for only a few weeks. There is a severe energy penalty for launching outside this "window."

at the orbit of Venus, at the point where the transfer orbit grazes Venus' orbit, at the same time that Venus does. Otherwise there will be no Venus flyby!

But transfers to Mars are even more complex to treat: the orbit of Mars is fairly eccentric, and the separation between the orbits of Earth and Mars varies from 0.37 to 0.67 AU. Thus, of the many times when the orbital positions allow us to enter a transfer orbit from Earth to go to Mars, only a few arrive at Mars when it is near perihelion. The amount of energy (propellant) needed for the best Mars opportunities is far less than the amount needed on other occasions.

It is not hard to use the data on the orbits of the planets to calculate the ideal launch times for flyby missions to Venus and Mars, and indeed to

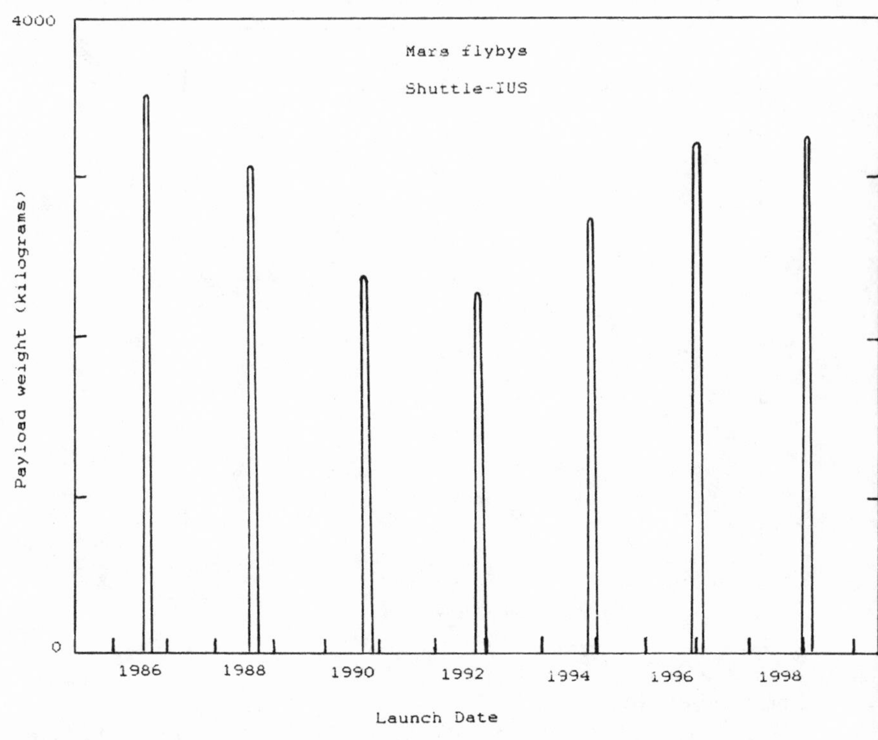

Figure 5.10
Launch Windows
The payload that can be carried by Mars flyby spacecraft launched from the Space Shuttle using the Inertial Upper Stage (IUS) is shown for a time period that includes seven Mars launch windows. Velocity change (delta V) requirements are so severe outside of narrow (about one month) launch windows that no mission is possible: even with zero payload weight, the booster could not reach Mars. Note the considerable range in the "optimum" payload weight that can be carried on missions that are launched in different launch windows. Generally, the largest payloads can be carried on missions that reach Mars near perihelion.

other patients as well. All Venus opportunities are rather similar in their energy requirements, but the Mars opportunities vary greatly in their attractiveness for the reason outlined above (fig. 5.10).

We have included a list of the dates for Hohmann least-energy transfers from Earth to Mars and Venus for the time period from 1960 to 2000. The best Mars opportunities are marked. The time interval (a week to a month) over which the launch energy is close to optimum is called a *launch window* (fig. 5.11).

Missions to Jupiter recur almost exactly every 13 months, and windows for direct flights to Saturn open every 12½ months. For the most distant planets, the opportunities repeat in intervals of just over one year. Sometimes nature is kind to us, and we find that we can fly by more than one planet on a single transfer orbit. Sometimes nature is doubly kind, and a flyby of one planet may use its gravitational pull to redirect the spacecraft, and even accelerate it, toward another target (see fig. 5.12). The Mariner 10 Venus-Mercury flyby and the Pioneer 10-11 and Voyager 1-2 missions to the outer planets are familiar examples of this technique. Every 170 years the jovian planets, Jupiter, Saturn, Uranus and Neptune, line up along a smooth curve, so that a single spacecraft can fly by all of them. This opportunity, called the "Grand Tour," was used by the Voyager spacecraft launched in 1977. The previous Grand Tour launch window was missed because of the low space priorities of Presidents Adams and Jefferson.

A very useful measure of how hard it is to get to a particular target is the minimum velocity change (delta V) needed to get to it. Since all launches to deep space have to pass through LEO along the way, and since future missions may be staged from American and Soviet space stations, the delta V needed for these missions is often given for the transfer from LEO to the target. This is especially appropriate if the mission is to return to LEO.

HOW TO GET AROUND THE SOLAR SYSTEM

The basic, familiar method of propelling a spacecraft is by means of rocket engines (fig. 5.13). The fundamental principle of a rocket is that it ejects mass at high speed in one direction. The force used to expel the exhaust accelerates the rocket in the opposite direction. All real rockets that you have ever seen in flight or on television are propelled by violent, heat-producing chemical reactions: combustion.

Although nature is very clever, people have a way of eventually catching on to Nature's tricks. Many other kinds of rockets have been designed, some have been built, and a few actually flown in space, some without recourse to combustion, and some without any kind of energy-producing chemical reactions. Even more interesting, some means of propulsion do not use the

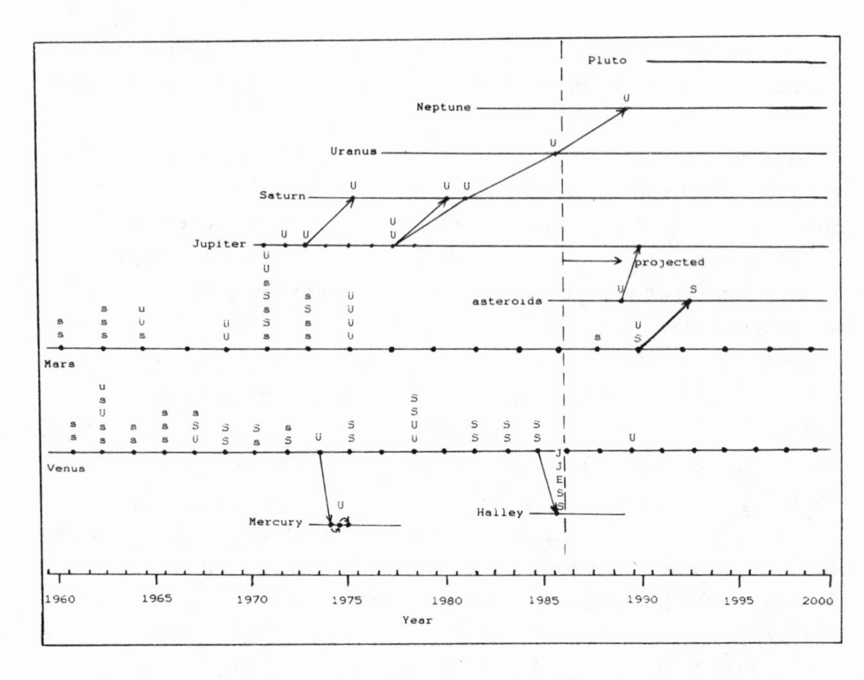

Figure 5.11
Planetary Launch Windows and Missions, 1960–2000
The launch windows for missions to Mars and Venus, and a few illustrative Jupiter launch windows, are shown as dots. The Jupiter launch windows recur at 13-month intervals, direct missions to Saturn recur at intervals of 12.5 months, and Uranus, Neptune and Pluto opportunities occur just over one year apart. In practice, missions to Saturn, Uranus, Neptune, and Pluto would normally be launched in Jupiter launch windows, and would take advantage of the gravitational slingshot effect (see chapter 5) to accelerate and redirect the spacecraft. Missions to Mercury or to certain nearby asteroids and comets would commonly use a Venus swingby to shorten the flight time.

Actual launch attempts are indicated with an S (for Soviet Union), U (United States), J (Japan), or E (ESA; European Space Agency). Note the far higher level of Soviet planetary activity and its strong orientation toward Venus, compared to the sparse but very broad coverage of the Solar System by American spacecraft. Successful missions (defined as those that arrive at their target and return a significant fraction of their intended scientific data quota) are indicated by upper-case letters, and unsuccessful missions by lower-case letters.

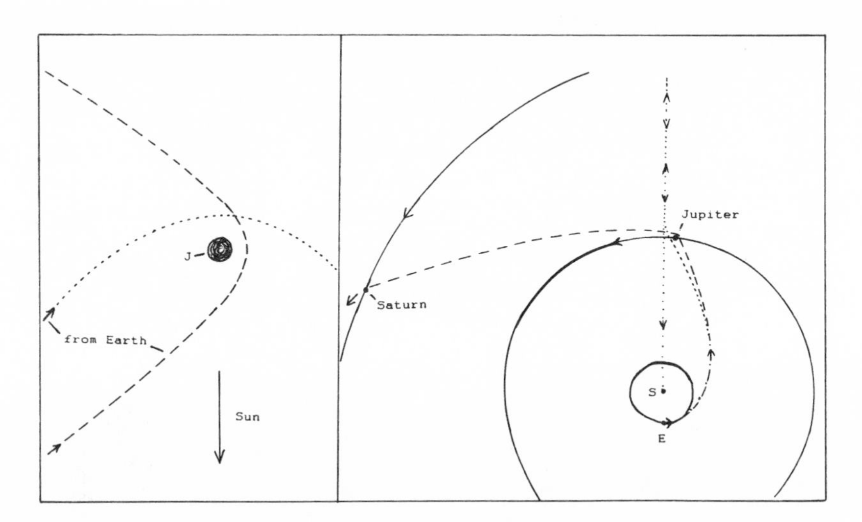

Figure 5.12
Gravitational Slingshots
A spacecraft that flies by a massive planet will be strongly affected by that planet's gravity. Imagine two spacecraft launched from Earth toward Jupiter (J) that arrive with similar speeds, one passing ahead of the planet and the other behind it. The left-hand panel shows these encounters as seen from a point that stays above the pole of Jupiter. The right panel shows the view as seen by an observer who stays above the pole of the Sun (and hence sees Jupiter moving). The dashed trajectory, followed by the spacecraft that passes behind Jupiter, is seen as following a hyperbolic, symmetrical trajectory by the observer above Jupiter: it departs with the same speed that it has when it arrives, but in a different direction. Relative to the Sun, the spacecraft speeds up enormously because it departs from Jupiter in the direction of Jupiter's motion. This gravitational "kick" can be used to accelerate and redirect the spacecraft to get it to a more distant target (here, Saturn) more quickly. The dotted path, followed by the spacecraft that passed ahead of Jupiter, also looks symmetrical from Jupiter's point of view, but, as seen from the Sun, it departs from Jupiter with its forward motion about the Sun removed. It travels straight outward, coasts to a stop, and then (when Jupiter has reached the far side of the Sun), the spacecraft falls into the Sun. From Earth, this is the easiest way to drop a probe into the Sun.

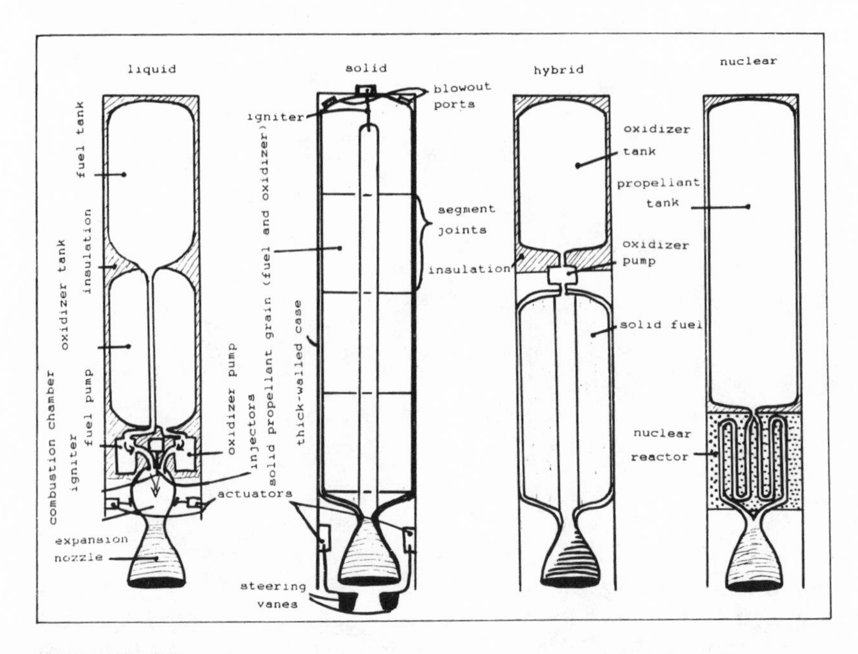

Figure 5.13
Four Types of Rocket Engines
A liquid-propellant rocket (left) burns a mixture of fuel plus oxidizer delivered to a combustion chamber by turbopumps. A solid-propellant rocket is very simple: it contains particles of both solid fuel and solid oxidizer, held together by a plastic or asphaltic "binder." A solid booster cannot be shut off once ignited: blowout ports are often provided at the front end of the rocket to give a means to limit the delta V provided by the rocket. The third type of rocket pictured is a hybrid, which has a solid fuel charge and a tank of liquid oxidizer. Such a rocket can be turned on and off like an all-liquid rocket, but also is about as complex as a liquid rocket. The fourth rocket type pictured is a nuclear rocket that uses the heat produced by a nuclear reactor to accelerate a working fluid (such as liquid hydrogen) to speeds unattainable in a chemical combustion reaction. All four rocket types work by expelling hot gases from a thrust chamber: the unbalanced force of the hot gases pressing in all directions pushes the rocket forward. Rockets do not need air to "push on." Quite the opposite, air pressure hinders the expansion and acceleration of the exhaust and reduces the performance of the rocket engine. Rocket engines perform more efficiently in space than in the atmosphere.

rocket principle. We shall briefly survey the main forms of propulsion, emphasizing promising new technologies for the next 25 years.

The principle of chemical propulsion, being most familiar, needs less introduction. A fuel and an oxidizer are mixed in a combustion chamber and ignited with a device that is essentially a fancy spark plug. The burning of the fuel releases a huge amount of heat, and the hot combustion gases are directed out a shaped rocket nozzle to give them the highest possible speed. The higher the speed of the exhaust, the better the performance of the rocket. A jet engine also burns fuel from its propellant tanks, but differs from a rocket in that jet engines gulp in huge quantities of air to supply the oxidizer (in this case, oxygen gas) needed for combustion. Jet engines therefore cannot operate in space where air is absent. Rockets certainly can!

Most rockets are of two basic types, liquid or solid. Liquid fuel rockets burn any of a number of different fuel-oxidizer combinations. The most common include kerosene-like hydrocarbons (such as aviation fuel) with liquid oxygen (LOX), hydrazine or hydrazine derivatives like MMH (monomethyl hydrazine), or UDMH (unsymmetrical dimethyl hydrazine), with liquid nitrogen tetroxide, UDMH with nitric acid, UDMH with liquid oxygen, and kerosene with nitric acid. The best liquid propellants are those that give off the most energy during combustion and have combustion gases with the highest possible temperature and lowest possible molecular weight. At the moment, the best chemical propellant available is the combination of liquid hydrogen with liquid oxygen.

Liquid propellants that boil far below room temperature (usually referred to as *cryogens*) are difficult to store for extended periods of time without heavy and expensive refrigeration equipment. The hardest of these propellants to handle is unquestionably liquid hydrogen: it boils at a temperature not far from absolute zero, has a very low density (and therefore requires very bulky tanks), and forms explosive mixtures with air. Other more friendly cryogens include liquid oxygen and nitrogen tetroxide. Propellants of lower volatility, not requiring refrigeration, include kerosene and nitric acid. These are called "storable" propellants.

The performance of rocket propellants is rated in a very simple way: what we really want is a propellant that produces the greatest possible engine thrust (force) compared with the propellant use rate. Imagine that one pound of a particular propellant combination, fuel and oxidizer, is burned at the exact rate needed to produce a constant thrust of one pound. Obviously, the best propellant is the one that will burn longest. Let us simply set up an engine, load the propellants, adjust the fuel pump speed until the measured thrust is exactly one pound, and time how long one pound of propellant lasts. Most storable liquid fuels will give 250 to 290 seconds of burn time. If LOX is used as the oxidizer, the efficiency increases, and burn times of about 300 to 340 seconds can be achieved. With the superb combination of liquid hydrogen and liquid oxygen, burn times of up to 420 seconds can be realized on Earth and up to 480 seconds in space.

Rather than saying "the amount of time that one pound of this fuel combination will burn while producing a constant thrust of one pound is 285 seconds" every time we want to rate a fuel, we abbreviate and say "this fuel has a *specific impulse* of 285 seconds."

The only way to beat the specific impulse of hydrogen-oxygen is to add liquid fluorine to the LOX, a procedure called "floxing." Unfortunately, fluorine is extremely corrosive, and the exhaust from this rocket contains huge amounts of hydrogen fluoride, which is very toxic and so corrosive that it will dissolve rocks (and, if you're not careful, rockets!). By comparison, the hydrogen-oxygen engine delivers nearly the same performance while producing benign exhaust gas that is essentially pure water vapor.

Almost all liquid propellant rocket engines use fuel from one tank and oxidizer from another. There are, however, examples of rockets that act upon a single propellant molecule which is itself unstable: if the molecule is heated sufficiently, the molecule reacts to form new chemical species and gives off heat. These single-propellant *(monopropellant)* systems are simpler than normal liquid rockets because they require only one set of pumps, but they are usually of rather low performance. The most common monopropellants are hydrazine (which decomposes to make hydrogen and nitrogen) and nitromethane. Hydrazine has often been used in low-thrust rockets that control the orientation and spin of satellites.

Solid rocket propellants are usually of relatively low performance, but they are simpler than liquid propellant rockets (no pumps, valves, insulation, or plumbing), can be made very large, and can be stored on Earth or in space for very long times before use, unlike high-performance liquid cryogens. Solids are likely, therefore, to be found on "both ends of the stack": the massive strap-on solid rocket boosters (SRBs) that lift the Space Shuttle off the launch pad and get it started on its way, or the small "kick stage" that rides with a spacecraft en route to Jupiter or Saturn and then, after one or two years of inert flight, ignites and places the spacecraft in orbit about the planet.

Solid rocket propellants usually consist of three parts, a fuel (tarry or rubbery organic material, aluminum powder, etc.), an oxidizer (such as ammonium perchlorate), and a binder to hold the powdered fuel and oxidizer together. This mixture, once ignited, is very hard to throttle or extinguish.

Some rocket engines contain a hollow tube of solid propellant which is made of fuel and binder only, with no oxidizer. A tank of liquid oxidizer is carried separately. When the rocket is to be fired, pumps force the oxidizer through the cavity in the solid fuel, and the engine burns nicely. When the desired burn time has been achieved, the pumps are shut off and the flame dies at once. These "hybrid" rockets are desirable for missions that require starting and stopping the engine several times, a feat that is quite difficult with "pure" solid-propellant rockets.

American satellite launchers of the recent past commonly employed first stages using kerosene-liquid oxygen (the Atlas ICBM and the first stages of the Saturn 1 and Saturn 5 superboosters) or hydrazine derivatives with nitrogen tetroxide (the Titan ICBM). The Space Shuttle vehicle is composed of several parts: the winged orbiter has three engines, totaling about 1.5 million pounds of thrust, which burn from launch until orbit insertion, using liquid hydrogen and liquid oxygen drawn from a huge external tank (the original ET) that attaches to its belly. The tank is normally jettisoned just before orbit insertion, and falls into the atmosphere to burn up over the Indian Ocean. At launch, two long solid-propellant rocket boosters (SRBs) attached to the ET help hurl the entire massive vehicle off the pad and up through the dense lower atmosphere.

Soviet boosters use a variety of liquid propellants. The A-class booster uses kerosene and LOX, the B1 has a first stage that uses UDMH and LOX and a second stage using kerosene and nitric acid, and the C1 and F1 apparently use storable UDMH with nitric acid. All major operational stages in Soviet launchers appear to use only conventional, low-energy liquid propellants. Soviet experimentation with hydrogen-oxygen propulsion goes back to at least the early 1960s, but we have no convincing evidence that such a stage has ever flown successfully in the Soviet space program. The *Kosmos 1767* satellite, placed in orbit on July 30, 1986, appears to have been carried by a new, experimental launch vehicle. It is possible that this vehicle may use H-O propulsion. Solid propellant rockets are also surprisingly absent from the Soviet space program, possibly because the main Soviet launch site, Plesetsk, gets so cold that solid propellants crack and fail to burn properly.

There are no major advances in sight that would permit significantly improved performance of chemical propulsion systems. The American space program has relied on high-energy hydrogen-oxygen engines since the Centaur flights of the early 1960s. The Apollo program and the Space Shuttle depend on H-O propulsion. There is no doubt that the Soviet Union could achieve great improvements in payload performance with existing boosters if they were to be successful in developing this technology.

NUCLEAR PROPULSION

Having reached the limit of the possible performance of chemical propulsion systems, it is natural to wonder if we are forever fated to be stuck at present levels of capability. Fortunately, the answer is a resounding "No!" The specific impulse of a rocket engine is proportional to the velocity of its exhaust. The limit reached by hydrogen-oxygen engines is determined by the amount of energy contained in the fuel mixture and the molecular weight of the exhaust. The lighter the molecules in the exhaust, the faster they move at a given temperature. Also, of course, the higher the temperature, the faster they move. An exhaust made of hydrogen molecules would move faster, and give a higher specific impulse, with the same engine temperatures—but how would we heat the exhaust? Atomic hydrogen would be even better. But best of all would be something even lighter than the lightest atom: light itself! But in every case, where do we get the energy to generate the exhaust if we don't use chemical reactions?

One possibility that we have mentioned is the nuclear rocket, in which the energy produced by a nuclear reactor is used to heat a stream of hydrogen to the highest temperature that the equipment can stand. The jet of hot hydrogen may have a specific impulse of about 1000 seconds. In the 1960s NASA was fully aware of the promise of nuclear propulsion: plans were openly discussed for dispatching a manned expedition to Mars in 1982 on a nuclear rocket. But the problems of the nuclear rocket are also well known: leakage of radioactivity; the heavy mass of shielding needed to make it safe to carry men; the possibility of a crash with a resulting major radiation release. The nuclear rocket is, for all practical purposes, banned from Earth.

The heart of the problem, the nuclear reactor, has by a curious twist of fate not been banned from space. The Soviet Union has launched a series of heavy ocean surveillance satellites that use nuclear reactors to power large synthetic aperture radar transmitters (SARs). Their main use is to keep track of American aircraft carriers and task forces. One of these reactor satellites, *Kosmos 954,* reentered the atmosphere over western Canada a few years ago, leaving a trail of radioactive debris over a lengthy swath of rugged wilderness. The satellite could just as well have dumped its radioactive load on Calgary—or Moscow, for that matter—if it had come down at a slightly different time.

ELECTRICAL PROPULSION

It is possible to use electrical energy to strip the electrons from a gas to make a mixture of positively charged ions and negatively charged electrons. The ions and electrons can then be separated by electric fields applied to a wire-mesh grid on the rear of the ionization chamber. The electron and ion beams are merged as they leave the rocket, so the net charge of the exhaust (and of the spacecraft) is zero. It is easy to apply electric fields so strong that the ions are ejected with speeds far greater than they could achieve in any chemical reaction. This is an example of electrical propulsion: the engine just described is often called an ion engine (fig. 5.14).

The ion engine can deliver specific impulses of 3000 to 4000 seconds, which is truly superior performance. However, the ion engine is not without its complications and drawbacks.

First, the ion engine works only on a very low-density gas. This means that, even though the engine is very efficient, the actual thrust is very low. A million-pound vehicle might reasonably deliver a thrust of 100 pounds, for a peak acceleration of 0.0001 gravities. Trips taken under such low accelerations are bound to be very lengthy.

The low thrust per pound of spacecraft is due in part to the necessity of supplying huge amounts of energy to the ion engine. While the vehicle needs to expel little propellant because of its high specific impulse, it nonetheless needs an uncommonly large mass of power-generating and power-conversion equipment. The source of energy could be a nuclear thermoelectric generator, of the sort flown on a number of satellites (not a large, complex reactor), or a large array of solar cells. Ion engines based on these two energy sources are called nuclear electric and solar electric propulsion systems, respectively.

A second problem is that the wire grid that pulls out the ions is bombarded by very fast ions and slowly erodes away. Thus very long missions

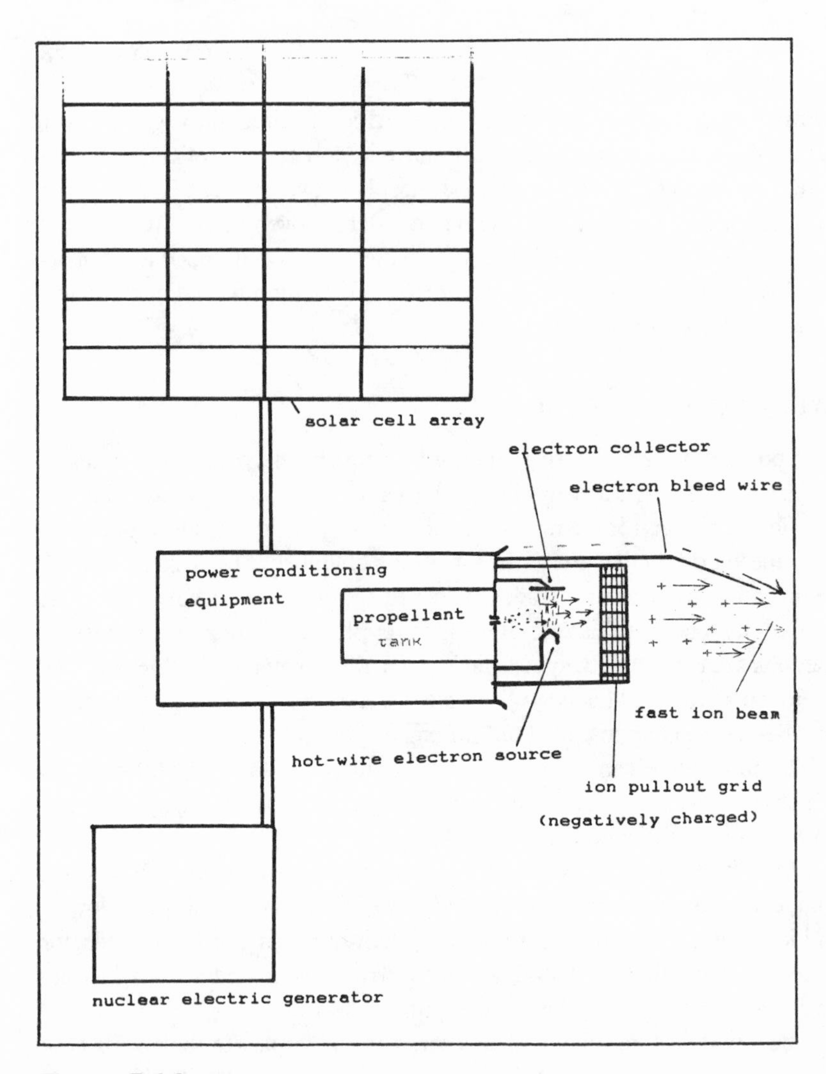

Figure 5.14
Electrical Propulsion
An electrical propulsion system requires very large amounts of electrical energy, supplied by either a solar-cell array (top) or a nuclear reactor (bottom). A propellant such as mercury, argon, oxygen, or cesium is sprayed in a diffuse cloud through a region being bombarded by an electron gun. The fast electrons from the gun knock off electrons from the target atoms, making them into positively charged ions. These ions are then accelerated by a large electrical potential energy difference and fired out the back of the spacecraft. The electrons that had been stripped off of these ions by the electron gun are bled back into the ion exhaust to preserve the electrical neutrality of the spacecraft.

must either have provisions for periodically replacing the grid, or operate at lower exhaust velocities (and hence lower specific impulses).

Small ion engines have been tested in space on a number of occasions since the 1960s, mostly for simple orbital station-keeping to counteract very small disturbing forces such as gas drag caused by the upper atmosphere. No mission has ever been dispatched from Earth to go elsewhere under ion propulsion. Advances in ion propulsion have been minimal and slow-paced for over 20 years.

SOLAR THERMAL PROPULSION

The main factor limiting the performance of electrical propulsion systems is the large mass of the electrical equipment, and the main factor limiting chemical propulsion is the large mass of propellant necessary to provide the needed thrust. It sounds at first as if there is no way to win.

Fortunately, it is possible to devise a scheme that uses solar power (and therefore need not carry a heavy reactor); it uses the solar energy directly, without conversion to electricity (and therefore does not need acres of solar cells and bulky power-conditioning equipment, nor does it suffer from the 20% efficiency of conversion of sunlight into electricity).

This scheme uses a large inflatable or deployable reflector made of very thin metal film to focus sunlight directly onto a blackened thrust chamber made of a material such as rhenium metal, which is extremely resistant to melting and evaporation: it is a *refractory* metal (fig. 5.15). Hydrogen is fed into the thrust chamber, where it is heated to temperatures of thousands of degrees by the absorbed solar energy. The jet of very hot hydrogen can have a specific impulse in the range of 1000 to 1200 seconds. Furthermore, sunlight provides more than 1000 watts of usable power per square meter. This means that it is easy to deliver megawatts of power to the thrust chamber with only a few kilograms of reflecting film. The thrust level may be quite high, and accelerations of 0.01 to 0.1 Earth gravities could be achieved. This "hot rod" delivers the specific impulse of a nuclear rocket, and it does so with clean solar power and without the enormous dead weight of a shielded nuclear reactor. Maintaining the acceleration of 0.01 gravities permits interplanetary trips of 50 million kilometers to be carried out in less than two weeks. At 0.1 Earth gravities, interplanetary trip times would be a few days.

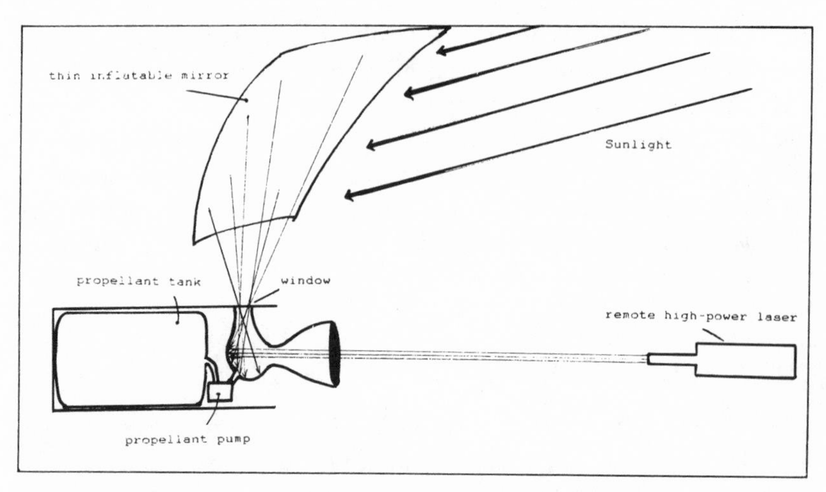

Figure 5.15
Solar and Laser Thermal Propulsion
The energy source for these thermal propulsion systems is either a large inflatable solar collector mirror that focuses sunlight through a window into the thrust chamber, or a remote laser that illuminates the inside of the thrust chamber with millions of watts of power. Propellant, pumped from a tank into the thrust chamber, is heated to temperatures higher than that of a chemical flame before exiting through the expansion nozzle.

Such fantastic performances, however, would consume vast amounts of propellant. Sending a 1000-ton payload to Mars by this means, at an acceleration of 0.01 gravities, would reach a maximum speed of 70 kilometers per second at the halfway point, and would use up 160 million tons of hydrogen. Surely solar thermal propulsion would be better used to replace chemical propulsion on shorter, lower-velocity trips. It is a very appealing alternative to nuclear propulsion for such trips, and delivers a performance far superior to any chemical propellants.

The use of sunlight causes this scheme to be very attractive to the inner Solar System, but much less desirable far from the Sun, where the intensity of sunlight is low.

LASER THERMAL PROPULSION

An alternative to sunlight as the source of energy to drive a thermal rocket is laser energy directed at the solar collector mirror from a remote transmitting station. The limitation on the use of a solar thermal system caused by the diminution of the intensity of sunlight does not affect a vehicle being driven by megawatts of laser power, broadcast in a beam that fits within the outline of the collector. Since the actual motor would work in the same way that a solar thermal rocket does, specific impulses of as high as 1200 seconds are possible. Again, it is the supply of propellant that limits the performance of laser thermal vehicles in interplanetary space.

There is also the possibility of using laser power to drive launch vehicles from the surface of the Earth to low Earth orbit. The "fuel tanks" of such a rocket may well be filled with water—the exhaust would be extremely hot stream. Obviously, use of large, flimsy collecting mirrors would not be possible during launch, but huge ground-based lasers could aim with sufficient precision to deposit their energy inside the rocket thrust chamber. The laser guiding accuracy necessary for this incredible feat is being developed as part of the Strategic Defense Initiative.

MASS DRIVERS AND RAIL GUNS

Not all schemes for electric propulsion expel diffuse streams of energetic ions. In fact, two potential forms of electromagnetic propulsion expel solid chunks of material with masses up to several kilograms. Because of this feature, both of these systems could also be used to send their "exhaust" to a desired destination while the "engine" itself remains stationary, on the surface of a large body such as the Moon.

The better known of these two types of electromagnetic accelerators, largely because of such popular books on space development and colonization as *High Frontier* (New York: Doubleday) by Gerard K. O'Neill of Princeton University, is the mass driver (see fig. 5.16). The mass driver, in its original conception, was effectively an extremely long solenoid coil that would work in the same way as the solenoids in electrically actuated switches. Electrical power, derived in O'Neill's example from large arrays of solar cells spread out on the lunar surface, would send pulses of power sequentially through a large number of loops of cable.

The material to be accelerated, for example, a kilogram chunk of iron, is set near to but "behind" the center of the first loop in the solenoid. A huge pulse of electric current is sent through the solenoid, which generates a very intense magnetic field. The metal slug is grabbed by the magnetic field and pulled rapidly forward toward the center of the loop. As it reaches the cen-

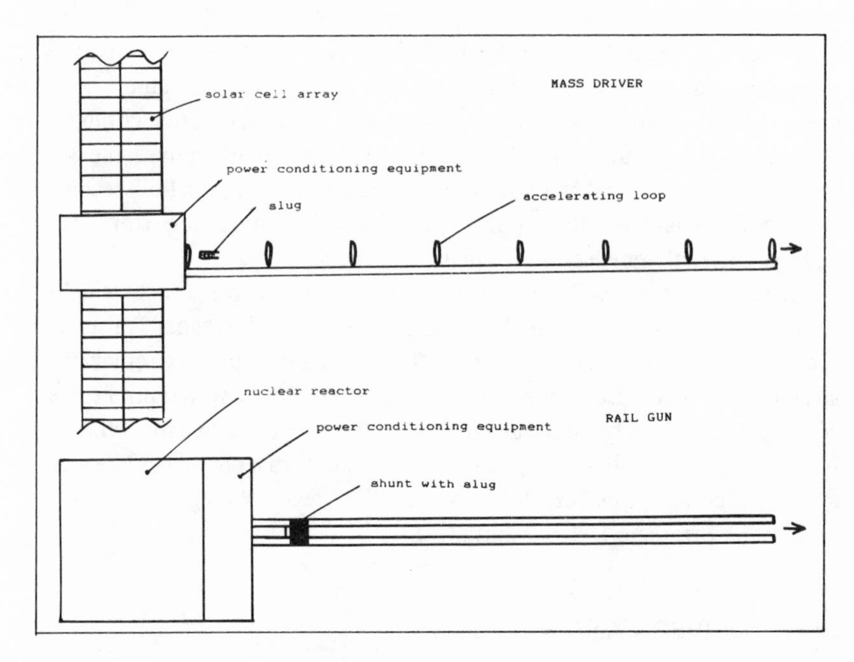

Figure 5.16
Mass Drivers and Rail Guns
A large source of electrical power, such as a solar cell array (top) or a nuclear power supply (bottom) is used to fire slugs of metal at very high speeds, perhaps up to tens of kilometers per second. The mass driver (top) drives large currents through a series of loops to kick the slug up to very high speeds. The rail gun (bottom) fires a metal "sled" by pouring an immense electric current through a shunt at the rear of the sled.

ter of the first loop, the electrical power to the first loop is turned off, and the slug continues on toward the second loop. As it approaches the second loop, a brief surge of power is sent through it, which again pulls the slug forward and accelerates it toward the third loop. The secret of high performance is twofold: the ability to tap very large amounts of stored electrical energy, and the ability to time the surges of power to each of the hundreds of coils. The slug may be accelerated to speeds of many kilometers per second, and hence may provide a specific impulse up to about 1000 seconds.

In this early version, the mass driver suffered from three main problems: first, it required a large mass of power-collection and power-storage equipment; second, it was many kilometers long, and had to be kept in excellent alignment, so that the mass of the structure also needed to be very large, and hence inflexible—it could not readily be aimed at more than one target; third, when used as a propulsion system it emitted streams of "bullets" into near-Earth space. Great care would have to be exercised to assure that the hypervelocity metal slugs were directed in such a way as to cause no hazard to anyone or anything in space.

Advances in power-storage technology, including promising applications of flywheels as energy storage systems, have reduced the concern about the first problem. Moreover, recent developments at Electromagnetic Launch Research in Cambridge, Massachusetts, have shown that mass driver "engines" can be made to deliver huge amounts of energy in very short distances. Thus mass drivers can now be designed to lengths of meters, not kilometers. The acceleration of the slug is, however, incomparably more violent than in the earlier concept. In fact, heating of the ejected metal by eddy currents induced in it by the mass driver can vaporize the slug. The net result is a form of mass driver that is very compact and low in weight, and that exhausts harmless vapors rather than deadly bullets. This is a recipe for a mass-driver engine for propelling bodies through space. It is obviously no way to shoot small payloads to their destination.

When used to deliver small payloads, the long, low-acceleration version of the mass driver is required. It is not necessary that all the ejected mass be made of metals. It is perfectly satisfactory to have a metal "bucket" full of dirt, which can be accelerated in the mass driver as a unit. It is even possible to use a long mass driver to accelerate the bucket full of dirt and then grab onto the bucket for reuse, letting only the dirt fly off into space.

The mass driver can easily achieve the escape velocity of the Moon. It therefore can be used to fire small packages of lunar material off the Moon toward destinations in near-Earth space. A lunar mass driver system, once installed, could deliver payloads into nearby space at much lower cost than any chemical rocket could hope to achieve. Further, it uses no fuel! Thus the need for large factories to squeeze the propellant lifeblood of rockets out of the lunar turnip would disappear. Instead, a mass driver could use a large solar cell array to provide its power. But then it would also need some form of energy storage device to turn the slow, steady trickle of power from the solar cells into the short, powerful bursts needed to shoot pellets.

On closer examination, other complications appear. Because of the slow rotation of the Moon on its axis, the Sun shines for about two weeks at a time, slowly drifting across the sky. Then comes a two-week night. There is no obvious way to store enough energy to run the mass driver through the night, when there is no solar energy reaching the solar cells. Also, in order to perform with high efficiency, the solar cell arrays should be steerable so that they can track the Sun as it moves across the daytime sky. This adds substantial weight and complexity to the solar power system. It may turn out that the best bet is to ignore the "free" solar energy and land a large nuclear power plant on the Moon. But these are very heavy and expensive.

On Earth, large power sources are relatively cheap. Why not use the mass driver to launch satellites from Earth? Unfortunately, the problems are so immense as to be insurmountable. First, satellites in orbit move horizontally. If we attempt to fire an electromagnetic launcher on the surface of the Earth nearly horizontally, then the slugs emerging from the mass driver would have to penetrate vast masses of atmosphere before exiting into space. Frictional drag would eat up more energy than would be saved. Second, most payloads (especially manned ones) are far too delicate to withstand the rigors of launch from a short mass driver: to get up to orbital velocity in a 10 kilometer mass driver would require an average acceleration of 320 gravities, enough to reduce an astronaut to a thin, very flat puddle. Once out of the muzzle of the mass driver, the payload would be subject to crushing deceleration while passing through the dense lower atmosphere: the direction of the acceleration would instantly reverse, and the puddle would slam into the ceiling.

If we desire to shoot through the least possible mass of atmosphere, then we must launch vertically. But we would not be able to build mass drivers that stand 10 kilometers tall. A practical limit would be about one kilometer, compared to the tallest structures ever built on Earth, which are about 0.4 kilometers high. This 300-story launch structure would have to accelerate its payloads at more than 3000 gravities to impart orbital speed to them, but even so they would be a total failure: the payload, emerging from the top of the atmosphere, would not be traveling horizontally in a stable orbit; it would coast upward a few thousand miles, come to a stop, and fall back to Earth. The size of the rocket motor it would need to carry to put itself in orbit would be nearly as large as that required to lift it from the ground—but this rocket would have to be strong enough and light enough to survive being fired from a gun!

Mass drivers may eventually see extensive use in space, but their use as satellite launch vehicles from Earth is clearly not feasible.

The second type of electromagnetic "pellet gun" is the rail gun (again, see fig. 5.16). As with the mass driver, a large source of energy and an energy storage system are needed. Solar cells or a nuclear power source would provide the energy for either kind of device. The rail gun has two long, parallel metallic rails that are aimed in the direction of firing. A huge electric voltage difference between the two rails is used to drive a powerful electric current arc across the backside of the pellet, which rides between the rails. The current induces an extremely strong magnetic field behind the pellet, which blasts the arc (and the slug ahead of it) down the track with very high acceleration. Rail guns have recently been tested at speeds up to 9 kilometers per second, and present planning calls for an upgrade to about 30 kilometers per second.

The rail gun is a nonnuclear weapon capable of destroying satellites or strategic missile warheads, and is therefore being developed principally with funding from the Strategic Defense Initiative. For our purposes, however, the rail gun is of interest primarily as a propulsion system that could deliver exhaust velocities of 10 to 30 kilometers per second: this translates into a specific impulse of 1000 to 3000 seconds.

All electric propulsion systems, whether solar- or nuclear-powered, require massive power sources and energy management systems (see fig. 5.17). All likewise require the expenditure of significant masses of propellant. Nonetheless, there is a wide range of circumstances in which these systems, because of their high specific impulse, can outperform any chemical rocket. They have the further advantage that the "propellant" can be almost anything that you might find lying around on an asteroid or natural satellite. Thus, for use off Earth and in locations where supplies of chemical propellants are absent (such as on a metallic asteroid), they may confer major operational advantages.

SOLAR LIGHT SAILS

Light (electromagnetic radiation) carries energy and momentum. If we could place a flashlight far out in space, away from any gravitating body, the flashlight would recoil very slightly as long as it was emitting light. By sending all of its stored chemical (battery) energy out as light, it would send out a certain amount of momentum in the same direction. The flashlight would end up with exactly the same amount of momentum, but would be moving in exactly the opposite direction. If we were to shine an intense

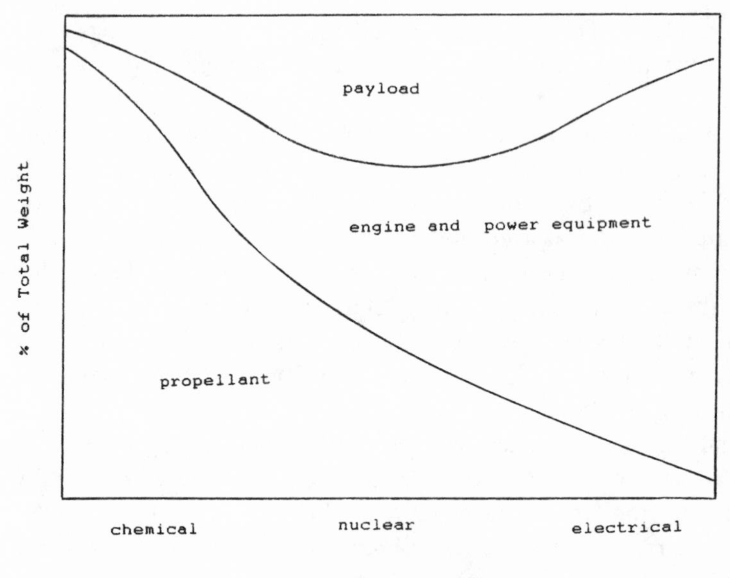

Figure 5.17
Tradeoffs in Propulsion Systems
This graph shows the weight breakdown of a space vehicle with a particular fixed mission (a Jupiter probe, Mercury orbiter, or the like). Use of chemical propulsion limits the vehicle to a low specific impulse, and requires that a huge mass of fuel be carried (left). Extremely efficient electrical propulsion techniques, with specific impulses of 10,000 seconds or so (right), use little fuel but require a great mass of electrical power supply and power-conditioning equipment. In either case, the payload is diminished. For any given mission with its associated delta V, there is some type of propulsion system that is optimal: the payload can be made largest at some intermediate specific impulse. For many planetary missions, nuclear or electrical propulsion systems with specific impulses of 1000 to 4000 seconds would be optimal. For missions with very small delta V requirements it does not make sense to carry the great weight of an electrical propulsion system, and a modest amount of chemical propellant gives the optimal payload weight.

beam of light from the flashlight onto a light, thin sheet, the sheet would recoil and be pushed away from the light source.

A dark sheet will absorb all the light hitting it, and thus it will pick up all the momentum carried by the light. If the sheet is thin enough so that both sides have the same temperature, and heat can be rapidly conducted between its front and back surfaces, then the sheet will quickly reach a balance in which the radiation absorbed by it will be exactly compensated by the heat radiation it gives off. Since the front and back of the thin sheet are at the same temperature, the amount of light (and momentum) given off in the two directions is the same, and exactly cancels out. Thus the sheet will accumulate momentum from the light beam with an efficiency of exactly 100 percent.

Now consider a perfectly reflective sheet, which absorbs no energy, but instead simply reflects all incident light back toward its source. This sheet experiences twice the recoil of the black sheet: it absorbs all the momentum of the light beam that strikes it, and then hurls it back toward its source, recoiling from that act as well. Thus, by reversing the direction of the light beam instead of absorbing it, the white sheet ends up with twice the momentum of the black sheet. It actually takes momentum from the light beam with an efficiency of 200 percent.

Several of the satellites that have already been placed in orbit were thin, inflated balloons. Their orbits were visibly altered by the pressure of sunlight on them: the centers of their orbits were actually pushed a kilometer or so away from the center of the Earth in the direction away from the Sun.

Balloon satellites that are carried up from Earth folded inside little canisters and then gently inflated in space, cannot be made very thin because they must be strong enough to survive the stress of folding, stowage, launch, and inflation. But metal films made in space can be much thinner. In space it may be feasible to make films so thin that the acceleration of the films under normal-intensity (Earth standard) sunlight could be made quite significant. It seems almost certain that thin metallic films capable of accelerating at 0.001 gravities could be made in space, and 0.01 gravities may be achievable with more advanced fabrication technology. These accelerations would permit trips between the inner planets to be carried out in a few weeks, rather than many months. Better yet, the propulsion system is very light (acres of thin metal film and guy wires). Even better, the "exhaust" travels at the speed of light. And best of all, it carries and uses absolutely no propellant! Maneuvering is done by tilting and twisting the sail by means of its guy wires, and one may tack on the light of the Sun just as a sailboat tacks on a breeze.

It is very important to keep in mind that this concept of "solar sailing," in which light sails are pushed about the inner Solar System by the light pressure of the Sun, has nothing to do with the solar wind. The latter is the stream of hot plasma, mostly made up of equal numbers of electrons and protons, emitted into space by the upper atmosphere of the Sun. It carries much too little energy to be of any use for propulsion purposes.

Despite the enormous promise of solar sailing, not even the simplest demonstration has ever been flown in space. Funding for this scheme is severely limited, and, with any practical demonstration of the technique dependent upon having several million dollars of funding, the prospects for an early flight test seem remote. But if we could sail even in a crude way on the light of the Sun, we could roam the inner Solar System at will, and even rendezvous with fast-moving comets. Testing truly high-performance light sails is not a very simple matter, since they are so sensitive to atmospheric drag that they cannot be used at altitudes below about 1000 kilometers. This rules out simply deploying them directly from the Space Shuttle in LEO. Nonetheless, the problem is readily solvable with a small kick stage. We could do a test in three years if we cared to try.

LASER-DRIVEN LIGHT SAILS

Solar sails are at the mercy of the light of the Sun. Since the intensity of sunlight drops off as the square of the distance from the Sun, the intensity of sunlight at Jupiter's orbit (5.2 Astronomical Units from the Sun) is less than 4 percent of the intensity at Earth's orbit (at 1 AU). Thus solar sails rapidly fade in usefulness beyond the inner edge of the asteroid belt.

However, a space-faring civilization with the ability to make Solar Power Satellites would be able to generate vast amounts of electricity. This electric power could be used to drive batteries of lasers directed at a light sail far from the Sun. Laser light beams are remarkable for their ability to stay in a very tight, very slowly spreading beam. It is entirely possible to imagine illuminating a light sail a billion kilometers away by laser light, and driving it with far more force than the feeble light of the Sun could exert at that distance. The entire solar system would then become accessible. Even interstellar probes could be dispatched with such high speeds that the duration of the journey to a nearby star could be cut down to a century or so. When we consider that the space age is already 0.29 centuries old, a century does not seem altogether ridiculous.

Although many other advanced propulsion techniques have been studied, the ones described in this chapter seem to span the range of propulsion methods that we are likely to use (or be capable of using) between now and 2010. It is likely that a number of these advanced ideas may eventually be used.

GRAVITATIONAL SLINGSHOTS

Propulsion advances are only one of the ways to get more performance per pound of propellant. Indeed, we have already encountered the idea of using the gravity of planets to hurl spacecraft off in new directions, and to increase or decrease their speeds. The *Mariner 10* mission in 1974 used the gravity of Venus to redirect the spacecraft toward Mercury, and the Soviet Vega missions to Halley's Comet used Venus in exactly the same way (Vega is short for Venera-Galley in Russian, meaning Venus-Halley).

The only four spacecraft ever sent to visit the outer planets, *Pioneers 10* and *11* and *Voyagers 1* and *2,* all used the gravity of Jupiter to accelerate and redirect the flyby probes. The most ambitious of these, *Voyager 2,* not only rode the Jupiter slingshot to Saturn, but also used Saturn to accelerate and redirect the probe to Uranus, which it reached in January 1986. It then used the gravity of Uranus to fling it on to the most distant planet from the Sun, Neptune. (Yes, Neptune: Pluto, near perihelion in its highly eccentric orbit, is currently vacationing inside Neptune's orbit, basking in the warming rays of the Sun. Not warm enough to melt nitrogen, of course, but warmer than usual!)

New applications of this technique may include the use of swingbys of the Moon to assist both outbound and inbound missions to asteroids and other planets, as was done in the low-budget mission of the American International Cometary Explorer (ICE) spacecraft in 1985.

AEROBRAKING

Another excellent example of how to carry out velocity changes without using propellants is the technique of using atmospheric drag to change a satellite's orbit. This technique, in its extreme form, uses the atmosphere to slow a payload down to subsonic speeds in preparation for landing on a planet. Atmospheric entry probes such as the Soviet Venera series, the Viking landers, the ill-fated *Mars 2, 6,* and *7* landers, and the Galileo Jupiter entry probe all use aerobraking to kill their large approach velocities.

In much milder form, the Pioneer Venus orbiter used atmospheric drag to make substantial changes in its orbit. Reducing the apoapsis altitude (the high point of its orbit about Venus), the eccentricity, and the orbital period is easy. But with aerodynamically shaped aerobrakes it is possible even to make significant changes in orbital inclination if the initial orbit is energetic enough. The ultimate in aerodynamical aerobrakes is the space plane, as exemplified by the Space Shuttle and the four model "Spaceski Shuttleski" craft that have been test-flown and recovered by the Soviet Union under cover of the Kosmos series.

Aerobraking is of great potential value for returning payloads to Low Earth Orbit. Consider a spacecraft returning with a cargo of minerals from an asteroid or the Moon, with the LEO Space Station as its destination. The spacecraft can be aimed just clear of the Earth's atmosphere, so that its point of closest approach to Earth is at the altitude of the Space Station. Because the returning spacecraft may begin with a significant velocity relative to Earth, and because it picks up enormous speed falling in the Earth's gravity well, it may be traveling 3 to 6 kilometers per second faster than the Space Station as it passes by. The conventional approach would be to fire a large rocket engine on the spacecraft to kill its excess speed, leaving it near the Space Station with about the same final speed. A tug from the Space Station could then pick it up. This approach would commonly require a mass of space-storable propellant that is 3 to 10 times the mass of the payload.

There is a much easier way to get rid of the excess speed. That is to equip the payload with a heat shield and to aim it to skim the top of the atmosphere instead of going directly for the altitude of the Space Station. The vehicle will make one pass through the upper atmosphere and then ride back out into space again on an orbit that is no longer hyperbolic relative to Earth: it has dissipated enough energy in its first encounter with the atmosphere to change the orbit to a highly eccentric ellipse. The amount of energy that needs to be dissipated is about the same as what the reusable Space Shuttle experiences during atmospheric entry. The spacecraft then takes a day (or a month, if desired) executing its eccentric orbit. It is completely cooled down to normal temperatures in a couple of hours. It coasts out to apogee, and then falls back for a second perigee passage in the upper atmosphere.

The second pass through the atmosphere (or the first for a vehicle returning from the Moon) can be used to put the vehicle in a nearly circular orbit with nearly the same period as that of the Space Station. A tiny engine burn at apogee will lift the perigee out of the atmosphere, and then it is a simple matter to rendezvous and dock with the Space Station.

This approach, which produces heating loads very similar to those encountered in routine spacecraft returns to Earth, is far less demanding than the Apollo or Zond return from the Moon. Apollo did not simply drop from about 11 to 8 kilometers per second in the atmosphere (from the eccentric lunar return orbit to circular orbit), but from 11 to 0. A heat shield capable of absorbing such punishment must be very heavy, roughly 25 percent of the total payload mass.

An even more interesting alternative would be to bring the return vehicle down from highly eccentric Earth orbit (HEEO) by a large number of very mild passes through the upper atmosphere, dissipating only a few percent of the excess energy on each pass. The heating would then be so mild that the return to LEO could be done with a very light heat shield, perhaps 1 or 2 percent of the payload weight. For return from an asteroid, a heavy heat shield could be used for the first, intense, perigee passage. The heat shield could then be detached and left parked in HEEO, there to await the next outbound expedition. Meanwhile, the payload, with a very light thermal shield, could slowly work its way down to LEO using a number of mild atmospheric passages. The next expedition would be spared the enormous cost of lifting the heavy main heat shield from LEO to escape velocity: it would pick up the heat shield on its way out, just before its escape from Earth.

TETHERS AND SKYHOOKS

There is an interesting low-tech way to move payloads around that can be practiced by a relatively massive satellite that is equipped with a long cable or tether (see fig. 5.18). A wide variety of neat tricks have already been discovered in which momentum is transferred back and forth between two tethered bodies in order to redirect one or the other. Vehicles traveling in the atmosphere at suborbital speeds may be "snatched" into space by a robust tether; small levels of artificial gravity may be maintained in-

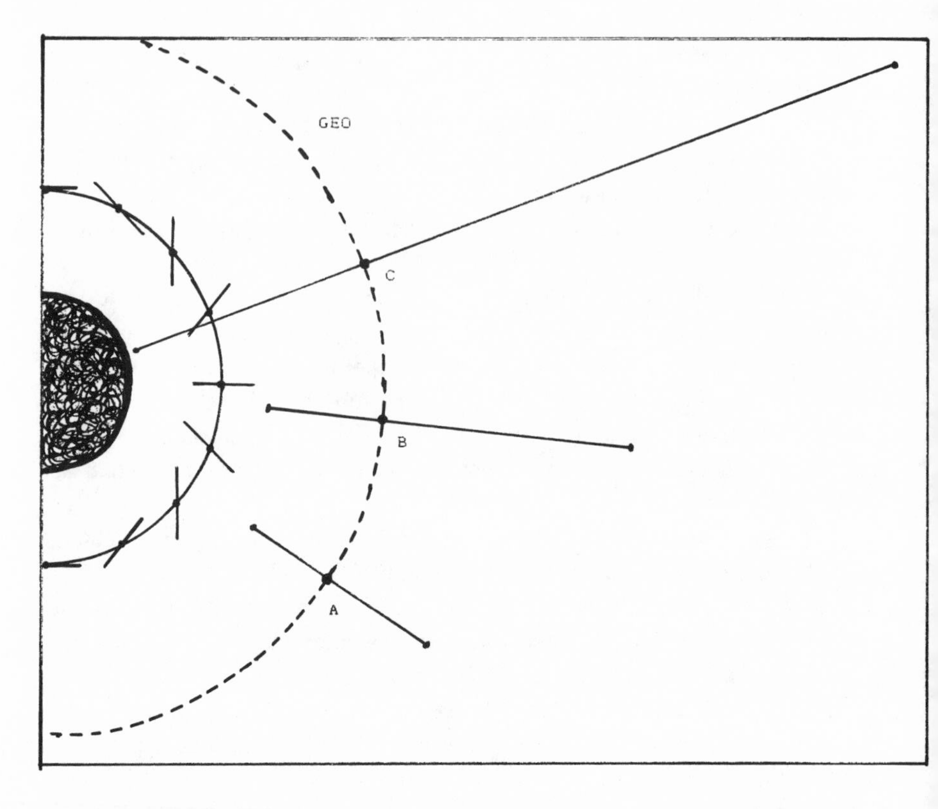

Figure 5.18
Tethers and Skyhooks
A satellite (A) may simultaneously reel out a line toward Earth and one away from Earth without changing the center of mass of the satellite or unbalancing the forces on it: the lower tether is held taut by the excess of Earth's gravitational force over centrifugal force, while the outer tether is held taut by its excess of centrifugal force over gravitational force. The tethers, if they are strong enough, may be paid out (B, C) until the lower end reaches the ground. If the satellite is in GEO, the lower end will come to rest at a spot on Earth's equator. Such a ground-to-space tether is called a skyhook. Tethers may also be made to rotate so as to provide the means to transfer satellites from one orbit to another (walking tethers), as in the example shown in a lower orbit.

definitely by connecting two satellites at different altitudes with a tether; high levels of artificial gravity can be achieved by spinning two tethered satellites at the ends of a tether like an Argentine bolo. Payloads may be lifted off the Moon or deposited on the Moon by "walking" tethers deployed from a Moon-orbiting satellite.

The ultimate use of tethers is for a satellite in geosynchronous orbit (forever orbiting precisely above a fixed point on the Equator) to lower one tether to the ground while simultaneously "extruding" one cable outward toward the Moon. Both cables would hang "downhill" from the geosynchronous satellite: the inner portion would feel an excess of gravity over centrifugal force, and hence be pulled taut by Earth's gravity. The outer portion would be traveling around the Earth at the speed of a geosynchronous satellite, and hence would be traveling faster than local circular orbital speed. Indeed, the tip speed of the tether may be well in excess of Earth's escape velocity! Thus it would experience an excess of centrifugal force over gravity, and also would pull upward on the satellite in GEO. The satellite would continue to orbit as before so long as these forces were kept in balance. This device, which permits one to simply "climb a beanstalk" from Earth's surface to escape velocity, is for obvious reasons called a "*skyhook*."

The first in-space use of tethers was in the Gemini program, when an Agena target vehicle was tethered to a Gemini capsule and the pair were spun at modest speeds. A joint Italian-American Space Shuttle mission is under development to use tethers up to 100 km in length to dangle an upper-atmosphere research payload into the atmosphere to explore the region too high for aerodynamic flight of winged vehicles, but too low for study by satellites. Genuine Earth skyhooks are, at best, very far in the future, but numerous exciting applications of shorter tethers will be well within reach in the near future. Sadly, materials strong enough to make Earth skyhooks are not yet available.

LAUNCH COSTS

One of the most striking illustrations of the gap between our expectations and our achievements in space has been the evolution of the cost of lifting payloads into orbit. In the 1960s it was common to project a steady, rapid decline of launching costs with the arrival of each new generation of launch vehicles. A typical illustration of these expectations was shown in Fig. 3.15.

The projected launch costs, in constant 1964 dollars, were to drop from $2000 per pound of payload for the Atlas booster used in the Mercury program to $1000 per pound for the Saturn 1 booster. The much larger Saturn 5, which was capable of orbiting five times as large a payload as the Saturn 1, was assigned a payload transportation cost of $250 per pound. This trend was to continue with the recoverable booster (now called the Space Shuttle), which was to deliver payload to LEO for $25 per pound. The shuttle was to be operational by 1977.

History has not been kind to these illusions. In figure 5.19 we show the actual costs of launching payloads by each of these systems. The trend is much simpler than Cole and Cox (or any of their contemporaries) anticipated. The real costs per pound of payload (again in constant 1964 dollars) were $1200 for the Saturn 1, $1400 for the Saturn 5, and $1400 for the Space Shuttle.

Perhaps we can squeeze a little understanding out of this story. Consider: the ultimate payload capacity of the Atlas was not realized by the *Mercury* 1-1/2 stage version, or even by the Atlas Able or Atlas Agena, but by the Atlas Centaur. The Atlas Centaur could lift about 5 tons at a cost of about $10 million (about $1000 per pound). The cost of each Saturn 1 launch was projected to be $40 million, each Saturn 5 launch was supposed to cost about $50 million, and each shuttle launch about $10 million. In other words, these wildly different vehicles with radically different sizes and hardware were all expected to cost roughly the same amount *per launch*. This is entirely unrealistic.

Second, we maintain that efficiency improvements are almost always the result of competition. The NASA booster development program always pitted each new booster against its predecessor. The only absolute cost criterion that had to be met was that each new booster should launch payload *no more expensively* than the last one. Accordingly, costs were free to rise unimpeded until they approached traditional operating costs. Then management got concerned. If there had been two competing development programs for a superbooster, one following a shuttle approach and the other directed toward some "big dumb booster" concept, much lower costs may have been achieved. Possibly, real launch costs would by now be down to $200 per pound. However, the naîve reaction of the critic of NASA is to describe this two-track competitive approach as "wasteful" because one of the competitors would never achieve operational status. Such people also must have difficulty understanding why consumers are so much better off in a

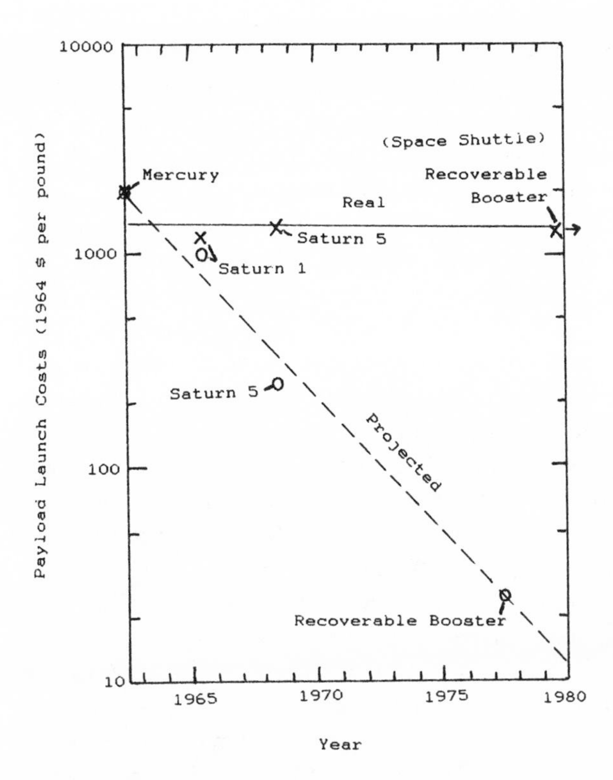

Figure 5.19
Actual Launch Costs
The circles show the projected decrease of launch costs expected in the early 1960s to result from the introduction of the Saturn boosters and the reusable Space Shuttle. The x marks show the actual costs realized by these systems. The projected launch cost of $25 per pound for the Space Shuttle compares very unfavorably with reality ($1400 per pound in 1964 dollars, or about $4000 per pound in 1985 dollars). No cost saving was realized by spending over $10 billion on the development of the Space Shuttle.

competitive capitalist economy where there is a "wasteful" diversity of choices.

If you want to fly a payload on the Shuttle, NASA doesn't charge you the real cost. You can rent a Shuttle flight for about $84 million, and fly up to 60,000 pounds into orbit. That comes to a launch charge of $1400 per pound. However, the actual cost of running the shuttle program is about

$250 million per flight. This means that NASA absorbs about two-thirds of the cost of each flight. This is regarded as a national service. In point of fact, it means that, if private industry can find a way to launch payloads for half the cost of the shuttle, they would have to charge $2000 per pound to break even. Because of the subsidy on the shuttle, no one would choose to fly the more efficient vehicle: NASA has in effect bribed users away by using tax dollars to subsidize the Shuttle and undercut private industry. Thus the subsidy has the effect of stifling competition *and hence preventing costs from coming down*.

Three separate attempts by American business to crash into the launch market have been stymied by the Shuttle subsidy. The Shuttle is now facing stiff competition from the European Space Agency's *Ariane 3* launcher, which is being similarly subsidized by ESA to assure launch charges lower than those charged by NASA. With the *Ariane 4* on the horizon, and with shuttle orbiters in very short supply since the loss of *Challenger,* capitalist economics may again assert itself in Washington. The plot is further thickened by the recent entry of the Soviet Union into the commercial launch business. At $25 million a shot, entrepreneurs can now buy a launch of the Soviet D1e heavy-lift launch vehicle, which can deliver 5 tons to GEO (or Venus, for that matter). The simpler D1, which is used for launches to LEO, would presumably cost less—dare we guess $20 million per shot? At 44,000 pounds per launch to LEO, that comes to about $450 per pound.

But that is not all: both Japan and the People's Republic of China are entering the market as well. Their prices will undercut both NASA and ESA. Asked how they could charge such low rates, a Chinese official responded with a single emphatic word: "Labor!" Chinese launches, following the Western example, will be insured by a new institution with the euphonious and paradoxical name of Chinese People's Insurance Company.

It is an amusing irony that the market forces of capitalist competition, suppressed by U.S. government subsidies, are now being unleashed by Soviet and Chinese competition! Don't we believe in our system? If so, let's let it work!

THE AEROSPACE PLANE

In full knowledge of the limitations of the Space Shuttle, and with the belief that launch costs could indeed be brought down sustantially, a number of studies have been undertaken to find the cheapest feasible means of lifting payloads into orbit. Present indications are that this vehicle may depart radically from existing rocket designs.

The general features of the emerging launch vehicle can be tentatively sketched out. It would consist of a large, vaguely Shuttle-like wheeled single-stage vehicle, somewhat like a cartoonist's rendering (or "artist's misconception") of a supersonic transport. The vehicle would use liquid hydrogen as its fuel, deriving almost all of its oxidizer from the atmosphere. It would take off from a long runway like an SST, accelerating with jet engines to high supersonic speeds.

At about Mach 3 the vehicle would switch over to a second set of engines that are optimized for operation at high altitude, and indeed can be ignited only at high supersonic speeds. These are called supersonic-combustion ramjets (scramjets). They would ingest huge amounts of hypersonic air, inject hydrogen, and accelerate to close to orbital speed, perhaps over Mach 20, at its optimum altitude. Then, when the vehicle is so high that its scramjet engines can no longer ingest enough oxygen to keep burning, it would pull up and coast to orbital altitudes, using rocket engines to give it the final kick to complete the transition to orbital flight.

After delivering its payload into low Earth orbit, it would reenter like the Space Shuttle and return to a runway landing at its base.

The vehicle could deliver Shuttle-like performance with a far smaller take off weight. The Shuttle carries all its own oxidizer, which, for hydrogen-oxygen engines, is the lion's share of the total weight: about 90 percent of the propellant mass used by H-O engines is oxygen. The aerospace plane, by comparison, would get almost all the oxygen it needs from the air. It would need to carry only enough liquid oxygen for the last rocket-propelled kick into orbit.

The very low density of liquid hydrogen presents some problems: even a 90 percent reduction in the amount of liquid oxygen still leaves a huge volume of tankage for the hydrogen. This may be somewhat alleviated by cooling the liquid hydrogen so that it solidifies partially to make a "hydrogen slush."

Many variations of this concept are under study at present. One would be an intercontinental Mach 15 civil transport plane, sometimes called the "Orient Express." Another would be a satellite launching vehicle, such as the British Aerospace HOTOL (Horizontal Take-Off and Landing) craft (fig. 5.20). The Strategic Defense Initiative is looking at an orbital version of the aerospace plane as a low-cost means of delivering heavy payloads to LEO.

Claims are circulating that the aerospace plane would permit the delivery of payloads to LEO at costs of 1 to 20 percent of present Space Shuttle costs. It is already clear that, even if real launch costs can be lowered only as far as $400 per pound (10% of present Shuttle costs), the

Figure 5.20
The British HOTOL Concept
The British conceptual design of a HOrizontal Take-Off and Landing (HOTOL) launch vehicle features a single-stage turbojet-scramjet-rocket combination fueled by liquid hydrogen. The vehicle takes off from a conventional runway, burns the hydrogen in its turbojet engine to accelerate to about Mach 5, and then uses its hydrogen-fueled supersonic-combustion ramjet (scramjet) engine to approach orbital altitudes and velocities. It uses a rocket engine fueled by hydrogen and oxygen for the final kick into orbit. After delivering its payload into orbit, HOTOL again fires its rocket engines to drop out of orbit. It reenters like the Space Shuttle, and lands on a runway. Given sufficient interest by the European Space Agency, HOTOL could be flying by the late 1990s.

program would be an enormous success. Equally as exciting (although somewhat less credible) are assertions that the first aerospace plane could be flying by 1995 if the program is initiated soon.

Consider the average tourist, weighing about 150 pounds. At launch costs of $400 per pound, a ticket to LEO would cost $60,000. At that level, many people would be willing and able to pay their own way. At $40 per pound, business would boom.

MASS PAYBACK

The large bulk of the mass actually placed in orbit is low-technology materials, mainly propellant. One crucial question is whether there might not be other sources of these materials, such as the Moon, the nearby asteroids, or the Martian moons Phobos and Deimos, that could provide these same materials in LEO at lower cost than lifting them from Earth.

Discussions of the relative costs of placing payload in LEO by Earth launch vs. return from asteroids or natural satellites often founder on a shoal of uncertainties about what launch services will be available in the future. It is an interesting fact of life in space that the cost of almost any near-term endeavor is dominated by the cost of launch from Earth. It is, of course, possible to make a useful estimate of what lunar or asteroidal material would sell for in LEO in 1995 if we assume that the shuttle will be back in service, with its present schedule of transportation costs, at that time. But what if some other booster, such as an unmanned launcher derived from the shuttle or an aerospace plane, is by then in use? Then *both* the cost of Earth and space materials will be reduced! Which, then, would be the better bargain?

There is one handy way of getting an estimate of relative costs that is not very sensitive to the actual launch costs. This is done by simply asking "how many pounds of material are made available in LEO *per pound of payload lifted off of Earth?*" This is called the mass payback ratio (MPBR).

Delivery of payload directly from Earth would then have an MPBR of 1.00. that's the number we need to beat. A straightforward Apollo-type scheme to fly a vehicle from LEO to the Moon with chemical propulsion, land the vehicle on the Moon, pick up cargo, and return the cargo to LEO would have a mass payback ratio near 0.1. I think we can agree that only a genuine nut would advocate doing this as a way to beat Earth launch costs.

The real killer is the fuel needed to take off from the Moon: if it all is carried up to the Moon from LEO, the amount of fuel needed to lift each ton of it (plus its tankage and insulation) from LEO to the Moon is about 5 tons.

If some precious material ("lunar diamonds") were found on the Moon, with a high enough Earthside market value to pay for the cost of its retrieval, then this might still be a conceivable scenario for economic reasons. No such commodity, except of course for limited masses of scientific samples, is known. What we must find is some *known or likely* lunar resource that *helps defray its own shipping cost or, even better, ships itself;* in other words, propellant. When we survey the resources of the Moon in chapters 6 and 7, the question of propellant availability will be foremost in our minds.

Return of material to LEO from a nearby asteroid is potentially much more interesting. The large expenditure of fuel needed to decelerate the vehicle and land it safely on the Moon without crashing, and the equally large amount needed to get off the Moon again (a delta V of about 3 kilometers per second for each) are not needed for landing on a small body that has almost no gravity.

The Moon's relatively large gravity has an interesting but unpleasant consequence: a high-thrust engine is required to land and take off. This rules out the use of engines with very low fuel consumptions and high efficiencies, which might otherwise promise to save much of the fuel weight through their higher performance. An ion engine, with a specific impulse of 4000 seconds and thrust-to-weight ratio of 0.0001, is capable of an acceleration of only 0.0001 gravities. It takes about 0.14 gravities just to hover near the lunar surface, and at least 0.3 gravities to land and take off safely. Thus these fuel-saving engines can be used only for the outbound leg from LEO to the vicinity of the Moon, accounting for only a third of the total velocity change needed for the round trip.

Consider the asteroid 1982DB. A near-Earth asteroid with both a low orbital eccentricity and inclination, it crosses Earth's orbit twice on each trip around the sun. Virtually the entire trip, from LEO to asteroid to LEO, can be conducted with low-thrust propulsion systems that are very sparing in their use of fuel. An asteroid mission with an ion engine could depart from LEO, spiral out to beyond the Moon's orbit, escape from Earth, fly to and rendezvous with the asteroid, land on it, load up with asteroidal material, take off, and return to Earth on an aerocapture trajectory to end up back in LEO. The total velocity change for the outbound trip to the asteroid would be about 4.5 kilometers per second (compared with 6 for the Moon), and the return to Earth would require a delta V of only 0.06 kilometers per second.

Imagine a return vehicle containing 1000 tons of ores sitting on the surface of the Moon. The mass of hydrogen and oxygen propellant required to lift it off the Moon and return it to LEO via aerocapture is about 1100 tons. The same 1100 ton mass of propellant on the asteroid 1982DB could return 55,000 tons of asteroidal resources to LEO even using low-performance space-storable propellants, or 80,000 tons using hydrogen and oxygen propellants.

Mission studies by Carolyn Meinel of Analytic Decisions Inc. and Kelly Parks of the Arizona Center for Space Resources show that it is easy to design missions to return asteroidal resources with a mass payback ratio greater than 10. If the resource returned to LEO is propellant (hydrogen and oxygen), then a small part of this propellant may be used for the next asteroidal mission, which removes the need to lift any more propellant from the surface of the Earth. This use of space-derived propellant for space missions to return more propellant is called *bootstrapping*. With bootstrapping, flights to the best asteroids could give MPBRs of about 100. Obviously, then, the availability of propellants on nearby bodies in space is of crucial importance.

The limit to the MPBR achievable with asteroidal resources would then be set by the need to lift the aerobrake heat shield into earth orbit: the large majority of the mass that must be carried to an asteroid by a mining vehicle is the return aerobrake. If some means could be found to make aerobrakes in space, MPBRs of at least 1000 could be achieved.

This means that the use of space resources can lower the cost of near-Earth space operations by very large factors, so large that the potential demand for facilities and materials in space may vastly exceed all current projections.

ECONOMIC PAYBACK

As the example of "lunar diamonds" above illustrates, mass payback and economic payback need not be the same. Economic payback can be better than mass payback if the returned commodity is "precious," which basically means that it is worth a lot more than propellant in LEO. That means that it must have a value of several thousand dollars per pound. Precious metals and gems would fall into this category. So would pharmaceuticals and *objets d'art*, such as paintings and integrated circuits (have you ever seen a photomicrograph of an integrated circuit? It *is* high art).

More likely, the materials returned to LEO will initially be low-tech items that happen to be needed in large quantities, such as propellants or metals. In that case, a mass payback ratio less than 1 would be intolerable. Because the mining and return operations have costs associated with them, MPBR's much greater than 1 will be required in order to break even economically. Dr. James R. Arnold, a respected expert on space resources and director of the California Space Institute in La Jolla, has said on numerous occasions that any scheme that returns ten pounds for one would be a sure economic winner.

Perhaps we could profitably state the crucial economic point: mission costs go up roughly *exponentially* with the required velocity change. Consider a mission that uses 1000 pounds of propellant to reach a speed of 10 kilometers per second. If we try to use the same type of propellant to achieve a ten-times larger velocity of 100 kilometers per second, the required mass of fuel increases not to 10,000 pounds, but to 2 million. This brutal fact of life of rocket propulsion forces us to search for the missions that require the least velocity change, and to fly only the very best of them. A corollary to this theorem is that, for a given velocity change, the option that has the least requirement for high-thrust (i.e., low-specific impulse) propulsion will use the least fuel, and therefore will be most attractive economically. The second corollary is that, for a given total velocity change, the missions that require the smallest proportion of the propulsion after loading the ore will be the most economical.

What are the easiest bodies to get to? Table 5.1 lists a number of presently known targets in descending order of attractiveness. It is instructive to see how the Moon fares in such a comparison.

The table also clearly shows why planets (especially Earth) are terrible sources of resources for operations in space. For us to try to carry out the exploration and exploitation of the Solar System by lifting everything from the ground would be like seventeenth-century Europeans trying to explore North America by forcing the explorers to take with them everything they would need until their return. Picture it: "Posted! No fishing, no hunting. Thou shalt not drink local water. Planting seeds strictly prohibited. No wood cutting or burning permitted." This policy, had it been followed, would have left the Western Hemisphere safe from European colonization forever.

Table 5.1
ΔV's and trip times between LEO and the surfaces of the Moon, selected asteroids, Mars, Phobos + Deimos*

	OUTBOUND		*INBOUND*	
Body	*ΔV, LEO→surface (m/sec)*	*time of flight (d)*	*ΔV, surface→LEO (m/sec)*	*time of flight (d)*
Asteroids:				
1982 DB	4450	210	60	480
1982 XB	5300	220	220	470
1982 HR	5300	180	260	320
1980 AA	5400	690	360	450
Anteros	5270	390	390	290
Phobos/Deimos	5600	270	1800	270
Moon	6000	3	3100	3
Mars	4800	270	5700	270

*All returns to LEO are via aerocapture. All arrivals in the Mars system are also via aerocapture in Mars' atmosphere.

The economic "bottom line" is still not known. Experts from the aerospace industry who have reviewed this issue believe that most of the cost is in the development of new technology at the beginning of the effort and direct launch costs. Since different space resource utilization schemes require dramatically different amounts of new technology, and since some use large numbers of men in remote locations while others do not, this issue must be worked out very carefully before useful cost estimates may be made. Any scheme that can minimize the amount of required new technology while satisfying the propulsion energy requirements would be especially desirable.

Nonetheless, it seems clearly feasible to reduce the costs of raw materials in near-Earth space by a factor of at least ten, and perhaps a thousand. It may mean that a cash investment that would buy a 100-ton space station in 1992 might buy a 100,000-ton space station a decade later. The impact of space resources on our future in space is not merely important; it is of paramount importance. Opening the vast resources of the Solar system to human use will make possible the industrialization and colonization of the Solar System.

NEEDS AND REQUIREMENTS IN LEO

It will be helpful to us as we survey the various potential targets of resource exploitation to have some idea of what materials will be in greatest demand for use in LEO and on Earth.

Two of these resources come to mind at once: water, which serves as an essential life-support fluid and is the source of hydrogen and oxygen for propellant use, and native metals, for use in space construction activities. NASA and other launch services carrying traffic to GEO will consume huge quantities of H-O propellant. If a shift to ion engines is made for the LEO-GEO transfer, then ion drives using oxygen as their reaction mass would be the logical choice.

Many manned activities in space, especially in high Earth orbit (HEO) and in interplanetary space, are very vulnerable to dangerous radiation from cosmic rays and especially from solar flares. Massive shielding (which may be asteroidal metals, lunar dirt, fuel, tanks of water—almost anything!) will be in great demand. The alternative is to accept severe radiation sickness as a common result of spaceflight. In the event of a major solar flare, radiation levels in and beyond GEO would be high enough to be fatal.

As our final example of potential demands for materials in LEO, we take one possible version of the Strategic Defense Initiative. In this version, no weapons are built: instead, the US and the USSR jointly embark on a program of armoring their essential communications, command, control, and intelligence (CCCI) satellites to make them virtually invulnerable to antisatellite measures. This would give both parties a guaranteed ability to detect launch activity worldwide. Thus a surprise first strike would no longer be possible. The nuclear arsenals of both nations (which are sized to permit a devastating response even *after* absorbing a surprise first strike that destroys 80 percent of the victim's strategic weapons) could immediately be reduced by a factor of about five. It is economically and technically impossible to launch the required million tons of armor and shielding from Earth, but it appears entirely feasible to bring it down from bodies in space.

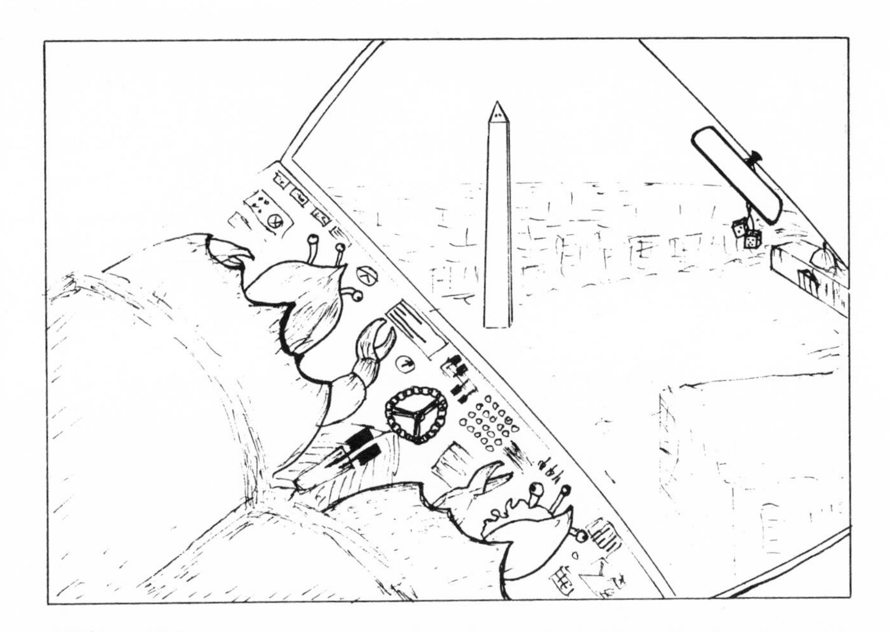

Figure 5.21
Advanced Propulsion Concepts
OK, time for another right-angle turn—you turn off the warp drive and I'll turn on the woof drive. 11, 10, 7, 6 . . .

6

LUNAR SCIENCE

What the Moon Has To Offer

The Moon is a mirror: everyone who looks at it sees the reflection of his own theories.—Harold C. Urey

The Moon, Earth's constant companion, is 3476 km in diameter, just a little over a quarter of Earth's size. It has a mass of only 1/81.3 that of Earth, and its density is markedly lower than those of the terrestrial planets, indicating a significant difference in composition. Indeed, the density of the moon is similar to that of Earth's rocky mantle. The dense metallic lunar core, if any, must be very small.

THE MOON BEFORE SPACECRAFT

The Moon orbits about Earth at a mean distance of 384,400 kilometers (238,300 miles), which is about 60 Earth radii out. The orbit is inclined about 5 degrees to the plane of Earth's orbit around the Sun. The axial tilt of Earth is large, however, close to 23 degrees. Because of the effects of the Sun's gravity on the Moon's orbit, the inclination of its orbit relative to Earth's equator can vary by $\pm$ 5 degrees, from about 18 to 28 degrees.

This seemingly obscure fact is of great importance to the people who launch spacecraft: the latitude of the Kennedy Space Center is 28.5 degrees, which permits NASA to launch spacecraft directly eastward, taking full advantage of the Earth's spin velocity, into orbits that lie almost exactly in the plane of the Moon's orbit. Soviet lunar and planetary probes, however, are launched from Tyuratam, which has a latitude of 45.6 degrees. The best that can be done from there is to launch into an orbit with an inclination of

45.6 degrees, which is inclined 19 degrees to the Moon's orbit even under the best of conditions. Soviet lunar flights must enter orbit, coast until they cross the plane of the Moon's orbit, and then execute a "dogleg" maneuver to change planes. This is very wasteful of fuel. In practice, however, the situation is much worse: any satellite vehicle launched from Tyuratam with an orbital inclination less than about 50 degrees would drop its upper stages onto Mongolia or, even worse, China. Thus orbital inclinations less than 50.5 degrees are never used for launches from Tyuratam (see fig. 6.1).

The orbit of the Moon is not perfectly circular, but has an eccentricity of about 0.055. This means that its actual distance can reach up to 5.5 percent above the average when the Moon is at apogee, and down to 5.5 percent below the average at perigee. Its distance from Earth thus varies from a minimum of about 363,300 km to a maximum of about 405,500 km (225,200–251,400 miles).

The rotation of the Moon is locked to Earth: the period of rotation on its axis is the same as the period of its orbital revolution about Earth. This means that the Moon always keeps the same side toward Earth. The rotation is at a perfectly constant rate, but because of the variation of its distance from Earth the Moon varies in its speed along its orbit. When it is near perigee [after falling some 42,200 kilometers (26,160 miles) closer to Earth and consequently speeding up] the Moon moves noticeably faster.

Because of the lack of perfect synchrony between the speeds of rotation and revolution, we sometimes see a little way around to the backside of the Moon, either the leading or trailing side. We actually get to see about 60 percent of the surface area of the Moon from Earth. The other 40 percent is "permanently" hidden from view of residents of Earth. This is sometimes naïvely called the "dark side" of the Moon, although it should be obvious that, at new Moon, when the Moon is between Earth and the Sun, the side toward us is dark and the side away from us is in full light. The proper name is "far side."

The Moon is roughly spherical in shape, but it is slightly elongated along the axis that points at Earth. The surface on its near side is marked with huge, dark, smooth-looking patches, many of which are nearly circular in shape. Because they have been known from antiquity, when no evidence of their true physical character was available, they long ago received the astonishingly inappropriate name of "seas" (*maria*, pronounced MAH-ree-ah in Latin; singular mare, pronounced MAH-ray). They are indeed low basins, but they are far drier than any terrestrial bone (see fig. 6.2).

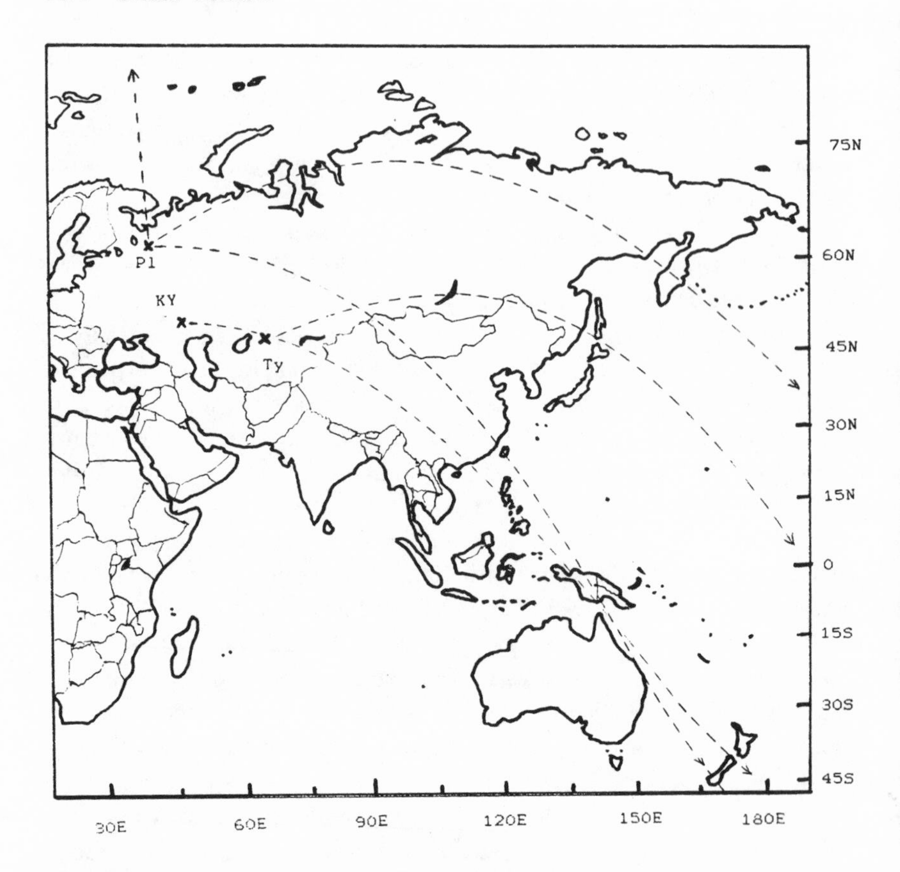

Figure 6.1
Launch Azimuths from the Main Soviet Space Centers
The Soviet satellite launch bases at Tyuratam (Ty), Kapustin Yar (KY), and Plesetsk (Pl) are marked. The closest of these to the Equator, Tyuratam, is used for all Soviet lunar and planetary missions. The launch direction used for these deep-space missions skims the northern boundary of China. The path is far from ideal for flights to the Moon and the planets, but is the best available to the Soviets from their far-northern location. The range of launch azimuths illustrated for the huge military facility of Plesetsk illustrates access to orbits with inclinations of 63, 73, and 98 degrees, the full range achieved from that base. The few satellites launched each year from KY are all placed in orbits inclined between 49 and 50.6 degrees.

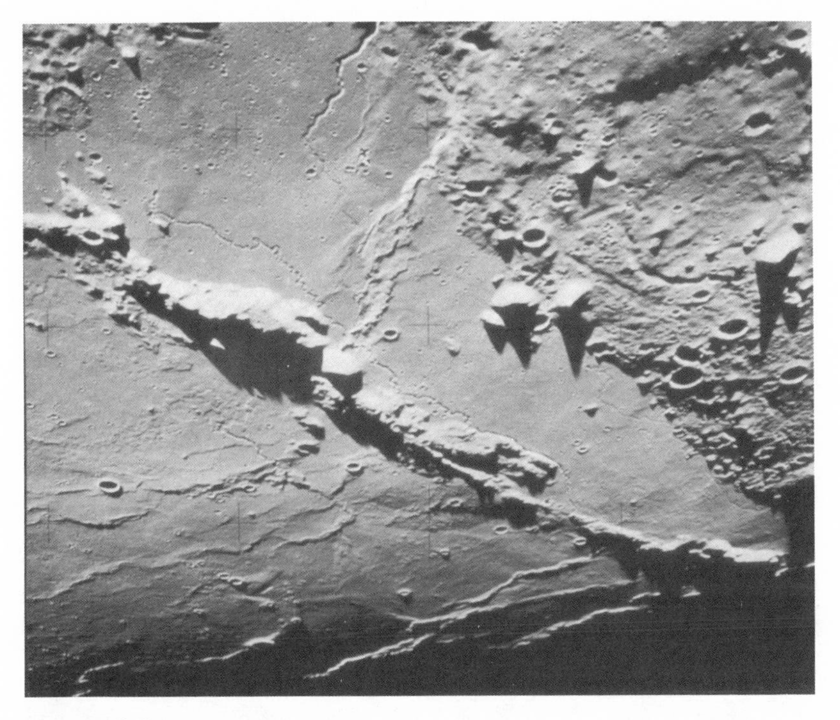

Figure 6.2
A Geological Sampler of the Moon
The relatively young mare floor at the bottom left is dominated by basalts similar to those in Earth's ocean basins. The wrinkle ridges are a relic of the lava-flow history of the basins. Note the small number of impact craters on the mare floor. The region in the upper right is older, with a higher crater density and a mantling of ejecta from distant impacts. The sharp, tall peaks (exaggerated here by the nearly horizontal lighting) are volcanic.

Most of the near side is covered with relatively bright, very heavily cratered "highlands," which evidently must be much older than the maria: the maria have not been around long enough to accumulate a dense coverage of large craters.

Pre-spacecraft studies of the Moon revealed very little of its chemical and physical nature. It was commonly believed that the lengthy history of shattering impacts would produce a deep layer of dust, dirt, and rubble (called "regolith" by planetary scientists). Indeed, one scientist, Thomas Gold of

Cornell University, believed that the mare basins were filled with a very deep layer of fluffy dust with so little strength that a spacecraft landing on it would instantly sink out of sight. Fortunately, Gold was wrong.

UNMANNED LUNAR PROBES

The birth of spacecraft study of the Moon can be traced to the *Luna 3* flight of 1959. *Luna 3* flew by the Moon in a highly eccentric Earth orbit and photographed the far side, the first time that it had been seen by any resident of Earth. The photographs were of very poor quality, badly overexposed, but they did show a few major features. Most prominent were the two large dark basins called the Sea of Moscow and the crater Tsiolkovsky. Most startling of the *Luna 3* findings was the extreme scarcity of large basalt-filled basins on the far side.

The Ranger program in the mid-1960s was devoted to crashing spacecraft with excellent camera systems into selected impact areas on the Moon. They struck the lunar surface at speeds in excess of 2 kilometers per second. The pictures obtained were transmitted back to Earth by what was for that time a very fast television link. The Ranger pictures clearly showed that the large craters seen from Earth were but the most prominent members of a distribution that extended on down at least to sizes of a few meters, and perhaps to much smaller sizes.

Both the Soviet Union and the United States had active programs for orbital photographic mapping of the Moon and for automatic surface landers. The first surviving lander out of six Soviet attempts, *Luna 9,* sent back photographic panoramas taken from a vantage point about one meter above the surface. Craters as small as one centimeter could be seen in the regolith. The Jodrell Bank radio telescope in England, having been given the transmission frequency of the Luna lander by the Soviets, tuned in on *Luna 9* and received some of the first pictures. Sir Bernard Lovell, the director of Jodrell Bank, did not know in advance how the data were encoded, but, upon seeing the data they received from *Luna 9,* he noticed that it looked like a wirephoto format. The data were fed into a standard wirephoto machine and out came a perfectly clear picture—which showed the surface of the Moon studded with rocks that were about seven times as high as they were wide. It looked like a lunar Stonehenge! The pictures were released by England, and the Soviet builders of the spacecraft were a bit piqued to have their data published first by others. They at least had the pleasure of pointing out that the Jodrell Bank images were vertically exaggerated by a factor of ten. I (JSL) was standing at the side of Dr. Harold C. Urey when he was handed the first *Luna 9* picture from Jodrell Bank, full of pillars, menhirs, stelae and telephone poles, by a San Diego television reporter. Urey's response was explosive: "Impossible! Somebody did something wrong!"

One other thing about Luna 9: despite its weight of over 100 kilograms, it definitely did not sink in the dust. So much for "Gold dust."

The American landers, starting with the first American landing attempt, *Surveyor 1,* returned tens of thousands of photographs with excellent resolution. Of even greater importance, however, were the first measurements of the composition of the lunar surface by *Surveyors 5, 6* and *7.*

A device called an alpha backscatter spectrometer, carrying a weakly radioactive source of alpha particles, measured the recoil of these particles from the surface. These particles bounced off the nuclei of atoms in the surface rocks, each particle recoiling with a speed that gave the weight of the atom it hit. The data could then be used to reconstruct how many atoms of each weight were present in the surface. From these atom abundances, excellent guesses could be made about what minerals were present and how abundant they were. From the calculated mineral abundances, the rock type could be determined.

The three Surveyors that carried the alpha backscatter instrument (numbers 5, 6 and 7) sampled widely separated areas of the surface and found significant differences in composition. *Surveyor 5,* in Mare Tranquilitatis, and *Surveyor 6* in the Sinus Medii, both found a basaltic composition. Basalt is the most common form of volcanic rock on Earth, dominating the ocean crust that covers three fourths of the Earth's surface. *Surveyor 7,* which landed in a highlands region near the large crater Tycho, found higher amounts of calcium, aluminum, and magnesium, and much less titanium and iron than in the mare sites.

In general, the analyses showed a surface that was clearly the product of at least one or two major episodes of melting and separation of materials according to their densities. The most dense materials, such as free metals and sulfides, settled to the center, while the least dense portion of the rocky material floated to the top to form a crust. This process is called *differentiation.* The mare basins were found to be holes in the low-density crust that had been flooded with kilometer-deep deposits of a relatively dense igneous rock, basalt. The basalts found in different basins were similar in many ways, but showed a significant range of compositions. One of the least expected findings was that some of the basalts were extraordinarily rich in titanium. The mineral ilmenite (the titanium-iron oxide $FeTiO_3$) is abundant in some mare basalts and in the accompanying mare soils, as analyzed by the Surveyor landers. Such high concentrations of titanium in a deposit on Earth would make it worthy of economic exploitation.

The next three Soviet lunar missions were the *Luna 10, 11* and *12* orbiters. *Luna 10* was heavily instrumented for geophysical and geochemical study of the Moon: magnetometers, radiation counters, gamma ray and X-ray spectrometers, meteoroid detectors, and an infrared radiometer. Very little was said about *Luna 11,* but it seems likely that it was quite similar in design to *Luna 10. Luna 12,* however, was of very different design, dedicated to photographic mapping of the lunar surface to help in the selection of landing sites for future missions. Very few pictures taken by *Luna 12* have been released, and the degree of success of the mission has been questioned.

A fringe benefit of Moon-orbiting spacecraft is that tracing of their orbital motions reveals much about the lunar gravity field. This is of great importance, because the gravity field near the lunar surface is very "lumpy" due to the dense basalt fill in the major mare basins. Since Soviet lunar landing craft are preprogrammed, with little ability to adapt to new information, it was necessary for the Soviets to have precise gravity-field data in order to compensate for the lumpiness of the gravity field in programming future landers. For reasons of safety, all early landing missions were targeted for smooth areas. But smooth areas tend to be in mare basins, which is precisely where the biggest irregularities in the gravity field are located.

The last Soviet lunar launch of 1966 was the *Luna 13* landing mission. Like all previous Soviet lunar craft, this lander was powered by chemical batteries. It differed from *Luna 9* by having two booms that extended out from the lander capsule to deploy instruments directly on the surface. One arm carried a small penetrator to measure directly the bearing strength of the surface, and the other carried a gamma-ray source that was used to determine how dense the surface was. A camera system similar to that on *Luna 9* was also carried. The net result of these experiments was to confirm the loose, granular nature of the surface, and to give assurance that the surface was strong enough to bear any landing craft with ease.

Also in 1966 and 1967, concurrent with the first five Surveyor flights, were the launchings of five Lunar Orbiter photographic mapping missions by NASA. Between them, these spacecraft mapped virtually all of the lunar surface, much of it with excellent resolution. The choice of landing sites for the Apollo program was made largely on the basis of the Lunar Orbiter maps.

By early 1968 the American Surveyor and Lunar Orbiter programs had been completed, and a 16-month gap in Soviet lunar activites had intervened after Luna 13. In April of 1968, *Luna 14* was placed in orbit about the Moon for the ostensible purpose of mapping the lunar gravity field. Again, very little information was released about this mission, and the treatment of it in the Soviet press was rather low-key. In retrospect, we may guess that it had more ambitious intentions, but we still do not know for certain whether it was in fact a sister craft of *Luna 12*. In any event, it was the last Soviet lunar mission launched by the A2e booster, and the last "second generation" Soviet Moon probe.

Soviet attention during this time interval was focused principally on the development of a manned lunar vehicle under the cover of the Kosmos and Zond programs (see chapter 3), which provided very little new insight into the Moon. A camera was flown aboard *Zond 5* to take photographs of Earth, but not of the Moon.

Two months later, *Zond 6* reflew the same basic mission but with a more complex and difficult return maneuver involving skipping off Earth's atmosphere over the North Pole on its way to a landing in the USSR. The photographic system flown earlier on *Zond 5* was used to examine the Moon during its close flyby, and some excellent film of the lunar surface was returned successfully to Earth. Soviet officials frankly and openly acknowledged for the first time that the Zond spacecraft was being prepared for manned lunar flyby missions.

The Zond series was interrupted at this time by the successful *Apollo 8* mission, in which three American astronauts not only flew by, but actually orbited, the Moon for ten revolutions. As recounted in chapter 3, the Soviet manned lunar flyby mission was then canceled. The remaining Zond spacecraft were expended during 1969 in unmemorable repeats of the earlier Zond missions, with negligible contribution to our understanding of the Moon.

Unmanned exploration of the Moon was abandoned by the United States with the advent of manned lunar flights. The Soviet program, having lost by a very thin margin the race to send men to the vicinity of the Moon, redirected its energies toward the use of advanced, heavy, unmanned craft, carried by the D1e heavy-lift booster. The lunar competition between the two nations switched to a race to recover the first lunar surface samples. The contestants were the manned American *Apollo 11* mission, scheduled for launch on July 16, 1969, and the unmanned *Luna 15* sample-return vehicle.

Luna 15 was launched just three days before the planned launch of *Apollo 11.* On July 17, now running only two days ahead of the Moon-bound *Apollo 11,* it entered orbit about the Moon. Then a most astonishing performance ensued. The craft remained in lunar orbit, twice changing its path about the Moon, until two days after the Apollo landing and the first Moon excursion by Neil Armstrong and Buzz Aldrin. *Luna 15* then fired its engines, dropped out of lunar orbit, and headed for the surface, where it ignominiously crashed amid a blizzard of attention and speculation.

It was not until September of 1970 that the first successful "third generation" Soviet probe was launched. *Luna 16,* the first admitted surface sample return attempt, was a stunning success. A sample of lunar soil weighing about 100 grams was returned to Earth in a small reentry capsule.

The 101-gram *Luna 16* surface sample was very similar in composition to the basalt-rich powdered regolith returned on the *Apollo 12* mission. The composition was dominated by silicates of iron, aluminum, calcium, and magnesium, and contained up to 4.9 percent titanium (in the larger rock chunks). Manganese and chromium were found to be present in small, uniform amounts, averaging about 0.2 and 0.3 percent respectively. Sodium and potassium, which are very common in the crust of Earth, were found only in minor amounts.

Luna 20 was the first attempt to return material from the lunar highlands. It returned somewhere between 50 and 100 grams of regolith with composition very different from the crushed basalt seen in the mare basins. The dominant rock type found was anorthosite, a rock dominated by the mineral anorthite, which is a calcium aluminosilicate. This rock type is typically light in color, in keeping with the observed contrast between the dark (basaltic) basins and the bright highlands. This also confirmed the compositional differences between mare basin fill and highlands that had originally been inferred from the Surveyor compositional measurements.

The target for the *Luna 15, 23* and *24* missions was the Mare Crisium basin, where one of the great "lumps" or mass concentrations in the Moon had been found from spacecraft tracking data. These mass concentrations (sometimes familiarly termed "mascons") were believed to represent thick lenses of dense basalt filling the major basins to great depth: one of the challenges of these missions was to drill deeply enough into the loose regolith to confirm the basaltic nature of the local "country rock." The damage suffered during landing by the drilling arm on *Luna 23* prevented it from taking samples, but *Luna 24* successfully drilled to a depth of about two meters, extracted the drill core, and returned it to Earth. As with *Luna 20,* the size of the returned sample was never made clear. A reasonable guess would be about 150 grams.

Because any sample of the regolith contains fragments of debris blasted out of a wide variety of nearby craters, it is possible for even a small sample to provide a wealth of information about chemical variations in the crust over a wide area. Further, small automated landers can be sent to landing sites that are too rough and hazardous to risk a manned landing. Finally, the cost of an unmanned sample return is only a tiny percentage of the cost of a manned landing. Thus the *Luna 16, 20,* and *24* samples were useful out of all proportion to their size and cost. Such samples are an excellent complement to detailed geological studies by manned missions.

The principal drawback to the Luna scheme for unmanned automatic sample return is that one can only select a region of interest, drop in the lander, and get a grab sample from wherever it happens to land. No discretion can be exercised about the choice of sample, and any material that happens to lie an inch away from the corer, even if it has an obviously very interesting appearance, would not be recognized and could not be collected. And of course, without mobility, there is no possibility of sampling nearby geological features. This scheme lacks selectivity, adaptability, and mobility.

The issue of mobility was very strikingly addressed by the *Luna 17* and *21* missions. These landers carried as their main payload small tanklike vehicles, named *Lunokhod 1* and *2* respectively (see fig. 6.3). The *Lunokhods* were designed to be driven slowly about on the lunar surface by direct radio control from Earth. Unfortunately, the total mass required to fly a Lunokhod and a sample return vehicle to the same landing site is far beyond the reach of the D1e "Proton" booster. Otherwise, it would have been very advantageous to use a roving vehicle, under radio control of geologists on Earth, to collect samples for return to Earth.

Lunokhod 1 was powered by a solar cell array, which lined the inside of a lid over the instrument compartment: if *Lunokhod* had been dependent on chemical storage batteries alone, it would have been unable to do any significant exploring. With the assistance of the solar cells, *Lunokhod* was

Figure 6.3
Lunokhod's Tracks in the Lunar Regolith
The lunar regolith is a poorly sorted loose, dry dirt with particle sizes ranging from ultrafine dust through sand, gravel, and rocks up to boulders the size of a large building. The lunar soil is pulverized by the violent impacts of comets and asteroids, which also leave their chemical signature in the regolith: about 1 percent of the typical sample of regolith is meteoritic debris, including both metal from the impact of asteroidal chondrites, irons, and stony irons and oxidized, relatively volatile-rich material from carbonaceous asteroids and comets. This picture was taken by one of the cameras on the Soviet *Lunokhod 1* roving vehicle.

kept in use for 11 months, greatly outliving any previous Soviet lunar spacecraft.

Because of the slow rotation of the Moon, each lunar day is two weeks long, as is each night. Lunokhod experienced a very wide temperature range between the hottest time of the lunar "day" (just after local noon) and the coldest time of night (just before local dawn). The thermal stress caused by these widely varying temperatures (from about −150 to +130°C—238–+266°F) is a serious hazard to delicate equipment, and of course ample opportunity exists for failures during the dark, cold night, when no solar power is available. A radioactive source was used to help heat the instrument compartment at night. Closing the lid at night assisted with thermal control of the instruments, and prolonged the lifetimes of both the internal instruments and the solar cells.

The 756-kilogram (1660 lb.) Lunokhod had four pairs of electrically driven wheels under a bathtub-like body that contained the instrumentation. It was navigated by means of a television camera, which returned its views to the "driver" on Earth. Because of the round-trip time of about 2.7 seconds for radio and television signals, it would be hazardous to run the lunar tank too fast: a picture showing a hazard ahead must be received, interpreted, and acted upon before Lunokhod has time to run into the obstacle. The chosen top speed was 0.1 kilometers per hour (.062 mph). The total distance covered over *Lunokhod's* career was about 10.5 kilometers (6.5 miles).

In addition to the equipment needed to rove about and return pictures of items of interest, Lunokhod also carried several scientific instruments. Among these were an X-ray spectrometer and a penetrometer, both advanced designs drawing upon *Luna 13* experience. A reflector target was supplied by French scientists. Laser pulses, transmitted to the Moon through terrestrial telescopes, were reflected back to Earth by the Target. Very precise measurements of the time taken for laser pulses to travel from Earth to the Moon and back were used to monitor the position and motions of the Lunokhod and of the Moon itself.

Lunokhod also carried instruments to measure radiation from cosmic rays and the Sun, and an X-ray telescope to map the distribution of X-ray sources in the sky.

Lunokhod 2, carried to the Moon by the *Luna 21* mission, weighed 840 kilograms (1850 lb.). Its instrumentation was similar to that on *Lunokhod 1,* except that a sky photometer, a magnetometer, and a third television camera were added. *Lunokhod 2,* which was now being driven by experienced teams of controllers, was permitted twice the top speed of *Lunokhod 1.* In its lifetime of just under five months it covered more than 36 kilometers (22 miles) and returned over 80,000 television pictures.

THE MOON AFTER APOLLO

The detailed history of the Apollo program has been the topic of several fine books, as well as a large part of chapter 3. The scientific results from Apollo, which are a proper part of this chapter, bulk large enough to fill about 100 volumes. There were six successful lunar landings in the Apollo program; 11, 12, and 14 through 17. Twelve American astronauts walked on the surface of the Moon. *Apollos 15, 16,* and *17* brought with them lunar roving vehicles that greatly extended the range of the astronauts on the lunar surface. *Apollo 12* landed near, visited, and brought back samples from the *Surveyor 3* lander. And *Apollo 13* survived a near-disastrous explosion that occurred when it was 330,000 kilometers (204,600 miles) from Earth. The crippled ship rounded the Moon and returned to Earth, with every operation improvised by the superbly competent Apollo team at Houston and at the Instrumentation Lab at MIT. Finally, after four days of heart-stopping suspense, it splashed down safely in the Pacific.

The scientific legacy of Apollo is hard to convey to anyone who was not aware of the paucity of our knowledge of the Moon in the early 1960s. The Moon changed in a few years from a remote, mysterious astronomical body to a familiar land about which vast amounts were known. Hundreds of millions of people watched the astronauts on the Moon via live color television. A world that had scarcely begun to accustom itself to communications satellites, that was used to receiving news from foreign lands via film reels flown across the oceans, saw the view of the Moon as the astronauts saw it. A billion souls watched with rapt attention the events on the Moon as they occurred 1.3 seconds earlier.

Only one thing has survived from our old view of the Moon: its origin is still a mystery.

Let us capsulize the Apollo discoveries that are of greatest significance for our task of identifying lunar resources. The vast body of detailed results in this and other areas cannot even be hinted at in so short a chapter as this.

First, the Moon is an ancient body that was once geologically active, but is now astonishingly inert. It was formed, along with the planets, meteorites, asteroids, comets, and the other planetary satellites about 4.5 to 4.6 billion years ago out of a vast swarm of small solid bodies called planetesimals. The formation of the planets, from a cloud of rocks and dust to full-sized moons and planets, occupied about 100 million years. The Moon may have been formed by the violent collision of another planet-sized body with Earth near the end of that era.

For about 200 million years the outer shell of the Moon was extensively melted, possibly even all at once (the "magma ocean" stage). As the supply of planetesimals was depleted by the formation of the planets, the original very intense bombardment decreased dramatically. A crust of anorthosite composition formed, probably with major variations in thickness, over the entire Moon. The molten rocks that gave rise to this crust were exposed directly to the hard vacuum of the lunar surface: all volatile elements, such as hydrogen, carbon, nitrogen, sulfur, chlorine, the rare gases, and even sodium and potassium, were lost into space, leaving a crust that is extremely deficient in volatiles compared with Earth's crust.

The decreasing bombardment flux continued after the formation of the crust, cratering it severely. As early as 4.1 billion years ago the lunar highlands looked very much as they do today. A few rare impacting bodies, presumably asteroids and comets, struck with so much energy that they punched huge craters completely through the anorthosite crust, exposing the dense, hot basaltic material on which the crust was floating. Relieved of the weight of the overlying crust, the basaltic magmas buckled upward, and filled the great basins with dark basaltic material, rich in iron and titanium. The era of basin formation ended about 3.9 billion years ago. The filling of the basins with basalt ceased to be an important process about 3.1 billion years ago, and basaltic volcanism finally ceased completely about 1 billion years ago.

The recent internal history of the Moon has been remarkably quiescent. Occasional large impacts superimpose fresh craters on the highlands (where they make little difference) and upon the mare basins. The mare surfaces were formed so much later than the highlands that they are generally very lightly cratered: these late impacts show up very clearly on them. The Moon today has virtually no seismic activity. Lunar earthquakes ("moonquakes") are for the most part caused by bodies striking the Moon, not by events within the Moon. For all practical purposes, the Moon is a dead planet.

It is of value to compare the evolution of the Moon and Earth. In both cases, extensive melting led to the upward migration of volatiles and of crustal rocks rich in aluminum, silicon, and alkali metals. In both cases, dense sulfides and free metals, notably iron and nickel, were efficiently separated out by sinking and collecting into a core (though the lunar core is very small). The resulting crustal rocks of both planets are therefore virtually devoid of free metals. Earth's crust retains huge amounts of volatiles:

since these molecules cannot readily escape from a body as massive as Earth, they remain today as the oceans and atmosphere. The Moon's volatiles, however, have been lost to space.

The structure of the outer portions of the Moon today, as we presently understand it, is basically composed of two layers. Anorthosite crust overlies a mantle rich in magnesium and iron silicates, out of which basaltic magmas were once "sweated" by heating and partial melting. Below a depth of about 300 kilometers (186 miles) the composition is not nearly so well known. The core, if it exists, is unlikely to make up more than 2 or 3 percent of the mass of the Moon.

Low-density rock types presumably derived rather directly from the original lunar crust include not only anorthosite, but also similar rock types called norite and troctolite. High-density rock types, basalts, show a significant range of composition from one location to another. An *Apollo 11* titanium-rich basalt sample, rock 10057, contained 11.44 percent by weight titanium dioxide and 39.8 percent silicon dioxide. An *Apollo 15* low-titanium basalt, rock 15076, by comparison contained only 2 percent titanium dioxide, but had over 48 percent silica.

Several kinds of chemical oddities on the Moon are of potential interest to us. One of these is a rare rock type called KREEP. The K is the chemical symbol for potassium, REE is short for rare earth elements, and P is the symbol for phosphorus. These rocks are rich in potassium and in phosphate minerals that contain very high concentrations of the radioactive elements uranium and thorium. KREEP has so far been found only as tiny rock chips in the regolith, ejected from distant cratering events. No KREEP bedrock was found by any lunar landing mission, either in the Surveyor, Luna, or Apollo series. One source region for KREEP has been located by measurements of the gamma ray emission of the lunar surface from orbiting Apollo spacecraft. This is a highland area between Mare Imbrium and Mare Serenitatis, too rough and mountainous to be a safe landing site for a manned mission.

The *Apollo 17* landing in the Taurus-Littrow valley discovered another very odd type of material, an orange soil composed largely of myriads of tiny orange glass beads and black particles of devitrified (recrystallized) glass (see fig. 6.4). A number of moderately volatile elements that are otherwise extremely rare on the Moon were found to be concentrated in these beads. These elements include sulfur, lead, zinc, and several other less common elements with similar volatility.

Figure 6.4
A Geologist Samples the Moon
Jack Schmitt, operating from the *Apollo 17* roving vehicle, packs up samples taken from rock outcroppings that protrude through the crushed regolith. The last man to set foot on the Moon, Schmitt was also the first geologist to fly in space. The soil in this area has an unusual orange tinge, suggestive of a novel chemistry and a unique history.

Among the similar materials that have been found on the Moon are green glass beads from the *Apollo 15* landing site at Hadley Rille. These beads appear to have formed in gas-driven fountaining of volcanic melts, out of a spray of droplets of liquid rock. The ultimate source of the volatiles that drove this fountaining may have been the deep interior of the Moon, or volatile-rich "foreign" material desposited in the regolith by impacts of carbonaceous asteroids or comets.

It has sometimes been suggested that there may be permanently shadowed craters in the polar regions of the Moon that have been cold enough to retain ice over the entire age of the Solar System. Water vapor released by ancient lunar volcanic activity, or by impacts of ice-rich comets or carbonaceous asteroids, would presumably have rapidly expanded to cover the entire surface of the Moon. Any water vapor reaching such an intensely cold "trap" would have condensed. At such low temperatures, ice has so low a vapor pressure that it could not evaporate fully even over billions of years. No Apollo or Luna sample-return missions have ever visited the regions where ice deposits might be found, and hence we cannot rule out this possibility with our available knowledge.

There are two problems with this view. First, local impact events would heat and splash such material widely, allowing the water to evaporate and be lost from the Moon. Second, major impact events can from time to time impart enough spin momentum to the Moon to break its stable lock on Earth. For a time, the Moon would then tumble, rotating about some wholly different pole, and the old polar regions would be exposed to sunlight. This would continue until the tidal forces of the Earth have had time to dissipate the Moon's spin energy and return it to a locked state, with one end of its long axis firmly aimed at Earth. Thus there has probably not been any location on the Moon that has been a "permanent cold trap," and storage of ice in crater bottoms near the present poles is very unlikely.

Nonetheless, we must be aware that there is no experimental evidence by which we can be certain that ice is absent. It is not difficult to design experiments, such as microwave sounders or gamma ray spectrometers, that can search for water ice down to depths of a meter or more in the polar regolith. A spacecraft in polar lunar orbit, carrying such instruments, could detect any massive ice deposit that may be present. This experiment has been technically feasible since the late 1960s, but has never been carried out. Despite the improbability that ice will be found, the consequences of finding it would be profound. In chapter 7 we shall discuss the great economic importance that would attach to any lunar ice deposit.

NITTY GRITTY DIRT

Virtually all of the surface of the Moon is covered by regolith, a mixture of rock dust, grains, glass beads, pebbles, and large rocks produced and mixed by impact events. The soil is of course dominated by the local rock types, but commonly contains a wide variety of "foreign" rock fragments hurled long distances from remote impact events. The regolith also contains a small trace of surviving, strongly heated—and even melted—debris from the impacting bodies themselves. Finally, the regolith is irradiated directly by the solar wind whenever the Sun is above the horizon, and small amounts of solar wind gases are implanted in the surfaces of regolith grains.

For these reasons, regolith composition varies from place to place on the Moon. The easy and safe landing places tend to be in the near-side mare basins, which are highly unrepresentative of the lunar surface, being dominated by basaltic debris instead of anorthosite.

Mare soils are dominated by four major minerals. These are pyroxene, plagioclase, olivine, and ilmenite. Pyroxenes are composed of three simple silicates of calcium, magnesium, and iron, with formulas $CaSiO_3$, $MgSiO_3$, and $FeSiO_3$ respectively. These three components join readily together in solid solutions, and lunar pyroxenes are combinations of all three of them. The dominant plagioclase mineral is a calcium aluminosilicate, $CaAl_2Si_2O_8$, which is our old friend anorthite: you will recall that anorthite is also by far the most important mineral in the highland anorthosite rocks. Olivine is composed of two iron and magnesium silicate components, Mg_2SiO_4 and Fe_2SiO_4, in solid solution. Ilmenite is iron titanate, $FeTiO_3$. These minerals are usually found in the regolith not as separate, distinct grains, but rather as little rock chips containing several minerals and as glassy "agglutinates" that have been welded together by severe transient heating during impacts.

All of these are stable minerals with large chemical binding energies. It is no simple matter to disassemble them into the elements of which they are made. For example, the prevalence of the aluminum mineral anorthite in the highlands does not by any means prove that aluminum metal would be easy to produce. An element, in order to be a useful resource, must not only be present (and useful!), but also readily extractable. This point will be forcibly brought home in chapter 7.

Surprisingly, the most abundant element in the Moon is oxygen. Oxide, silicate, titanate, and aluminate minerals all are very rich in oxygen, and mare soils, which are composed almost entirely of such minerals, contain about 40 percent by weight oxygen. The only other abundant elements are silicon (up to 20%), iron in the moderately oxidized ferrous state (Fe^{++}; about 12%), and magnesium and titanium (about 6% each). Titanium, however, is highly variable from place to place.

The volatile elements carbon, hydrogen, nitrogen, and sulfur are in general very rare in the Moon, especially in rocks. However, all four elements have been found in low concentrations in the regolith. The observed carbon abundances range from as low as 5 to as high as 280 parts per million (ppm; grams per metric ton). Typical concentrations are near 100 ppm, and most samples have between 50 and 150 ppm carbon.

Hydrogen, like carbon a gas implanted in lunar regolith by the solar wind, usually ranges between about 2 and 60 ppm in abundance. Typical concentrations are near 40 ppm.

Nitrogen abundances have been found to range from about 50 to 150 ppm, with 100 ppm most common, Sulfur, which is present in minor amount in lunar rocks as sulfide minerals, is not largely derived from the solar wind. Its concentration in the regolith ranges from about 400 to 1300 ppm.

All of these elements may be brought into the regolith in part by impacts of volatile-rich asteroids and comets. Such impacts are, however, very inefficient at depositing volatiles because of the great ease with which gases may escape from the Moon: the fiery explosive impact of a large body on the Moon effectively drives off almost all of the volatiles.

Edward Anders of the University of Chicago has measured the abundances of a large number of elements in the lunar regolith and compared them to the abundances found in "native" lunar rocks and in "foreign" meteorites. He has found that the lunar regolith contains about 1 to 2 percent by weight of asteroidal debris from impacting bodies. This meteorite-type source may contribute importantly to the sulfur, carbon, and nitrogen abundances in the regolith.

One interesting result of the impact of asteroids with the Moon is that asteroidal native metal particles may be placed in the regolith. The lunar rocks themselves have undergone melting and solidification in the Moon's gravity field, and any free metal that they may have once contained has sunk out of sight into the deep interior. Thus, if it were not for later asteroid impacts, the lunar surface would be devoid of free metal. The same is true, to a more limited extent, of sulfur in sulfides. Sulfides, however, do have a slight solubility in molten silicates, and hence are not completely absent in solid igneous rocks.

The bodies impacting the Moon, like meteorites, contain an average of about 5 percent free metal. Since the regolith is about 2 percent meteoritic material, this means that the concentration of metal grains in the regolith should average around 1000 ppm (one kilogram per metric ton; two pounds per ton). This is in fact a typical observed value. In some places, near the impact points of metal-rich projectiles, concentrations up to a few tenths of a percent are possible. A portion of this metal is native to the Moon.

EVOLUTION OF EARTH AND MOON

Earth, initially nearly homogeneous in composition, melted and differentiated into layers of different density and composition at a very early date. Melting was almost certainly triggered by the violent accretion of large bodies. Their impacts drove powerful shock waves through the growing Earth, depositing huge amounts of energy in the deep interior. The densest components of Earth's material, metallic iron-nickel alloy and iron sulfides, together formed a dense, fluid melt and sank to form the core. The large bulk of the rest of Earth's material after formation of the core was silicates of magnesium and iron, which form moderately dense minerals that are rather hard to melt. A few other elements, such as calcium, can to some limited degree substitute for the atoms in these "ferromagnesian" minerals, and enter into them in low concentrations.

Elements that cannot substitute for ferrous iron, magnesium, or silicon are numerous: they are the elements that either have the wrong electrical charge or size to fit neatly in their places. All elements that have ionic charges of +1, especially the abundant elements sodium and potassium, are incompatible with the structures of the ferromagnesian silicates and are excluded from them. These materials generally have a rather low melting point, and hence can exist as melts in contact with dense, solid magnesium and ferrous iron silicates.

The elements that prefer an ionic charge of over 2, such as aluminum, ferric iron (Fe^{+++}), titanium, and the rare earth elements (REE) likewise are not compatible with the structure of the ferromagnesian silicates and are excluded from them into the melt. The light elements in the melt, such as sodium, are vastly more abundant than the heavy rare earths: for this reason, the density of the melt is low, and it rises out of the zone where the ferromagnesian silicates accumulate. All volatiles, such as water, carbon dioxide, nitrogen, and the rare gases are excluded in the same way, and join with the rising melt.

The vast reservoir of iron and magnesium minerals that comes to dominate the mantle of Earth thus rejects a rising melt that contains all the ingredients of the future crust, oceans, and atmosphere. Earth assumes a layered structure, determined by density and chemical affinities, with a core, mantle, crust, oceans, and an outer tenuous envelope of atmospheric gases.

Much of the early history of the Moon is obscure: we do not know for certain whether it was captured from orbit about the Sun, whether it accreted in orbit about Earth, or whether it was once part of Earth. We have our own preferences, but we need not (and will not) trouble with them here. Nonetheless, it is certain that the Moon underwent melting and thorough density-dependent differentiation during its very early history.

Lunar differentiation did an excellent job of hiding free metals and sulfides in a core that is totally inaccessible to mining operations. Just as on Earth, the elements that readily form free metals and sulfides were efficiently extracted by the core-forming melt and permanently buried at great depths. These elements include not only the obvious major elements iron, nickel, and sulfur, but also a host of other minor and trace elements including cobalt, platinum, iridium, osmium, palladium, gold, silver, arsenic, gallium, antimony, bismuth, copper, selenium, germanium, and so on. Many of these elements, although present in significant amounts in meteorites and asteroids, were so badly depleted in the outer regions of Earth and the Moon by core formation that they are very valuable resources today. Almost any meteorite or asteroid contains concentrations of these elements that are similar to those in their best known terrestrial ores.

The problem of the supply of these elements on Earth, however, is far worse than even this sad tale would suggest: these elements are not only strongly fractionated into the core, but the small amount of them remaining in the crust is very inequitably distributed as well. South Africa has 96 percent of the world's platinum metal reserves. Chromium is also derived predominantly from South Africa, as are diamonds. Cobalt comes mostly from Zaïre. The industrial countries of the Northern Hemisphere are largely deficient in the strategic metals required by their industries.

Thus on Earth's surface, water is virtually free and platinum metals sell for about $20,000 per kilogram.

On the Moon, something similar happened—with one major difference. The free metals were indeed extracted from the outer regions of the Moon, and the volatiles did indeed rise toward the surface: but as volcanic activity spewed forth crustal magmas onto the hard vacuum and low gravity of the lunar surface, everything volatile escaped. Even rock-forming elements like sodium and potassium, which play an immensely important role in Earth's continents, were boiled away into space and largely lost from the Moon. The remaining lunar crust was the involatile residue from the mess of volatile and incompatible elements rejected by the lunar mantle. It ended up very rich in aluminum, calcium, and rare earth elements—and produced anorthosite, norite, troctolite, and KREEP rocks.

Thus, on the lunar surface, platinum metals might well sell for $20,000 per kilogram, for the same reasons as on Earth. But, on the Moon, water would also sell for that price.

The most attractive resources on the lunar surface are the tiny admixtures of free metals and volatiles supplied by asteroids and by the solar wind. The only native lunar ore of immediate interest is ilmenite, which is a carrier of extractable oxygen and iron. It is ironic to imagine going to the Moon to mine its 100 ppm trace of volatiles or the 1000 ppm trace of free metals when we consider that they are not even lunar materials. They are, in fact, asteroidal, cometary, and solar materials. Any random asteroid is likely to contain 10 or 20 percent free metals, and any volatile-rich asteroid may contain 10 to 20 percent easily extractable water.

Clearly the viability of the Moon as a source of resources must be studied with some care. This we shall do in chapter 7. Equally clearly, metals and volatiles wanted for near-Earth space development may be much more readily obtainable from asteroids, where they are very abundant, than from the Moon. Thus we must also examine the chemical and physical nature of the asteroids and assess their desirability as sources of these same materials. This survey will occupy chapters 8 and 9.

7

LUNAR RESOURCE EXPLOITATION

Many lunar materials have been proposed as raw materials for space manufacture. Only those that are abundant and extractable by relatively simple means may be feasible for use.—Larry A. Haskin

Much has been said and written about the uses of lunar resources. We shall make no attempt here to "use everything except the squeal." Instead, we shall concentrate on those resources that are easiest to extract, most useful, and most likely to be used before the year 2010.

KEEPING IT SIMPLE

The products most in demand, mainly propellants, free metals, and bulk shielding, direct our attention to a few specific lunar materials. The lunar resources that we shall consider for early use begin, modestly enough, with unprocessed regolith. Extraction of free metal grains and volatiles, and one-step processing of the regolith (such as melting it to make glass), also seem worthy of consideration. There is a strong incentive to find a source of rocket propellants on the Moon: we will need to look at processes for extracting oxygen from abundant crustal minerals. Finally, some more demanding or less familiar processes, such as distillation of rocks by solar furnaces or decomposition of molten rock by electric currents (electrolysis) may become competitive within the time interval that we consider.

It is, of course, a complete mystery to us what facilities will be available on the Moon in 2010. We do not know how many people, how many tons of equipment, or how many megawatts of electrical power the lunar factories will have at their disposal. If we simply were to assume that the answer to all three questions is "zero," then we could end this chapter here. Yet the ability to operate factories on the Moon is contingent upon the presence of a variety of goods and services. The cost of building and operating a manned lunar base, irrespective of its relationship to the use of lunar resources, is probably in the range of $10 billion to $100 billion. Such an investment almost certainly would not be made for the sole or even the principal purpose of mining the Moon: here we tiptoe about the impenetrable thicket of political decision-making, and leave prognostications to others.

Still, it is perfectly possible to ask what could be done with lunar resources if, *for any reason whatsoever,* there is a manned presence on the Moon. We shall therefore assume, for our present purposes, that some form of lunar base exists. It is not our purpose at this point to judge whether a lunar base is a good or bad idea, or whether it is likely or unlikely to be built. The question before us is whether, given such a foothold on the Moon, the addition of a limited mining and processing operation could confer some new abilities on human operations on the Moon and elsewhere in space.

Because of the moderately deep gravity well of the Moon, it is not a foregone conclusion that any lunar resource may be exported economically. Nonetheless, exportable resources, especially propellants, are potentially of enormous economic importance in the exploitation of space. We must search diligently for them. If we do find resources that might be profitably exported, then we still need to compare the relative merits of the Moon and other possible sources of the same resources, such as nearby asteroids.

But the basic rationale for use of lunar resources is even more direct: if any future manned activities are planned for the lunar surface, there is a very real possibility of using lunar materials in support of them. It seems to us that any long-term, expensive, complex schemes for eventual use of a wide range of lunar resources must begin with and evolve from a modest scheme to use materials readily available on the Moon, subjecting them to a bare minimum of highly automated processing, using the least practical mass of equipment and process energy, and yielding useful products of high mass and low sophistication. From such a stance, an evolutionary, stepwise growth to more complex processes and a wider range of products would be natural. To begin operations with a highly complex processing scheme could well fatally tax the ability of a lunar base to support these activities.

In all of this it is best to recall that the lunar environment lies well outside the range of normal industrial experience here on Earth. The differences in temperature, pressure, gravity, etc. may seem at first glance to be complicating and disturbing factors. However, as our familiarity with the lunar environment grows, we will find ways to take advantage of local conditions and develop literally un-Earthly processing schemes. What is simple and inexpensive to do on Earth is not necessarily so on the Moon—and vice versa!

DIRT, DIRT, GLORIOUS DIRT

The curtain rises on Act II, Scene I of the dramatic presentation, Life on the Moon. After decades of neglect, the exploration of the Moon has resumed. An expedition from Earth has just landed, carrying everything needed to construct and begin operation of a small lunar base.

The pressurized modules of the base are lowered to the lunar surface, and the use of lunar resources begins at once. A small electric bulldozer digs a trench two or three meters deep in the lunar regolith, and a crane carefully lowers the base modules into the trench, where they are then fastened together. The bulldozer then covers the entire structure, except a single airlock, with a thick layer of lunar dirt. The thick blanket of dirt serves as thermal insulation and as radiation shielding, protecting against cosmic rays and solar flares. Solar cell arrays or nuclear power generators are deployed on the surface and plugged into the base, and it is now open for business, just in time for the two-week lunar night.

The alternatives to this scenario are most unattractive. If we leave the base unshielded, the time that astronauts may spend there will be severely restricted by their radiation exposure. In fact, the chances of debilitating injury or death by solar flare radiation are quite large. If we try to bring the massive shielding all the way from Earth, the cost of carrying the shielding will be immense. A 20-ton base would need 500 to 100 tons of shielding, which, if shipped from Earth at a cost of $20,000 per pound, would cost $20 to $40 billion delivered to Luna City. The conclusion is very simple: a lunar base *must* use lunar material for shielding. It would be crazy not to.

The same safety considerations apply to the use of shielding at manned installations in Earth orbit. Solar flares and galactic cosmic rays present a serious hazard to anyone in space, and every manned station would at the least need a heavily shielded "storm shelter" in which the residents could ride out severe solar flare events. The shielding need not be any special material—all that is needed is mass. If the economics of lifting material off the Moon were found to be sufficiently attractive, the shielding needed by space stations may be more easily supplied from the Moon than from Earth. This would open up a vast new market for lunar materials, which could help fund the growth and diversification of a lunar base. But can these materials be exported economically from the Moon? We shall return to this issue later in this chapter.

We can see that there is a very strong incentive to begin the use of lunar regolith immediately upon the arrival of the first lunar base. Then, given the ability to move hundreds of tons of regolith, it is reasonable to search for other potential uses of unprocessed lunar dirt.

Unfortunately, finding such uses is not an easy task. The dirt is a loose powder, with particles ranging in size from very fine dust to boulders as big as an office building. Any significant handling of the dirt would involve screening out excessively large particles. If we wish to make use of the regolith as a structural material, some means must be found to cement or weld the powder and seal it against gas leakage through its pores.

Three techniques for direct use of the regolith come to mind. The first, and classical, method is melting the soil to make a magma, which is solidified in molds to make basalt glass bricks. These bricks will be almost perfectly opaque because of the iron and titanium in them, and will be very brittle unless cooled extremely slowly. What one does with basalt bricks is a neat question, one that we have been unable to answer. Direct casting of large structures of basalt glass is technically very demanding because of the problem of thermal stress during cooling: like large glass telescope mirrors, cooling rates much less than one degree per day may be required to prevent the structure from cracking during cooling. Also, once finished, the resulting glass house would be vulnerable to breakage. Making the glass thick enough to be nearly immune to breakage would make thermal stress cracking an insuperable problem. Brick construction requires water-tight sealants.

The second technique would be to make concrete out of the regolith. Unfortunately, bulk mare regolith contains far too little calcium to make a good cement. Possibly mineral separation schemes, or use of calcium-rich highland soil, could provide a cement with suitable chemistry. Then comes the biggest problem: where do we get the water to make concrete? Concretes generally require at least several percent water. Bringing large amounts of water from Earth is utterly out of the question. The best solution to this problem so far suggested by concrete advocates is to ship tanks of liquid hydrogen from Earth and combine it with lunar rocks to make the

Figure 7.1
The Homey Art of Microwave Processing
Microwave heating of lunar regolith can sinter (spotweld) the grains in it into a porous, strong material somewhat similar to Space Shuttle thermal protection tiles. The sintering device pictured here is a flight prototype now undergoing extensive testing in laboratories and kitchens on Earth.

necessary water. This would reduce the mass supplied from Earth to about 0.5% of the mass of the finished concrete.

The third technique for using regolith, microwave sintering, appears to be the most promising, but is the least understood. The device used is like a souped-up microwave oven, generating several kilowatts of microwave power compared with the 0.5 kilowatts used by the typical home oven unit (see fig. 7.1). Almost all of the grains in the regolith are electrical insulators. No electricity flows through their interiors when they are illuminated with microwaves. However, the surfaces of the grains are full of defects, holes, irregularities, and impurities. As the microwave radiation's electric field oscillates back and forth, electricity is carried through a very thin layer on the surfaces of the grains. At the points where these angular grains contact each other, electric current arcs from one grain to another and welds them together. The product is porous, light, and strong, bearing a striking similarity to the Space Shuttle's insulating tiles. Unfortunately, the porosity permits gases and liquids to flow readily through the sintered blocks. Some means would have to be found to seal them against leakage, or they could not be used for construction.

Note that all three of the schemes for direct use of processed regolith—melting, concrete-making, and microwave sintering—give products that seem useful only as construction materials. Note also that all three processes yield materials that have serious drawbacks when used for construction. Clearly much more work will need to be done to demonstrate the utility or cost-effectiveness of any of these schemes. For now, we may be justified in searching for other processes and other products to support construction on the Moon. For example, lunar metals may be far more tractable and economical as lunar construction materials than processed regolith of any sort.

A different class of products becomes possible if we are able to process the regolith in a simple way. One such technique would heat the regolith for the purpose of extracting its content of volatile elements. An electrical power supply would be used to warm the regolith up enough to release most of the solar wind gases and other volatiles in them. Since the concentrations of hydrogen and carbon average about 40 and 100 ppm respectively, vast masses of regolith must be heated to secure any significant mass of products. Consider a rocket that requires a fuel load of 50 tons of liquid methane and 200 tons of liquid oxygen to depart from the Moon. Releasing enough carbon and oxygen to make the fuel requires heating 350,000 tons of regolith to about 1000 degrees Celsius! This is a major logistic feat and requires a staggering amount of heat. It also leaves us with the problem of making the liquid oxygen.

As a crude measure of the energy cost of these schemes, let us calculate how much energy is needed to make each gram of useful product. The winner is microwave sintering: it uses only 10 to 20 calories per gram of product. The cost of heating regolith up to temperatures near the melting point is about 300 calories per gram, and the cost of melting the material to make basalt glass is about 1300 calories per gram. If the regolith is heated up to near melting (250–300 calories per gram), but the useful product is the 140 ppm of volatiles released, then the cost of the product is 1,700,000 calories per gram of volatiles! This scheme would make absolutely no sense unless the regolith material itself were melted and made into some useful product; that is, the products would be mainly glass, plus 140 ppm of released volatiles, for an overall cost of 1300 calories per gram. Thus the economic feasibility of extracting volatiles from the regolith probably hangs on the discovery of some desirable use for strongly heated bulk regolith: the solar wind gases may never be more than a useful byproduct of some other, as yet unknown, process. We need to search further for a practical source of volatiles on the Moon.

MAKING OXYGEN ON THE MOON

Although oxygen is the most abundant element on the lunar surface, most of it is tied up in compounds so stable and so strongly bonded that they are very difficult to decompose. Fortunately, one oxide in the lunar regolith is far less stable than the others, making it a very attractive target for extraction. That component is iron oxide, FeO. The highest concentrations of iron oxide are found in the mineral ilmenite, $FeTiO_3$, which is abundant in the regolith in most mare locations.

With the use of some simple technique for separating minerals, ilmenite may be isolated with reasonable degree of purity from the lunar regolith. Electrostatic separation is under active investigation for this task. The ilmenite may then be heated with hydrogen gas (brought up from Earth) at a temperature of about 1000°C (see fig. 7.2). The products are water vapor and an intimate, probably sintered, mixture of solid metallic iron with titanium dioxide (the mineral rutile). The water vapor is condensed and decomposed by an electric current. The products of electrolysis are oxygen, which is liquefied for use as a propellant, and hydrogen, which is recycled through the ilmenite reactor.

Some hydrogen will be lost in each cycle, but some implanted solar wind hydrogen will also be released from the heated ilmenite. With some luck, it may be possible to make up the hydrogen losses without having to ship more hydrogen up from Earth on a regular basis. The energy requirements are about 500 calories per gram of ilmenite-rich solids, or nearly 5000 calories per gram of oxygen. This is a fair return, since oxygen is very useful.

Another technique for extracting oxygen from the regolith is to melt a kettle full of regolith and stick two electrodes in it (see fig. 7.3). D. J. Lindstrom and Larry A. Haskin of Washington University in St. Louis have shown that an electric current will then liberate oxygen gas at the anode. The cathode becomes plated with an iron-rich metal alloy. Although it is not a useful material in its own right, this metal could then be purified into

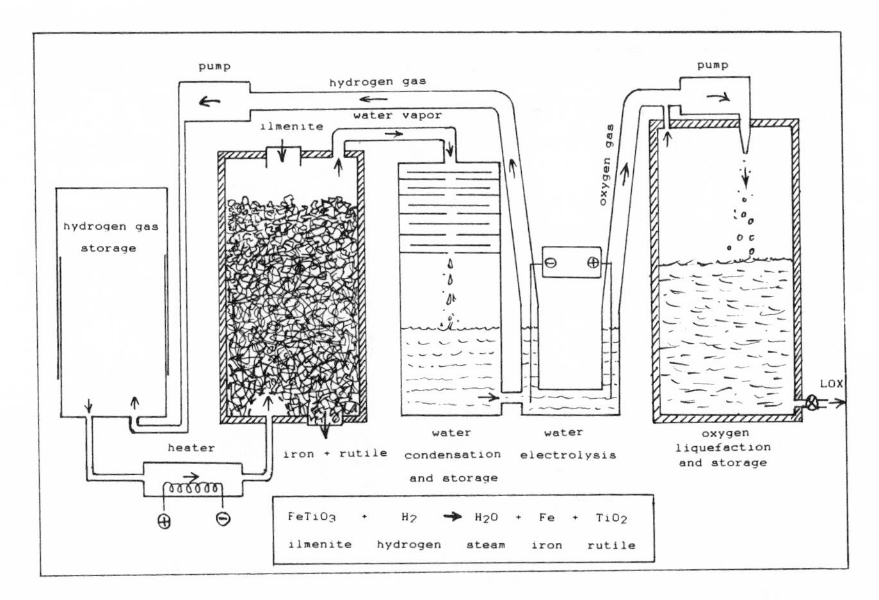

Figure 7.2
The Lunar Ilmenite Process for Oxygen
Hydrogen gas brought to the Moon from Earth is reacted with the lunar mineral ilmenite at a temperature of about 1000°C to make water vapor. The water vapor is condensed and electrolyzed into hydrogen and oxygen gases. The oxygen is liquefied and drawn off for use in rocket engines, and the hydrogen is recycled. The byproducts are metallic iron and rutile (titanium dioxide). Both heating the ilmenite charge and electrolyzing the water consume large amounts of energy. The heating could be done inexpensively with a solar furnace, but decomposing the water requires a large amount of electric energy from a solar cell array or a nuclear reactor.

its components (iron, chromium, magnanese, titanium, and even some silicon) if an appropriately simple and efficient scheme could be devised. This process uses about 1500 calories of thermal energy and a similar amount of electrical energy per gram of solids. For reasonable yields of oxygen (10% of the mass of the solids) the cost is about 30,000 calories per gram of oxygen.

It is too early to be certain what the chosen process will be, but the attractions of oxygen are so great, and there is so good a prospect for finding simple processes, that it would be rash to criticize the idea too strongly. At the

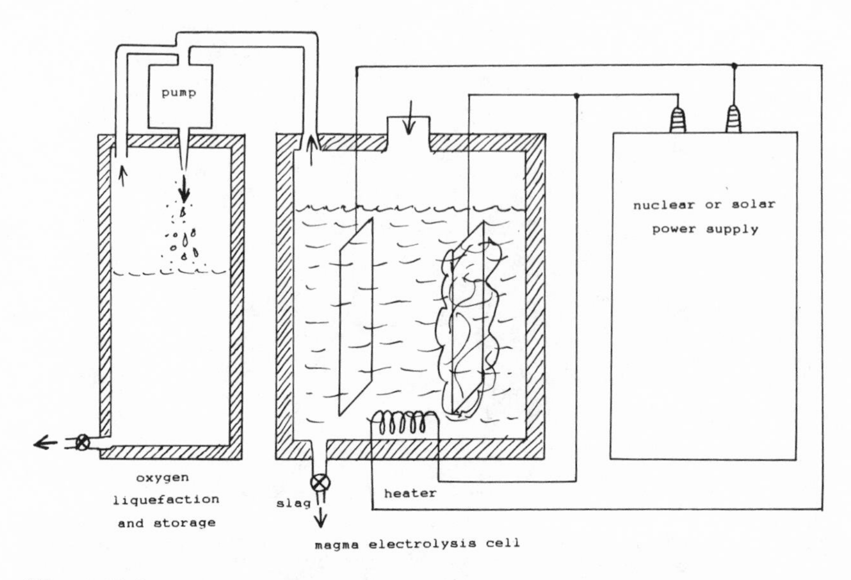

Figure 7.3
The Lunar Magma Electrolysis Process for Oxygen
Lunar material is strongly heated and completely melted in a furnace and is electrolyzed by means of a powerful electric current. Oxygen gas is released, and is liquefied for commercial use. A massive metallic electrode deposit composed of iron, nickel, chromium, manganese, silicon, and possibly other metals is produced. This metal deposit is not of direct commercial use, but provides a metallic concentrate of a number of elements of economic and industrial interest. The magma electrolysis scheme uses very large amounts of electric power.

moment, however, the ilmenite process looks stronger. It operates at lower temperatures, it does not require massive sources of electrical power, all the process energy could be provided by a solar furnace, it is far more energy-efficient than electrolysis, and molten silicates need not be handled. The melt-electrolysis technique, however, may possibly produce both a free metal product and a silicate residue product with useful properties. If future research shows that to be the case, the additional expense and complexity required may be justified.

Both the ilmenite and melt-electrolysis process produce free metals in a form that is not directly useful, but so highly concentrated that they are very attractive targets for extraction and purification. The very pure iron left by the extraction of ilmenite, like the complex iron-manganese-chromium-etc. alloy from melt electrolysis, needs only such a separation scheme to make it an extremely useful resource.

METALS FROM THE MOON

One very promising technique for extracting and purifying iron alloys is the gaseous carbonyl process, invented by Ludwig Mond a century ago. The epithet "Mond-Prozess" is exceptionally appropriate, since Mond is also the German word for Moon.

In Mond's scheme, metallic iron, nickel, and certain other metals can be reacted with carbon monoxide gas at 100 to 200°C and 10 to 100 atmospheres pressure (see fig. 7.4). These metals react spontaneously and exothermically with the gas to make gaseous compounds such as nickel tetracarbonyl and iron pentacarbonyl. These carbonyls are clear liquids at room temperature, with vapor pressures similar to that of water. They may readily be separated by distillation, or they may be selectively decomposed at 200 to 300 degrees Celsius at normal atmospheric pressure to produce extremely pure iron and nickel. Purities of 99.97% can be achieved in a single step.

The Mond process has been in use for a century on Earth, mainly by International Nickel. It has long been the source of most of the Free World's supply of nickel. The thermodynamics and kinetics of the process are very well understood.

An interesting fringe benefit of the proces is that nickel carbonyl may be used, in a process called chemical vapor deposition (CVD), to fill molds of very complex shape with stong, dense, crystalline nickel. Optical-quality hardware can be "grown" in molds in a few days out of a stream of nickel carbonyl vapor. Vaporform Products of New Kensington, Pennsylvania, and Formative Products Co. of Troy, Michigan, have pioneered this technique for making complex, high-quality parts in a single step: no labor-intensive and time-consuming machining, milling, drilling, or polishing is needed. Further, Susan D. Allen of the Center for Laser Studies at USC has shown that very thin, tough metal films of controllable and uniform thickness can be made by a related technique called laser chemical vapor deposition (LCVD).

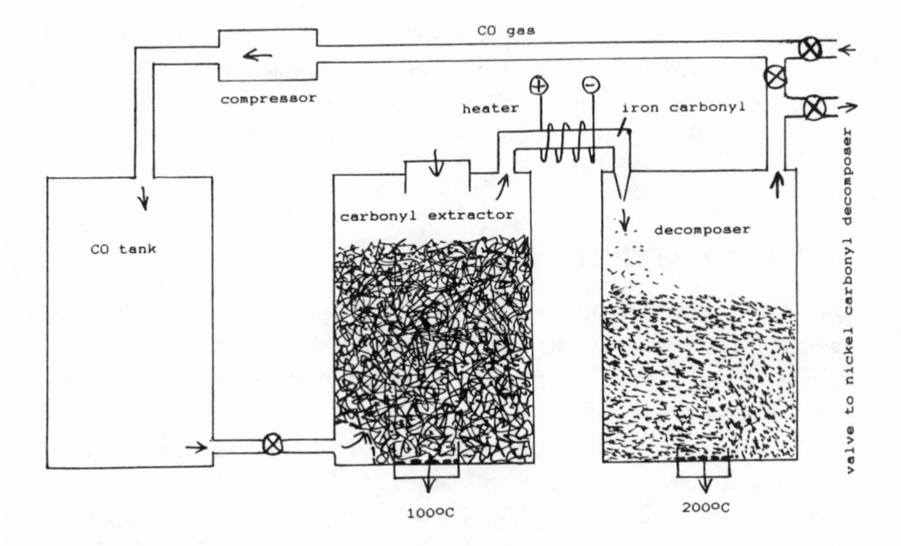

Figure 7.4
The Gaseous Carbonyl Process for Metals
In the gaseous carbonyl (Mond) process, native metal alloys (or magma electrolysis electrode metal deposits, or iron waste from the ilmenite process) are treated with warm carbon monoxide at a pressure of 1 to 10 atmospheres. Iron and nickel are readily volatilized as their gaseous carbonyls, and cobalt can be volatilized with somewhat higher pressures. The gas stream of metal carbonyls can be decomposed readily by gentle heating, precipitating iron and nickel (together or separately) and releasing carbon monoxide gas, which is recycled. All the modest process heat can be supplied by passive thermal control and small solar collectors. Virtually no electricity is required.

The energy required by the Mond process is small and easily supplied by solar heating. The peak temperatures needed are rarely higher than 120°C during volatilization of the metal, or 300°C during deposition of the metal. Since the main reactants are gases, cycling and reuse of heat used in the process is relatively easy. The energy required to initiate the formation of the gaseous carbonyls is essentially zero at normal daytime lunar surface temperatures. Thereafter, carbonyl formation is spontaneous.

The full strength of the Mond process is realized when it is used to treat native metals rather than ores or complex minerals. A simple magnetic rake would extract rather pure metallic iron from the lunar regolith, with probably less than 10% of contaminants such as agglutinates. This metal is a meteoritic alloy, similar in composition to iron meteorites: it is mostly iron, with a nickel content ranging between 7 and 60 percent and a cobalt content of 0.5 to about 4 percent. The total concentration of the platinum metals would be 30 to 250 ppm.

A second potential use for the Mond process would be to extract the iron from the sintered mess left over after the extraction of oxygen from ilmenite. This process would not only extract all the metallic iron in a single step, and provide extremely high-purity iron as a product, but would also leave as its sole byproduct a supply of very pure titanium dioxide, an excellent refractory and a promising starting point for the production of titanium.

High-purity iron is of considerable interest: it is very stong and highly resistant to corrosion even in the moist oxygen-rich atmosphere in a habitat. The Mond process is an excellent means of separating and handling this "poor man's steel."

Finally, the puzzle of what to do with the weird iron-rich cathode deposit from the electrolysis of molten silicates may be solved by application of the Mond process. Extraction of iron, manganese, chromium, nickel, and cobalt from the cathode alloy can be done directly by means of gaseous carbonyls. Depending on the composition of the regolith material used, the residual element after carbonyl extraction can be either silicon or titanium.

DOUBLE, DOUBLE, TOIL AND TROUBLE

Elbert A. King of the University of Houston has carried out several provacative experiments on the effects of heating meteorite samples up to very high temperatures in a solar furnace (see fig. 7.5). The meteorite samples can actually be boiled away, leaving a highly refractory residue very rich in calcium, aluminum, and titanium oxides. Controlled distillation of lunar materials, with slowly increasing temperatures, may permit the separation of quite a number of useful products.

To date, no such experiments have been conducted on lunar samples or lunar simulant mixtures. This would be a highly desirable project.

One of the greatest attractions of solar furnaces is the stark simplicity of the hardware needed to generate extremely high temperatures and heating rates. A reflector 30 meters square and weighing only a few kilograms could

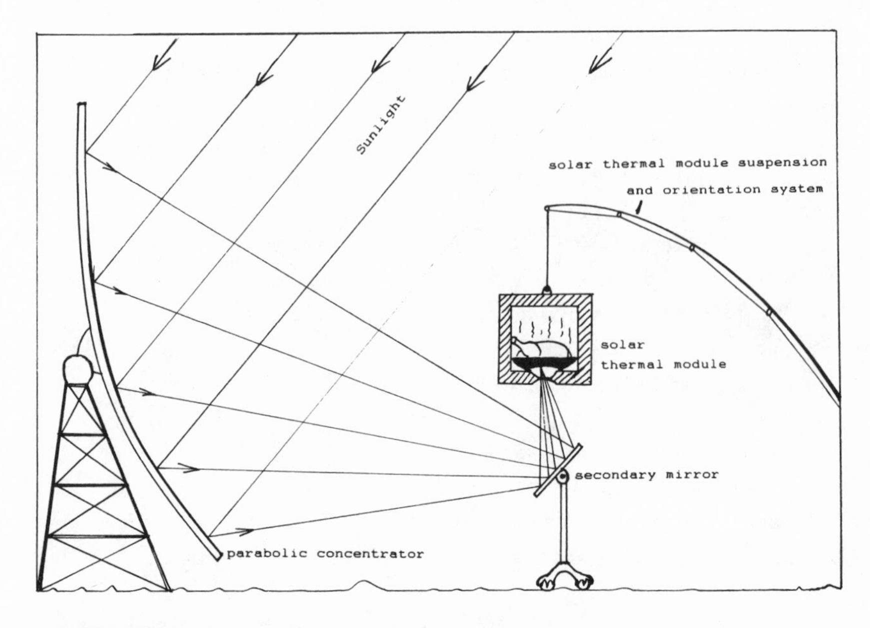

Figure 7.5
The R. Goldberg Memorial Solar Furnace
The efficiency and simplicity of the solar furnace strongly recommend it for use in space. Each square meter of solar collector on the Moon or in near-Earth space can provide over 1000 watts of thermal power. The Sun illuminates low-latitude sites on the Moon for two weeks at a time, and then sets for two weeks, stopping all non-nuclear power generation. Also, solar furnaces on the lunar surface must track the Sun as it moves across the sky. These are classic limitations inherent to planetary surfaces, but no such limitation occurs out in open space.

collect over a million watts of solar power. (Turning this energy into electricity can be done with an efficiency no better than 20 percent, so the simple reflector would have to be replaced by a solar cell array with five times its area. The equipment needed to regulate and condition this power is heavy and complex. It would seem wise to avoid using megawatts of electrical power if an alternative process using only thermal energy is available.) Tests of solar furnaces with a wide variety of meteoritic and lunar materials would be easy. Somewhat more demanding, but still clearly feasible within a few years, is the actual use of a small solar furnace in space, perhaps deployed from the Space Shuttle payload bay.

ECONOMICS OF EXPORT FROM THE MOON

The principal difficulty faced by schemes to export materials from the Moon is the 3.1 kilometers per second velocity change required to return to LEO. Using the highest performance practically attainable from chemical propellants (420 seconds specific impulse, from a hydrogen-oxygen engine), the mass of propellant needed to lift a 100-ton payload off the Moon is almost exactly 100 tons. Adding the tankage, insulation, and engine weight (about 12 tons), the total liftoff weight from the Moon would be 212 tons. For maximum performance, the hydrogen and oxygen should be in exactly the correct proportions for total combustion of both (one oxygen atom for every two hydrogen atoms; this comes to 88.9 tons of oxygen and 11.1 tons of hydrogen). However, the hydrogen must be brought from Earth, and is terribly expensive, perhaps $20,000 per kilogram. The oxygen may be made locally: if this is done, it will be because doing so is cheaper than bringing it from Earth. It would then make sense to burn a fuel mixture that has been beefed up with a surplus of lunar oxygen and a slight deficit of terrestrial hydrogen. The efficiency of the engine would be somewhat decreased, but the cost would be less.

The main factor in the cost of exporting materials *from* the Moon may well turn out to be the cost of carrying liquid hydrogen from Earth *to* the Moon!

To carry 11.1 tons of liquid hydrogen, plus its tank and insulation, from LEO to the Moon requires the presence of a 12-ton payload unit in LEO with an attached propulsion stage (more likely two stages, a translunar injection stage and a lunar landing stage). The total velocity change is about 6 kilometers per second, and the total mass of the unit departing from LEO is 53.8 tons. The direct transportation cost of lifting this assembly into orbit is $500 million. It would require two shuttle flights, each lifting 27 tons. This is about two tons more per flight than the shuttle has ever lifted, but within its projected payload capacity.

If we utterly ignore the cost of developing and deploying all the new space and lunar facilities needed for this scheme, and count only the transportation costs, the cost of the 100 tons of lunar product delivered back to LEO is then $500 million. This is $5 million per ton, or $2500 per pound, which is less than the $4000 per pound cost of launching the same mass from Earth. This means that, if development and operational costs can be contained, and if the transfer vehicles used to shuttle payloads from LEO to the Moon's orbit, to the lunar surface, and back to LEO are sufficiently reliable and reusable, it may be economically feasible for a lunar source of liquid oxygen to be delivered at competitive prices to LEO.

Clearly, however, no huge profits are to be made if the program must amortize $100 billion for the initial cost of the facilities. The profit margin cannot possibly surpass $1500 per pound, and is probably far less than $1000 per pound (and may be negative!). At $500 per pound, amortizing $100 billion of investment would require delivering 200 million pounds (100,000 tons) of lunar oxygen to LEO. If the interest burden on the initial investment is added in, the break-even point would not be reached until at least 250,000 tons of lunar oxygen had been delivered. Up to that point, prices for lunar oxygen in LEO would be competitive with Earth-launch costs only if they were subsidized.

What would happen if a new Heavy Lift Launch Vehicle were to arrive on the scene? If lift costs to LEO could be cut to $1000 per pound, wouldn't this instantly doom lunar oxygen at $2500 per pound? Not at all, because the main cost of returning lunar oxygen is the cost of lifting the necessary equipment and hydrogen from Earth. Consider orbiting the 53.8 tons in the previous example, but using the HLLV instead of the Space Shuttle. The costs to LEO would then be $107.6 million. The 100 tons of lunar oxygen returned to LEO would then have a transportation cost of $1.08 million per ton, or $538 per pound. If amortized startup costs and other operating expenses could be kept less than $400 per pound, the lunar oxygen could compete successfully in the market at LEO.

The secret of the commercial viability of lunar resources in near-Earth space is the containment of costs. If the transportation system has to carry the overhead of men and life-support equipment on every flight, this added expense alone would doom the project. If the full cost of a lunar base and the requisite mining and processing equipment had to be borne by the sale of liquid oxygen, that also might prove a killing handicap. But, to the best of our present ability to prognosticate, there seems to be no clear proof that the scheme is uneconomical. Certainly enough is known to justify the most careful technical and economic assessment of its chances.

Nonetheless, to predicate our future plans for the Moon on the assumption that lunar oxygen will pay for a lunar base is truly foolish. We must assure that our best-designed schemes for extraction of lunar oxygen are tested on a small scale, preferably on the Moon and with real lunar materials and operating conditions, at the earliest possible date. This will enable us to make informed decisions based on real operational experience.

If the scheme is found to be economical based on small-scale trials, then it would be stupid not to use it on larger scale. In fact, even if it can only match the costs of lifting oxygen to LEO from Earth, it would be worth doing for the boost it would give to our ability to run complex operations in space. It would confer on mankind new abilities to carry out industrial operations in space, and would do so for no additional cost beyond what we would have spent without the space operations.

Still, the deep gravitational well of the Moon is a serious problem. Every time any vehicle lands on the Moon or takes off from it, a substantial fuel penalty must be paid. It would be better if we could find a source of extractable oxygen that was energetically nearer to LEO than the lunar surface is. And it would be *enormously* better if we could find a source of hydrogen anywhere in nearby space, on the Moon or elsewhere!

LAYING THE GROUNDWORK

We have much unfinished business on the Moon. The complete cessation of the exploration of the Moon after the end of the Apollo program left us with no source of new lunar data. Within a short time, in fact, we had discarded all of the hardware and capabilities that we had paid for so dearly during the Apollo era.

Our knowledge of the composition of the lunar surface is excellent for tiny areas surrounding the points from which samples have been returned to Earth. Earth-based astronomers have studied the visible and infrared reflectivity of the entire near side, and have produced maps of the distribution of a few major minerals in the regolith. A limited area near the equator has been compositionally mapped from lunar orbit by gamma ray spectrometers. These spectrometers have very poor resolution: they measure the average composition of every point on the lunar surface that can be seen from the spacecraft at a given time, an area that is well over 100 kilometers in diameter. Further, their coverage of the surface is limited to low latitudes by the low orbital inclination of the Apollo orbits. It is from these orbital measurements that the only known concentrations of KREEP were found.

There is a great need for a complete compositional map of the lunar surface. This has been an achievable goal since the early 1970s. Unfortunately, NASA requests for a lunar polar-orbiting mapper have been greeted by the OMB with nearly complete incomprehension. Their usual response has been, "What do you want to do that for? You've been to the Moon already!"

The very idea—that tens of millions of dollars spent on chemical mapping by a small unmanned spacecraft could tell us a vast amount that the $24 billion of Apollo didn't—is beyond their belief. They have failed to recognize the true nature of Apollo: it was a political gesture, undertaken for reasons of national prestige. The science that it spawned was not the reason for the program, but a low-cost byproduct of the public-relations effort. Of the twelve Americans who walked on the Moon, only one was a trained geologist. He was Jack Schmitt, the last man to set foot on the Moon, the scientific afterthought personified.

An unmanned lunar polar-orbiting satellite designed to produce maps of the chemical and physical properties of the surface is an absolute minimum requirement for any future program, whether that program is oriented toward basic science or resource use.

However, remote sensing is not the whole story. Many elements and minerals of potential economic importance are not detectable from orbit by any known technique. We need to know the composition of the surface in detail not attainable from orbit. This requires the ability to return samples from the lunar surface to Earth. The ability to return small samples from a wide variety of interesting (and possibly hazardous) sites is of great value in assessing resources. Further, we need to be able to place roving vehicles on the lunar surface to traverse hundreds of kilometers, visit a wide range of sites with different chemical and physical properties, and pick up small samples on command. The rover must carry a sufficient variety of instruments so that the choice of samples can be made intelligently.

The separate problems of operation of automated sample return vehicles and remote-controlled rovers were addressed and solved by the Soviet Union in the early 1970s. However, there is a vast advantage to combining the abilities of these two types of vehicle. Operating such a combination would also be enormously enhanced by using orbital chemical and physical mapping data to plan the sampling program and the routes followed by the rovers.

Finally, we must begin to experiment with the processing of lunar materials. Much progress can be made in our laboratories on Earth. We can practice our techniques on artificial lunar regolith and rocks to work most of the problems out of the most promising processes. We would then be ready to try doing the same thing on the Moon on a small scale.

It is still much too early to decide how much processing could be done by a fully automated system. At some level of complexity, the presence and occasional active intervention of men would be required for efficient operation. But once the expense of a manned operation has been incurred, there is a serious question what the optimum use of the base personnel would be. Any base that keeps its human staff busy only a small part of the time is too small for efficient use of its most expensive asset, the crew.

It should be borne in mind that manned operations on the Moon are expensive. Even a minimal lunar camp, manned by two or three astronauts from time to time, would cost about $10 billion up to the end of its first decade of use. A large, permanently manned base with a staff of several dozen would have a runout cost close to $100 billion.

We believe that it may be possible to develop a strong rationale for a minimal lunar base, but we cannot conceive of the circumstances under which an early commitment to a $100 billion lunar base program could be made. If an early lunar camp succeeds in developing any product for export, then expansion of the base may well make economic sense. And if there are any lunar products that seem simple enough and useful enough to merit export, lunar oxygen, and bulk, unprocessed regolith seem the most promising. But no export from the Moon makes sense without lunar production of liquid oxygen.

8

THE EMERGENCE OF NEAR-EARTH ASTEROIDS

Because of (their) infrequent repeat performances, it would be unlikely that we would be aware of the existence of small Earth-orbit-group planetoids, and we should not be surprised if some were discovered in a determined search. Of all the possible groups, these would be the easiest to visit and to capture.—Dandridge M. Cole and Donald W. Cox, 1964

EARTH'S CATASTROPHIC PAST

One of the most profound effects of our exploration of the Solar System has been the dawning awareness that Earth really *is* a planet, subject to the same laws and processes as the Moon or Venus. Earth-based telescopic studies of the Moon showed it to be covered with craters ranging from hundreds of kilometers in diameter down to a few kilometers. When the early Ranger probes returned pictures of the surface of the Moon they revealed craters ranging from kilometers in diameter down to meters. When *Luna 9* and *Surveyor 1* photographed the view from the lunar surface, they found craters ranging down to millimeters in size. Particles of lunar soil returned by the Apollo and Luna landers displayed craters measured in micrometers. At the same time, the Lunar Orbiter mapping missions clearly showed previously unknown giant impact features.

Mariner 10, which flew by Mercury three times, found immense impact basins and a heavily cratered, depressingly lunar landscape on Mercury (see fig. 8.1). *Mariner 9* skimmed the surfaces of the moons of Mars and found them pocked with craters almost as big as the moons themselves (see fig.

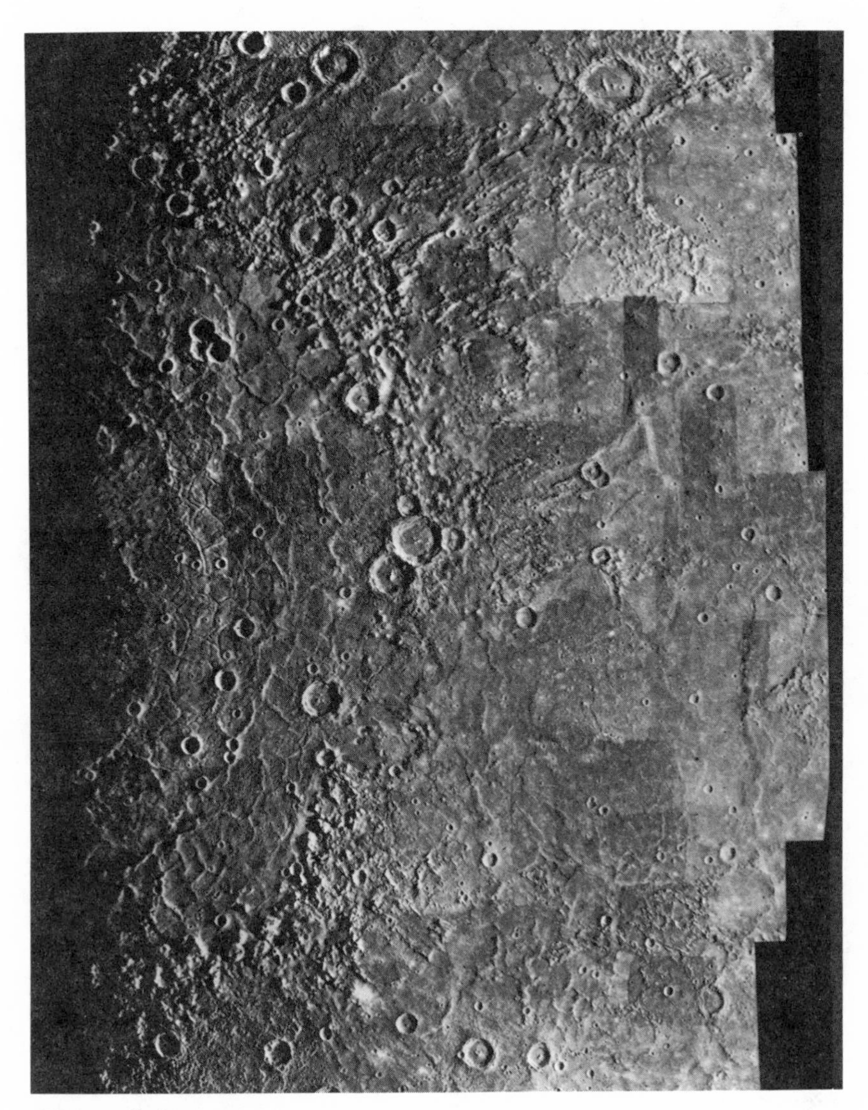

Figure 8.1
Craters on Mercury

This *Mariner 10* photographic mosaic of Mercury shows the edge of the gigantic Caloris Basin impact complex. The entire surface of Mercury is, like the Moon, heavily cratered by impacting comets and asteroids. Measuring the abundances of craters of each size on several planets, combined with absolute age determinations for mare basin and highland rocks on the Moon, permits an excellent evaluation of the impact frequency of these bodies and a very good estimate of the population of asteroidal and cometary bodies in earth-crossing orbits. Compare this photograph with the Lunar Orbiter view of the Mare Orientale basin on the Moon (Fig. 3.1).

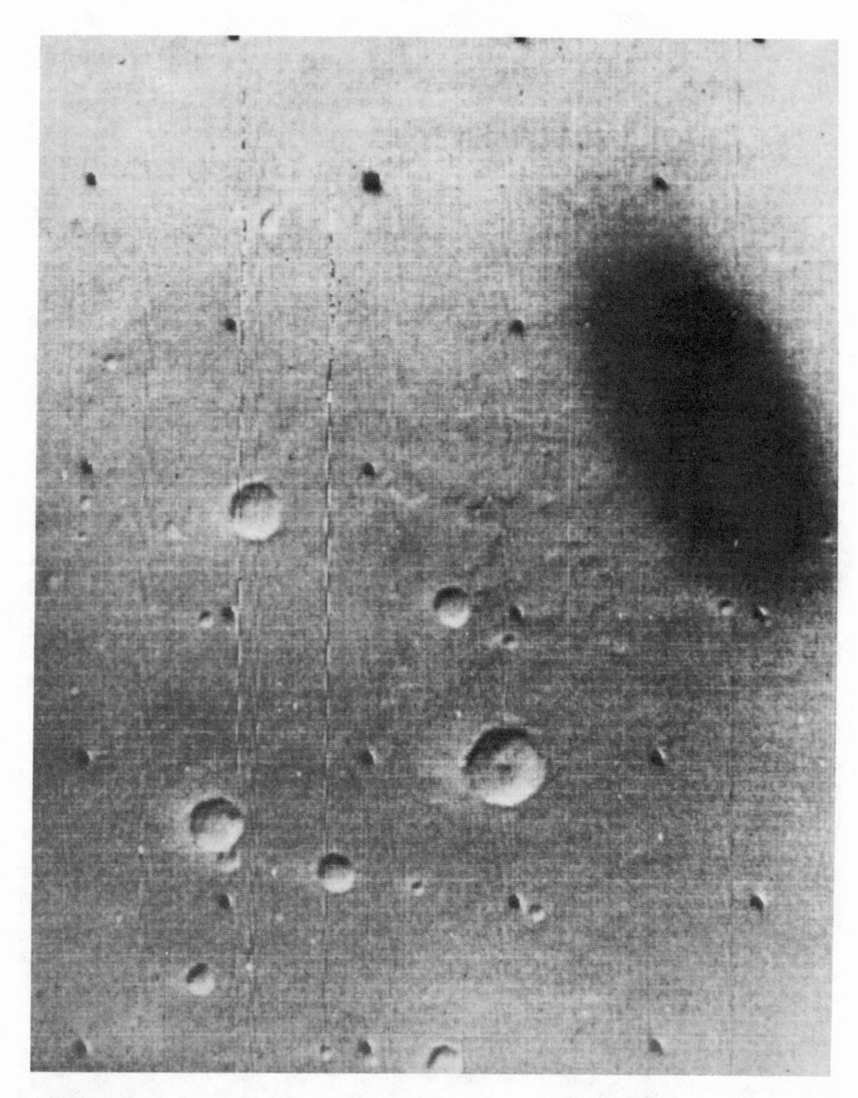

Figure 8.2
Craters on Mars
This *Mariner 9* photograph shows impact craters on a typical Martian terrain. The regular lines of dots are not craters, but are calibration marks in the optical system of the camera for use in correcting distortion of the images. The large black ellipse is the shadow of the inner Martian moon, Phobos. The dark area inside Phobos' shadow is experiencing a total eclipse of the Sun. These craters help us estimate the number of asteroids and comets with orbits that cross the orbit of Mars.

8.2). Powerful Earth-based radars penetrated the thick global cloud cover on Venus and found huge, ancient impact craters. And photographs of the Earth taken from space showed that the ancient cores of the continents, called *cratons,* were scarred with huge, shallow craters that had survived billions of years of weathering and scouring by glaciers. The oldest part of North America, the Canadian Shield, revealed to the eyes of satellites huge basins that had never been identified from the ground.

The surfaces of the planets and their satellites bear testimony to bombardment by cosmic projectiles, comets and asteroids, and vast numbers of smaller bodies. Since we have returned samples from several of the mare basins on the Moon to terrestrial laboratories, we have been able to determine the ages of those basins using radioactive dating techniques. We then can compare how many craters of each size have accumulated on targets of different ages. From that, it is a simple step to calculate the frequency with which such impact events have occurred over the last four billion years. We can then estimate the relative impact rates on small vs. large bodies by taking into account the gravitational enhancement of the impact rate on large bodies.

When we do this calculation, we find that Earth has indeed suffered a bombardment like that which scarred the Moon and Mars. In fact, Earth must have been even more intensely bombarded than the Moon: not only does Earth's strong gravity attract and collect passing bodies that would have ignored the blandishments of the little Moon, but Earth's gravity accelerates them greatly as they fall toward the surface. Even an asteroid that starts out moving very slowly with respect to the Earth when it is several million kilometers away will be accelerated by Earth's gravity to a speed of more than 11.2 kilometers per second by the time it strikes the Earth. The amount of kinetic energy carried by a moving body is proportional to the *square* of its velocity—in this case, the speeds and amount of energy are far outside the range of normal experience. A homey example would be a nice, slow rifle bullet, traveling at a mere 400 meters per second. A slow asteroid, impacting at only 12 kilometers per second, is traveling 30 times as fast. Thus each piece of the asteroid with the mass of a rifle bullet will carry 30 × 30, or 900 times as much impact energy as the bullet. An asteroid 10 kilometers in diameter also has quite a bit more mass than a rifle bullet, about 1,500,000,000,000,000,000 grams instead of the bullet's 10 grams.

The impact of such an asteroid, which happens on Earth about once every 100 million years, would cause a global catastrophe that would wipe out most forms of life on Earth. In fact, as early as 1973 the eminent space scientist and Nobel Laureate Harold C. Urey suggested that such impact events might mark the boundaries between the geological ages, the points in the sedimentary record where one finds sudden global changes in sedimentation behavior and in the fossil species preserved in the rocks.

The planetary scientists quickly explained this behavior to their satisfaction: large numbers of massive bodies had indeed struck the Earth within the period of time covered by the geological record, that is, within the last four billion years. Earth, however, is a very active heat engine, with constant internal turmoil and a steadily moving crust. At many places on the surface of the Earth, such as under the western rim of the Pacific Ocean and under the Andes, vast slabs of ocean floor rocks with their burdens of oceanic sediments are being shoved under the continents and down into the hot mantle, where they melt, release their volatiles, and create earthquakes and volcanic eruptions. So active is this process that the ocean floor rocks are recycled through the mantle every 100 million years or so. Hardly any of the ocean floor is as old as 200 million years, and most is less than 60 million years old. The ocean floor, which covers nearly three quarters of the surface of our planet, is much too young to bear a record of ancient impacts. Further, much of the present surface area of the continents is made of sediments laid down within the last few hundred million years, and those sediments were in turn made by by weathering away and grinding up incredible quantities of older crust.

In short, the "recording medium" in the Earth's geological tape recorder, the crust, is constantly being erased and re-recorded. So pervasive are these processes that only a handful of cratons older than 3 billion years are known. The only surviving sample of rocks older than 3.8 billion years on Earth today is a small chunk of the west coast of Greenland. It is hard to find areas that are old enough and large enough to have accumulated a lot of craters, but have survived this vigorous recycling program.

Geologists had long ago come to grips with the issue of whether the geological history of the Earth was placid or cataclysmic. In the eighteenth century the two schools of thought were cleanly polarized; so cleanly, indeed, that any modern student of human thought would be inclined to doubt them both. The *catastrophic* school held that the thick layers of grossly uniform sediments so commonly found in the geological record were produced instantaneously by titanic cataclysms, in which the entire face of the planet was rearranged. By this means, the antiquity of the Earth could be brought into serious question; indeed, many catastrophists were motivated by extremely literal readings of the book of Genesis, in which the "days" of creation were assigned a duration of 24 hours. *Uniformitarians,* on the other hand, held that all the work of geology was done by the kind of slow, mundane processes of weathering, transport, sedimentation, etc., that we can observe casually on a stroll through the countryside today.

While the overt debate concerned matters geological, the covert debate concerned convictions theological. The gut issue was whether the Earth was created in 4004 B.C., or was shockingly old, perhaps millions of years. The uniformitarian school invented a mythology of mountian building to keep their machine running smoothly (one that we find quite incredible today), and with their arguments, marshalled elegantly by James Hutton in his *Theory of the Earth* (1785), the uniformitarians carried the day.

With the advent of Darwinism eight decades later, it became widely accepted that the origin of new species occurred as a result of very slow, gradual change: rare random mutations would from time to time confer some characteristic on the offspring that would give them a competitive advantage, and this favorable trait would, because of its survival value, slowly spread through the gene pool until it became common. This factor, combined with chance geographical isolation of different breeding stocks from an initially homogeneous population, made it possible for two or more species to descend from a common ancestral species. The demise of the outmoded species was not viewed as a problem: the old species often did *not* die out, it simply was coopted and pervaded by the better gene. In light of the tediousness of this mechanism, it seemed evident that the time needed for biological evolution to run its course must be many hundreds of millions of years. Hence, uniformitarian geology seemed to be on very secure grounds. Unfortunately, not content with getting the better of their opponents, most uniformitarians concluded that there were not, and never had been, any catastrophes.

In the 1960s, when the first lunar and planetary probes were launched, geologists were wrapped up wholly in a fascinating debate concerning the nature of *orogeny,* the building of mountains. Spearheaded by the prescient work of Harry Hess at Princeton, the ideas of seafloor spreading and global tectonics were gaining currency. Geologists were entirely tied up in debating the issues of whether continents moved, what causes midocean ridges, why the heat flow rate through the ocean floors is so high, and why deep trenches are found at many ocean margins. All of this was deeply laced with marine geophysics, and increasingly with marine geochemistry as well. They were looking down, not up, for the answers to the big questions of geology. Thus as late as 1980 few geologists had come to grips with the alien concepts of planetary evolution being imported from alien, heavily cratered planets.

It was in 1980 that a team head by Nobel Laureate Luis Alvarez first presented geochemical evidence that a massive extraterrestrial body, an asteroid or comet, had impacted the Earth precisely at the end of the Cretaceous era. At that time, 65 million years ago, the sedimentary rocks over the entire globe are marked with a thin layer of dust containing heavily shocked particles of minerals and large amounts of elements such as iridium, which are extremely rare in the crust and oceans, but much more abundant in extraterrestrial materials such as meteorites. At exactly the same point in the sedimentary record, the overwhelming majority of all the species and genera present in the late Cretaceous disappeared without a trace.

Journalists often make a great issue of the "extinction of the dinosaurs" at this point in history. This is probably a red herring, since there were only a handful of species of dinosaurs left at that time, and they had clearly been going out of fashion for millions of years as more and more species found it expedient to be mammals. Further, the exact times of disappearance of these several dinosaur species are only poorly known because the large dinosaurs are only sparsely represented in the fossil record: even if they did go out with a bang on the same afternoon, it would be very hard to prove it from scattered samples separated by hundreds of thousands or millions of years.

Nonetheless, the second most severe biological extinction of all time took place at the end of the Cretaceous, in a time interval that has been estimated by some sedimentologists and paleontologists as being less than 100 years, and possibly "instantaneous."

The amount of dust in the iridium-rich layer at the end of the Cretaceous would be provided by a typical asteroidal body six kilometers in diameter. If the body had a higher iridium content, like an iron meteorite, it could have been several times less massive, and if it had the composition of one of the icy satellites of Jupiter (about half meteoritic rock and half ice), then it might have been several times more massive. We simply do not know enough about the compositions of comets to tell whether the impactor was an asteroid or a comet.

WHODUNIT?

When the general features of Earth's bombardment history became clear in the early 1960s, it was natural to begin to search for the bodies causing the large impacts. The culprits had to be reasonably large, numerous, and in orbits that crossed the orbit of Earth. The most obvious class of bodies then known were the long-period comets. These bodies pursue extremely eccentric elliptical orbits about the Sun, with perihelion distances often inside Mercury's orbit and aphelion distances far outside the realm of the planets, perhaps 10 percent of the way to the nearest stars.

Further, these orbits are almost randomly distributed over the celestial sphere: about half of them are traveling about the Sun in a direction opposite to the orbital motions of the planets. This is called *retrograde* motion. The long axes of their elliptical orbits also point every which way, and there is no tendency for these orbits to lie in the *ecliptic,* the plane near which all the planets orbit the Sun. Because of the immense size of their orbits, these comets often have orbital periods of millions or tens of millions of years. This means that not one of them has ever been observed more than once (the earliest astronomical records on Earth are only a few thousand years old).

Collisions of long-period comets can be satisfyingly violent, since head-on collisions with Earth can occur. Consider Earth, innocently plodding along its nearly circular orbit at 30 kilometers per second, encountering a comet heading the opposite way. The long-period comet has dropped in from virtually infinite distance, and will thus be traveling at a speed equal to the escape velocity of the Sun from a point on Earth's orbit. By a simple and basic law of celestial mechanics known since the 1600s, the escape velocity from a given point is the circular orbital velocity multiplied by the square root of 2, or 42.4 kilometers per second (km/s). In a head-on collision, the impact speed is the sum of the speeds of the two bodies, which is 72.4 km/s: New York to Los Angeles in one minute! Recalling our earlier example, this is 72.4/0.4 or 181 times as fast as the rifle bullet: each bullet-sized chunk of the comet will carry a kinetic energy that is 181 × 181 or 32,761 times as much energy as a rifle bullet of equal mass. Such impacts would be devastating.

Unfortunately, there are not nearly enough long-period comets to make craters at the rate that we see. The only other candidates are the short-period comets and the asteroids.

However, most asteroids come nowhere near earth (fig. 8.3). Almost all of the asteroids, the "vermin of the Solar System," reside in the wide gap between the orbits of Mars and Jupiter, in a wide band usually referred to as the *asteroid belt.* These bodies are very numerous, and almost all that we can see from our distant perspective are large enough to make quite handsome holes in planets. However, again, their orbits rarely depart from the belt. Until very recently, only a few tiny asteroids were known to venture across the orbits of Mars, Earth, and Venus, and these bodies were extremely elusive.

Astronomers are normally interested in stars and galaxies, not planets, and certainly not asteroids. Those few astronomers with a perverse fascination with vermin were accustomed to searching for new asteroids by taking long time exposures with telescopic cameras aimed to move along relative to the background stars at the same rate than an asteroid in the heart of the belt would move. This causes the light from the asteroid to be concentrated in a very short trail on the photographic plate, while the background stars leave long trails. This makes it possible to detect asteroids that would have been too faint to observe if the camera were fixed on the distant stars, with the asteroid's feeble light smeared out along a lengthy trail. Because of the nature of this technique, even intentional asteroid searches discriminated strongly against asteroids in nearby orbits, which move very rapidly against the sky. Thus the few known nearby asteroids, the ones that cross Earth's orbit and are therefore candidates for collisions with Earth, were discovered accidentally.

The absence of intentional searches for Earth-crossing asteroids was easily rationalized by reference to the statistics: since so few were known, there must not be many of them, and so they are not important, and so there is no point wasting valuable telescope time looking for them.

The short-period comets seemed to present a rather different set of problems. These comets typically have orbits that lie wholly within the realm of the planets, with orbital periods ranging from about 100 years down to as little as three years. Their orbits also lie relatively close to the plane of the ecliptic, and none of them are retrograde. Because of the short orbital periods, many of these comets have been observed through large numbers of perihelion passages: Halley's comet has been observed at each of its apparitions for over 2000 years. Since they pass close to the Sun so frequently, the volatile ices in the nuclei of these comets are quite rapidly depleted. Thus these bodies may evaporate away in a few millenia, leaving behind

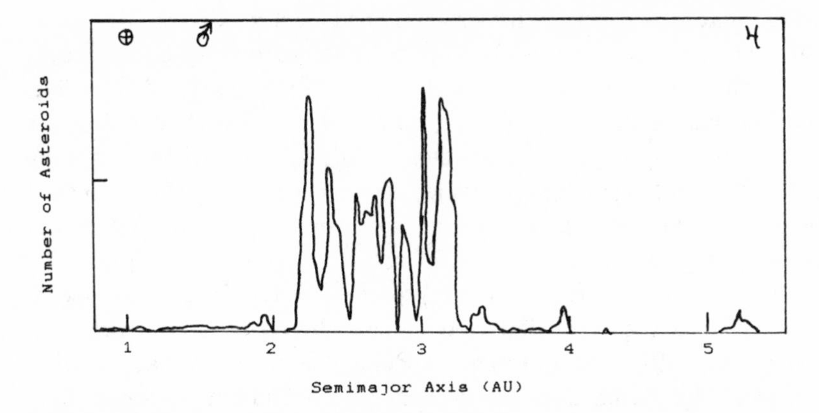

Figure 8.3
Distance Distribution of Asteroid Orbits
The mean distances from the Sun of over 1000 asteroids are plotted here to show the location and general features of the Asteroid Belt. The densely populated region from about 2.2 to 3.5 Astronomical Units is the Belt. The positions of Earth (1.0 AU), Mars (1.5 AU), and Jupiter (5.2 AU) are marked at the top. The deep gaps in the Belt are produced by Jupiter's gravitational perturbations. All of the gaps correspond to asteroids that would have orbital periods that are commensurate with Jupiter's orbital period. The Apollo, Amor, and Aten asteroids are at the far left, the group near 1.9 AU are the Hungarias, the group near 3.9 AU are the Hildas, and the Trojan asteroids are located on Jupiter's orbit.

vast swarms of involatile particles of dust. These swarms are responsible for the spectacular meteor showers that from time to time brighten the night sky, when Earth happens to be passing close to the orbit once pursued by the vanished (or vanishing) comet. Every October 24 we see meteors that are dust grains shed centuries ago by Halley's comet.

Some comets, however, if they avoid very violent heating by keeping a safe distance from the Sun, may not be active enough to blow off their dust as their ices slowly sublime into space, and may accumulate thick insulating blankets of fluffy, ice-free dust, hiding their icy inner hearts from prying eyes. Such comets would have very faint or even invisible comas and tails, and may thus be indistinguishable from Earth-crossing asteroids. Indeed, the "asteroid" 1983 TB is associated with a known meteor stream, and was almost certainly an active comet at one time.

Fortunately, astronomers have recently had much success in discovering and studying these two classes of bodies. The progress in finding Earth-crossing asteroids is to a large extent due to the efforts of asteroid expert Eugene Shoemaker of the US Geological Survey and astronomer Eleanor Helin of the Jet Propulsion Laboratory. With only limited funding, they have been searching for "fast-moving objects" since 1973. Helin has indeed found a number of nearby asteroids, some of which even have orbital periods less than one year.

The orbits of the known Earth-approaching asteroids are presented in table 8.1, ranked by perihelion. Note the large number that have been discovered since 1973, many of them by Helin. The asteroids that penetrate the inner Solar System are for convenience divided into three families. The closest to the Sun are the *Aten* asteroids, with mean distances from the Sun

Table 8.1
Orbits of Earth-Crossing Asteroids

Name	*Discovered*	*Perihelion*	*Aphelion*	*a*	*e*	*i*
*1983 TB	1983	0.14	2.47	1.30	0.894	22.8
1566 Icarus	1949	0.19	1.97	1.08	0.827	23.0
2212 Hephaistos	1978	0.36	3.97	2.16	0.835	11.9
*1974 MA	1974	0.42	3.13	1.78	0.762	37.8
2101 Adonis	1936	0.44	3.30	1.87	0.764	1.4
2340 Hathor	1976	0.46	1.22	0.84	0.450	5.9
2100 Ra-Shalom	1978	0.47	1.20	0.83	0.437	15.8
*1954 XA	1954	0.51	1.05	0.78	0.345	3.9
*1984 QA	1984	0.52	1.46	0.99	0.474	10.1
*1984 KB	1984	0.53	3.88	2.21	0.760	4.6
*1982 TA	1982	0.53	4.07	2.30	0.769	12.1
1864 Daedalus	1971	0.56	2.36	1.46	0.615	22.1
1865 Cerberus	1971	0.58	1.58	1.06	0.467	16.1
Hermes	1937	0.62	2.66	1.64	0.624	6.2
1981 Midas	1973	0.62	2.93	1.78	0.650	39.8
2201 Oljato	1947	0.63	3.72	2.17	0.712	2.5
*1981 VA	1981	0.63	4.22	2.46	0.744	22.0
1862 Apollo	1932	0.65	2.29	1.47	0.560	6.4
*1979 XB	1979	0.65	3.88	2.26	0.713	24.9
2063 Bacchus	1977	0.70	1.45	1.08	0.349	9.4
1685 Toro	1948	0.77	1.96	1.37	0.436	9.4
*1983 LC	1983	0.77	4.50	2.63	0.711	1.5
2062 Aten	1976	0.79	1.14	0.97	0.182	18.9
2135 Aristaeus	1977	0.79	2.40	1.60	0.503	23.0
*1983 VA	1983	0.81	3.67	2.24	0.636	15.4

Table 8.1 ***(continued)***

Name	*Discovered*	*Perihelion*	*Aphelion*	*a*	*e*	*i*
*1982 HR	1982	0.82	1.60	1.21	0.322	2.7
6743 P/L	1960	0.82	2.42	1.62	0.493	7.3
*1983 TF2	1983	0.82	3.62	2.61	0.387	7.8
2329 Orthos	1976	0.82	3.99	2.40	0.658	24.4
1620 Geographos	1951	0.83	1.66	1.24	0.335	13.3
*1959 LM	1959	0.83	1.85	1.34	0.379	3.3
*1950 DA	1950	0.84	2.53	1.68	0.502	12.1
1866 Sisyphus	1972	0.87	2.92	1.89	0.540	41.1
*1978 CA	1978	0.88	1.37	1.12	0.215	26.1
*1973 NA	1973	0.88	4.04	2.46	0.642	68.1
1863 Antinous	1948	0.89	3.63	2.26	0.606	18.4
2101 Tantalus	1975	0.91	1.67	1.29	0.298	64.0
*1982 BB	1982	0.91	1.91	1.41	0.355	20.9
6344 P/L	1960	0.94	4.21	2.58	0.635	4.6
*1982 DB	1982	0.95	2.02	1.49	0.360	1.4
*1979 VA	1979	0.98	4.29	2.64	0.627	2.8
*1984 KD	1984	1.01	3.41	2.21	0.544	13.7
*1982 XB	1982	1.01	2.70	1.86	0.454	3.9
*1981 ET3	1981	1.02	2.52	1.77	0.422	22.2
2608 Seneca	1978	1.02	3.93	2.48	0.587	15.6
*1980 PA	1980	1.04	2.82	1.93	0.459	2.2
*1980 AA	1980	1.05	2.73	1.89	0.444	4.2
2061 Anza	1960	1.05	3.48	2.26	0.537	3.7
1915 Quetzalcoatl	1953	1.05	3.99	2.52	0.583	20.5
1943 Anteros	1973	1.06	1.80	1.43	0.256	6.7
1917 Cuyo	1968	1.06	3.23	2.15	0.505	24.0
*1983 RD	1983	1.07	3.12	2.10	0.488	9.5
1221 Amor	1932	1.08	2.76	1.92	0.436	11.9
*1980 WF	1980	1.08	3.38	2.23	0.514	6.4
*1981 QB	1981	1.08	3.39	2.24	0.518	37.1
*1983 RB	1983	1.09	3.35	2.22	0.490	18.0
*1982 DV	1982	1.10	2.96	2.03	0.457	5.9
*1982 YA	1982	1.11	5.09	3.10	0.641	33.2
1627 Ivar	1929	1.12	2.60	1.86	0.397	8.4
1580 Betulia	1950	1.12	3.27	2.20	0.490	52.0
2202 Pele	1972	1.12	3.46	2.29	0.510	8.8
433 Eros	1898	1.13	1.78	1.46	0.223	10.8
887 Alinda	1916	1.15	3.88	2.52	0.544	9.1

Notes: The catalogue numbers and names of most asteroids are given in form "1221 Amor." The newest (unnamed) asteroids are temporarily known by their year and order of discovery, such as "1984 KB" and are set off with an asterisk. a is the semimajor axis, e is the orbital eccentricity, and i is the orbital inclination. Perihelion and aphelion distances and a are given in units of Astronomical Units (AU). Inclination is in degrees. The most accessible asteroids are those which cross Earth's orbit, and which have low orbital inclinations and eccentricities. These data were compiled by Eleanor Helin, who also discovered many of the asteroids on this list.

that are less than one Astronomical Unit (1 AU is the mean distance of Earth from the Sun, approximately 150 million kilometers). They thus also have orbital periods less than one year.

Next come the *Apollo* asteroids, which have periods longer than one year and which, at their point of closest approach to the Sun (perihelion) approach to within 1.017 AU of the Sun. That is the Earth's greatest distance from the Sun (aphelion distance). This means that Apollo asteroids all cross the Earth's orbit every time they circle the Sun. (However, it makes no guarantee about how close the body actually gets to Earth: the asteroid's orbit will not generally lie in the same plane as Earth's, and hence the asteroid, when it crosses Earth's orbit, may be significantly "above" or "below" the Earth's path.)

The third and most distant of the near-Earth asteroid families, the *Amors*, also have orbital periods longer than one year and spend most of their time outside Earth's orbit. In fact, they are defined as having perihelion distances between 1.017 and 1.3 AU, so that they do not at the present time actually cross Earth's orbit at all. However, fully half of the Amor asteroids are from time to time perturbed by the gravitational effects of the planets into orbits that do dip inside Earth's aphelion distance. A little reflection reveals that an Amor suffering such a perturbation would immediately become an Apollo asteroid. Conversely, a mild perturbation of an Apollo object might well turn it into an Amor. Thus the group names have no particular long-term significance: they merely denote whether the present orbit crosses Earth's. Aten and Apollo group members are called Earth-crossing asteroids. Atens, Apollos and Amors are collectively called Earth-approaching or near-Earth asteroids.

There is little doubt that increased funding for the search for near-Earth asteroids would greatly accelerate the rate of discovery. Helin's work has been the beneficiary of generous support from the World Space Foundation, a private space advocacy group headquartered in South Pasadena, CA. A new asteroid search program, Spacewatch, headed by Tom Gehrels of the University of Arizona, is also extensively supported by private donations.

The task yet to be accomplished is large and exciting. From a knowledge of the completeness of the telescopic search so far conducted, estimates can be made of the total numbers of Atens, Apollos, and Amors. The cratering statistics on the terrestrial planets and their moons can also be used to make an estimate of the total number of Earth-crossing bodies. The asteroid numbers estimated in these two ways are in reasonably good agreement. We shall include in chapter 9 a table based on Shoemaker's study of this problem. We note here only that he predicts that there are hundreds of thousands of asteroids larger than 100 meters in diameter in Earth-crossing orbits.

Another great contribution to the study of asteroids and comets has been made by the Infrared Astronomy Satellite (IRAS), which was launched on January 26, 1983 from the Western Test Range aboard a Delta rocket. From its vantage point in a 900 kilometer high polar orbit, IRAS mapped the distribution of a vast number of sources of infrared (heat) radiation in the sky. IRAS made over ten thousand observations of asteroids and comets, so many that the first catalogue of observations did not appear until 1986. Because only a tiny portion of the asteroid data have yet been analyzed, we know only that IRAS discovered a number of comets and great numbers of asteroids. Because of the sky-survey use of IRAS (dictated by stellar and galactic astronomers who vastly outnumber those with interests in asteroids), almost all of the previously unreported asteroids seen by IRAS were observed only once. It is impossible to determine the orbit of a body from fewer than three well-separated observations, and it appears that the principal contribution of IRAS to asteroid science will be a statistical understanding of their infrared properties (temperature and some spectral information), not revelations about particular identified asteroids. There is a vast amount of valuable science that could be done by flying another satellite like IRAS, but one dedicated to discovery and orbit characterization of asteroids rather than deep-sky mapping.

WHAT ARE ASTEROIDS LIKE?

No spacecraft has yet visited an asteroid. The *Mariner 9* Mars orbiter of 1971 and the two Viking orbiters of 1975 did survey the two natural satellites of Mars, Phobos, and Deimos, which are small, irregularly shaped bodies similar in their properties to many asteroids. Also, NASA made a decision in 1984 to use all future missions to the outer Solar system as probes of the asteroid belt while they are en route to their ultimate targets. The first fruit of this sensible policy was the decision to use the *Galileo* spacecraft, scheduled for launch toward Jupiter in May of 1986, to execute a high-speed flyby of the main-belt asteroid Amphitrite (Am phih TRY tee) in December of 1986.

These plans were shattered by the explosion of the Space Shuttle

Challenger in January of 1986. Pending the conclusion of inquiries into the cause of the explosion, the remaining shuttle fleet (*Columbia, Atlantis,* and *Discovery*) was grounded, with no hope of returning to service before January 1988. Four of the five major space science payloads that had been scheduled for 1986 were among the first five flights affected. The two Jupiter-bound payloads, the European Space Agency's *Ulysses* solar probe and the *Galileo* mission, were both indefinitely postponed. The two other science missions scheduled for early 1986 were instrument packages dedicated to observing Halley's Comet (from low Earth orbit) during the time that the flotilla of Soviet, Japanese, and European probes were actually flying by the comet at close range: one of these was lost with *Challenger,* and the other was canceled.

The Halley missions have been irretrievably lost. The comet, of course, will not return until the next century is nearly two-thirds over. The *Ulysses* and *Galileo* missions will be rescheduled for 1989 or later. It is still much too early to say what asteroid observations might be made by *Galileo* when it traverses the Belt on its new trajectory.

Despite our lack of *in situ* asteroid data, we do have a wealth of astronomical data that reveals much about their composition. Almost all we know about the composition of individual asteroids comes from measuring their reflectivity over a range of wavelengths spanning visible light from the ultraviolet (wavelengths near 0.3 micrometers, or 3000 Angstroms) through the red (0.8 micrometers, 8000 Angstroms) and into the somewhat longer wavelengths of the infrared (usually 0.8–2.5 micrometers, but sometimes farther). The quality of the information depends on how much light we receive from the asteroid: a large, bright asteroid in the heart of the belt will give us far more light than a small, dark asteroid near the outer edge. Very crude information on the colors of the brighter asteroids has been available for several decades, but the state of detector technology at the time was such that the received light could not be subdivided into a large number of wavelength intervals. Instead, the light was lumped into a handful of very broad wavelength bands. The most common kind of observation was to measure the amount of light transmitted through each of three widely spaced broad-band filters.

Earth's atmosphere absorbs all incoming light with wavelengths less than about 0.3 micrometers (ultraviolet), thus making observations from Earth-surface observatories impossible. The shortest wavelengths reaching the ground were isolated and measured using a filter that absorbs visible light and infrared, but transmits ultraviolet light to the detector. This was called a U (for ultraviolet) filter. The second filter passed only light centered on the blue (B) part of the spectrum (near 0.45 micrometers), and the third broad filter was centered on yellow light, near 0.6 micrometers. This third broad channel covered green, yellow, and orange light, which constitutes most of the visible part of the spectrum. Accordingly, it was called the V (visible) filter.

These filters were chosen for their usefulness in the study of stars, and were routinely used to extract information on the color, and hence the temperature, of stars. Almost nothing was then known about the colors of asteroidal minerals, and no attempt was made for many years to ask what wavelengths carried the most information about the composition and abundance of likely minerals. Thus the technique of measuring brightnesses through these U, B, and V filters (called *UBV photometry*) provided limited compositional information, much of it needlessly ambiguous.

The crudeness of the asteroid data was a source of endless anguish for meteoriticists. Dozens of classes of meteorites had been identified through careful laboratory analysis and microscopic study, and much had been deduced about the conditions of temperature and pressure under which these classes of meteorites formed. Some meteorites are chemically reduced (rich in metallic iron alloys) and very dry. Others are highly oxidized (free of native metal), ande rich in volatile materials such as water and organic matter (*carbonaceous*). Most are "primitive" objects called *chondrites:* they are made of a mixture of all the rock-forming elements in about the same proportions that they are found in the cosmos, and have never been subjected to melting and separation according to density. Some are igneous rocks, crystallized from magmas of silicate liquid (*achondrites*) or molten iron (*irons*), or mixtures of the two (*stony-irons*) (see figs. 8.4–8.6).

Unfortunately, none of these meteorites carry labels that say "Made on Ceres" or "Produkt von Pallas." It is as if we were given hundreds of still photographs taken by dozens of different photographers in various locations during World War II. Imagine further that these photographs had been shuffled to scramble their order completely before they were given to us, so that they are not geographically or thematically grouped, and are in random time sequence. Some of the photographs immortalize famous and crucial events; most show representative shots of the day-to-day horrors of war. From this evidence, we are called upon to reconstruct the history of World War II.

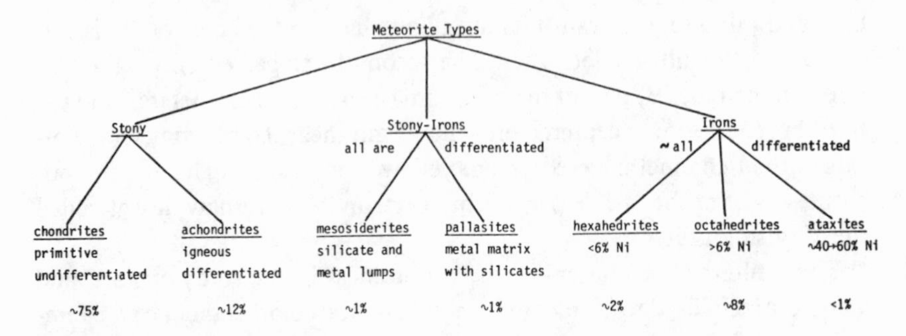

Figure 8.4
Meteorite Types
The known meteorites are broken down according to type. The percentages at the bottom show clearly that three quarters of the known meteorites are chondrites. In order to become a known meteorite, an asteroidal fragment must enter an Earth-crossing orbit, collide with Earth, survive atmospheric entry, be found, and be recognized as a meteorite. This means that our meteorite collections are strongly biased in favor of debris from Earth-crossing asteroids.

Some of the questions that occur to us as we contemplate this problem are: Is there any way to put the events in chronological order? Is there any way to sort them geographically and thematically, so as to let us isolate subplots of the greater drama? Is there any way to tell whether we have a reasonably complete and representative sample of the events we wish to understand?

Fortunately, meteorite compositions and textures permit us to associate them into thematic groups, and even to establish genetic connections between some of these groups. Even better, good techniques are available for dating the ages of meteorites. Unfortunately, their "geography" presents very difficult problems. Using our photographs, we have reconstructed World War II to the best of our ability, even to the point of giving the absolute chronology of events and reconstructing some of the better-documented battles—but we do not yet know where the war was fought!

In the late 1960s, Thomas B. McCord, a student working with astronomer James A. Westphal at the California Institute of Technology, carried out a program of observations of the Moon with a number of narrowband filters, chosen to span the entire accessible ultraviolet, visible, and

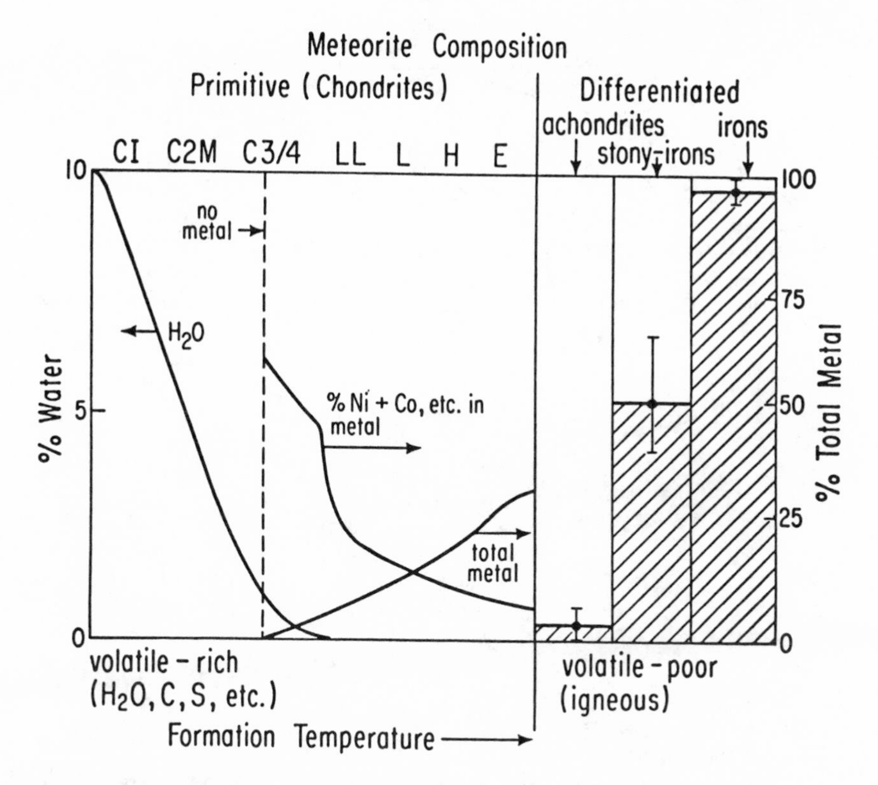

Figure 8.5
Meteorite Composition
The abundances of the economically important metal and water components of meteorites are collected here for quick reference. The objects with lowest formation temperatures (left) are most rich in volatiles such as water. The objects that have been heated strongly enough to melt them and permit density-dependent differentiation form the achondrite, stony-iron, and iron classes of meteorites.

near infrared spectrum. Later, as an assistant professor at MIT, and in collaboration with John B. Adams and Torrence V. Johnson, McCord made the first narrow-band filter observations of an asteroid.

As one might guess, McCord chose as his first target a large, bright main-belt asteroid, 4 Vesta. (All asteroids with known orbits are assigned catalog numbers in their order of discovery. Over 3000 asteroids have been catalogued. Vesta's number, 4, indicated that it must be extraordinarily

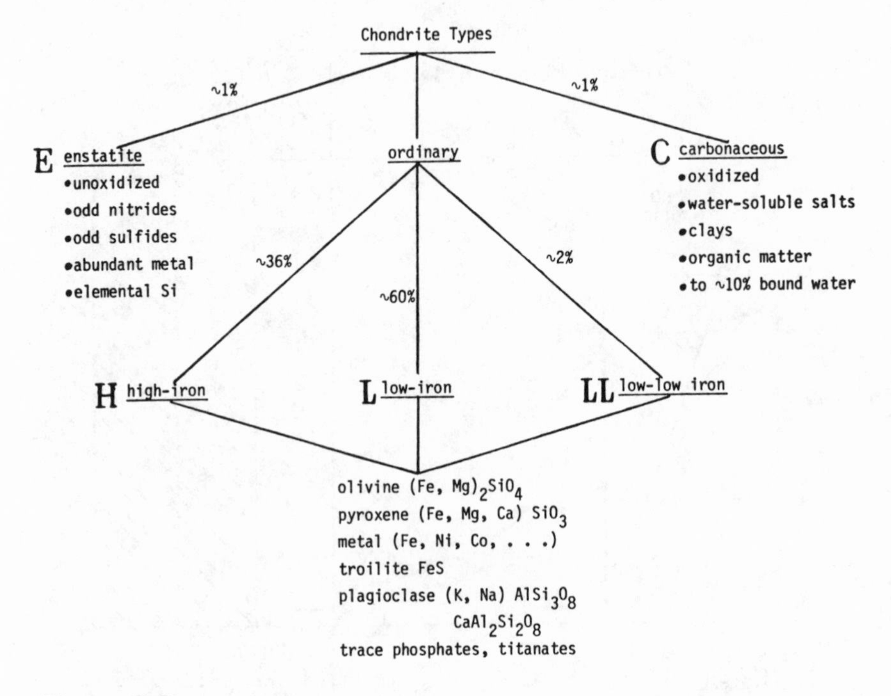

Figure 8.6
Chondrite Types
This chart continues the breakdown of the abundant chondritic meteorites into their five compositional classes. Note the enormous compositional diversity seen within the chondrites.

bright because it tells us that it was the fourth one discovered.) The results were almost astonishingly good. Vesta's brightness varied strikingly from one wavelength (color) to the next. The asteroid was very dark in ultraviolet light. Its reflectivity rose rapidly to a broad maximum in the visible, and dropped precipitously to a deep minimum in the near infrared, near a wavelength of 0.9 micrometers. The reflectivity then rose again into the middle infrared. The region of strong absorption near 0.9 microns, the most striking feature of the spectrum, is characteristic of minerals that contain ferrous oxide, FeO. The exact wavelength of the center of the absorption feature tells us what mineral the FeO resides in, the depth of the absorption feature tells us how much FeO is present, and the shape of the absorption band (i.e., the precise way the intensity of reflected light varies with wavelength) tells us whether we have a mixture of two or more FeO-bearing minerals.

The results for Vesta were compared to laboratory reflection spectra of a variety of minerals and a wide variety of real meteorite samples. McCord's team was able to conclude that Vesta's spectrum was a near-perfect match for a class of meteorites called basaltic achondrites. The significance of their work lies not so much in that one identification (although it is of considerable interest), as in demonstrating the principle that familiar meteorite types on Earth might be traced back to one or more parent asteroids by astronomical observational techniques. Thus the immensely rich body of data on meteorite compositions could at last be tied down to particular locations in the Solar System. We finally had found a way to tell where many of the battles of the war had been fought (fig. 8.7).

In the years since 1970, over 600 asteroids have been observed with the purpose of collecting information about their compositions. A very wide sampling of the meteorites that we have in our collections has been studied with the same techniques to make certain that we would recognize their spectral signatures if we saw them on asteroids. The general features of the results of these studies are fascinating. First, the highly processed (igneous) types are strongly concentrated near the inner edge of the belt, closest to the orbit of Mars. Both metallic (M) and achondritic or stony-iron (S) asteroids are most abundant there, as are the less common E and A classes, both of which appear to be types of achondrites. The heart of the belt is dominated by very dark (C) asteroids with reflection spectra indistinguishable from those of carbonaceous chondrites.

Larry A. Lebofsky of the University of Arizona has extended the spectral data on some of the largest and nearest C asteroids to longer infrared wavelengths, where absorption bands from water are found. Several of these asteroids show strong, clear absorption from chemically bound water, just like laboratory specimens of carbonaceous chondrites. Some of the asteroid spectra even show absorption features from organic matter!

The outer reaches of the belt and the even more distant families of the Hilda and Trojan groups are dominated by the "supercarbonaceous" D and P classes. These seem to be made of the same mixture of materials as the C asteroids, but with even higher proportions of volatiles.

One of the greatest puzzles in associating asteroids with meteorites is that the most common classes of meteorites known on Earth, the metal-bearing "ordinary" chondrites, are found to be rare or even wholly absent

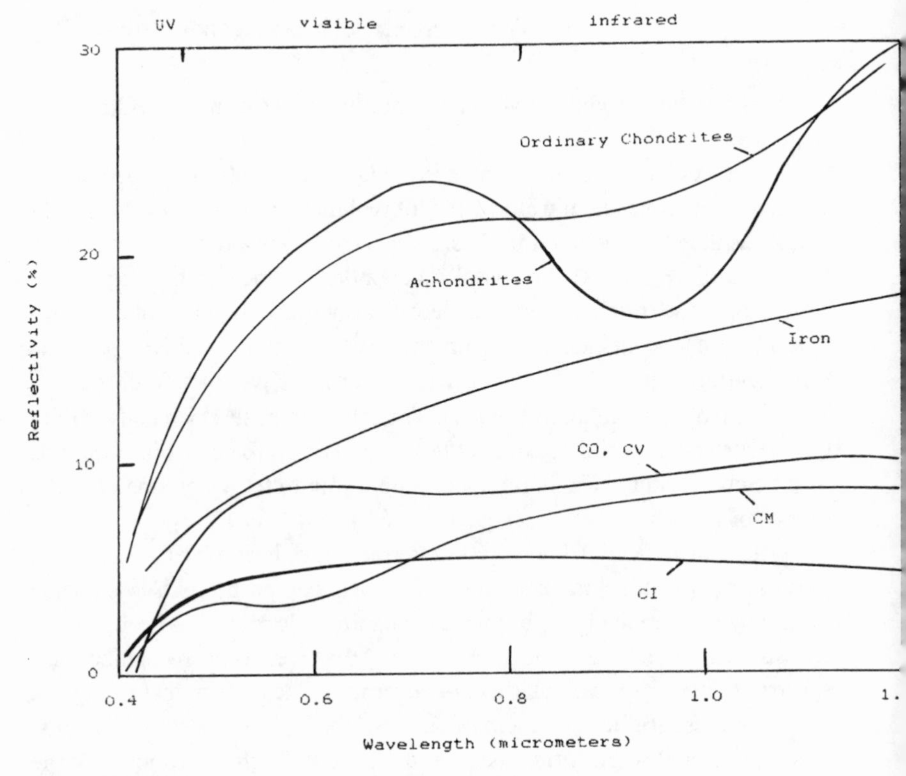

Figure 8.7
Reflection Spectra of Major Asteroid Types
The reflectivity (in percent of incident light) of several different asteroid classes is shown. White objects would have flat reflection spectra at levels near 100 percent, while very black objects (soot; black velvet) would have flat reflection spectra at a level of under 10 percent. The carbonaceous CI and CM chondrites are rich in "soot," magnetite, and dark clay minerals. The ordinary chondrites contain almost none of these darkening agents, and are relatively bright at all wavelengths in the visible and infrared. Metallic iron is moderately dark and reflects less light toward the blue end of the spectrum: its presence in an asteroid (as in the ordinary chondrites) lends a slope to the spectrum that makes the object look slightly red. The CO and CV carbonaceous chondrites are intermediate in composition, color, and reflectivity between the ordinary chondrites and the very dark CI and CM families. S-type asteroids have spectra like the two at the top of the diagram. The dip in their spectra near 0.9 micrometers wavelength is due to the presence of ferrous iron minerals such as olivine or pyroxene. C-type asteroids have reflectivities and colors similar to the C meteorite classes. The M-type asteroids have spectra dominated by metals. D and P asteroids are very dark and appear to be related to the C asteroids, but more volatile-rich. The E asteroids (not pictured) contain no ferrous iron and little or no free metal: they are similar to the enstatite achondrites.

in the asteroid belt. Why should such a rare class of asteroid give rise to the large majority of meteorite falls on Earth?

The best way to approach this problem is to ask what kind of orbit an object must be in so that it will fall to Earth. Phrased this way, the answer is obvious: anything that hits Earth must be in an Earth-crossing orbit. Computer simulations have been carried out in which small bodies are placed in a very wide variety of Earth-crossing orbits. Sure enough, most of the bodies hit Earth or Venus. In fact, they do so in a time ranging from a few tens of millions of years to a few hundred million years, depending on the initial orbit. Thus, from an orbital point of view, Earth-crossing asteroids are excellent candidates as sources for ordinary chrondrites. But Earth-crossing asteroids are rare birds. None of the Belt asteroids come anywhere near Earth (see fig. 8.8).

Since most of the near-Earth asteroids are very small, and since many have them have been known for only a few years, we have very little compositonal information on them (table 8.2). Lucy Ann McFadden, in her dissertation research done while working with Michael Gaffey at the University of Hawaii, compiled all of the photometric observations of the near-Earth asteroids done by herself and other astronomers. These observations cover only 18 of the 64 known Earth-approaching asteroids. Many spectral classes are found: C asteroids make up about 20 percent of her list, even though there are very strong biases against them. Very black, carbonaceous objects are so faint in visible light that they are much harder to find than similar-sized stony objects. S type asteroids are also found in the sample, as well as at least two bodies with reflection spectra indistinguishable from ordinary chondrites. The D and P asteroids are not represented, but this is hardly suprising: they are composed of material so volatile-rich that they probably would outgas like comets if moved into near-Earth orbits. After a brief "drying-out" period, they might well appear to be C asteroids. Similarly, the absence of M asteroids and the even more exotic E and A classes among the small near-Earth sample is scarcely surprising, since they are so rare that a random sampling of 18 asteroids from the belt would probably not find any of them either.

The other important class of bodies in Earth-crossing orbits is the short-period comet family. They are, from a dynamic point of view, conceivable sources of meteorites. There are other problems, however. First, genuine cometary material is probably much too delicate to survive blazing entry into Earth's atmosphere. The constituent ices would melt, boil, and crush

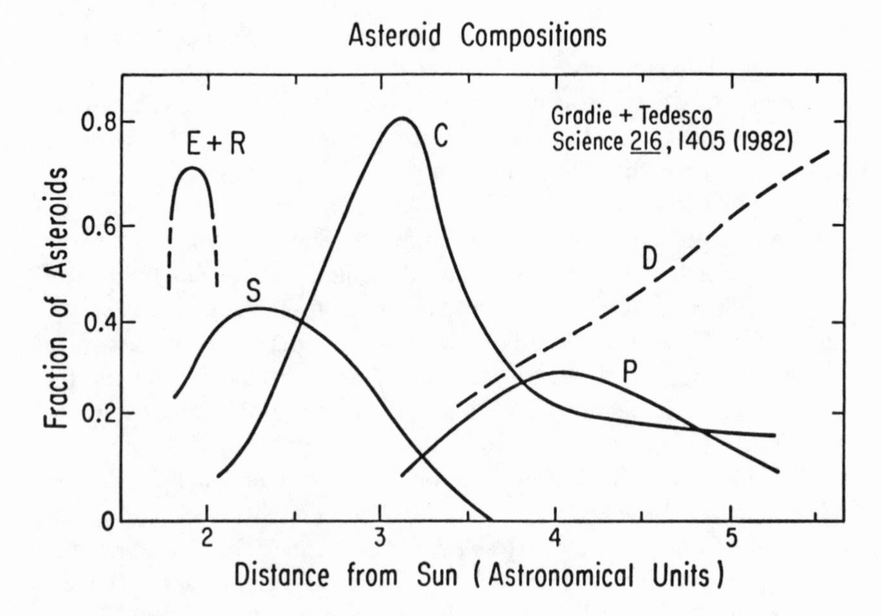

Figure 8.8
Asteroid Compositions
The major types of asteroids found in the Belt have very different distance distributions. The high-temperature E, R, and S asteroids, which have spectra similar to irons, stony-irons, and achondrites, are strongly concentrated near the inner edge of the belt. The C asteroids dominate the heart of the belt, and the very dark P and D asteroids are important mainly beyond the outer edge of the main belt. This is a simplified summary of work done by Jonathan Gradie and Ed Tedesco.

so easily that no meteorite-sized chunk of comet nucleus could possibly hope to make it to the ground intact. Long-period comets (see fig. 8.9) are even less likely sources of meteorites because their material would be at least as weak and volatile-rich as that of the short-period comets, and they would encounter Earth at such immense speeds that the shock of collision with Earth's atmosphere would crush any reasonable material into dust. The energy of motion of the long-period material would provide up to 1000 times the amount of energy needed to vaporize it completely.

We unfortunately have virtually no direct information on the composition of short-period comets. Spectroscopic studies of the gases in comet

Table 8.2
Near-Earth Asteroid Compositions

Asteroid	*Diameter (km)*	*Classification/Composition Data*
433 Eros	16	S type with metal, olivine (ol), pyroxene (px)
887 Alinda	4	dark S, similar to C3 chondrites, with ol and some carbon
1036 Ganymed		S type with FeO-bearing minerals and metal
1566 Icarus	1.4	FeO- and MgO-bearing (mafic) minerals
1580 Betulia	6.5	C type like CI or CM chondrites
1620 Geographos	2.4	S with metal and silicates
1627 Ivar		S with metal and mafic silicates
1685 Toro	7.6	S with mafic silicates and metal
1862 Apollo	3?	Px, ol, and metal
1863 Antinous		SU: ol, px, metal but not like any known meteorite types
1864 Daedalus		SO
1865 Cerberus	1.6?	S type, similar to C3 (CV or CO) chondrites
1915 Quetzalcoatl		Px: possible basaltic achondrite
1916 Bore		S type
1943 Anteros	2.3	S or E or M: CV or CO?
1980 Texcatlipoca		SU (stone-like unknown)
2062 Aten		S
2100 Ra-Shalom		C
2212 Hephaistos		S?
2340 Hathor		CSU (stony/carbonaceous unknown)
2368 Belt		S
*1977 VA		C-like
*1978 CA		S
*1979 VA	5	C (CI or CM?)
*1980 WF		unknown type
*1980 AA	0.5?	mafic silicates (and metal?)
*1982 DV		S
*1982 XB		S

We are indebted to Dr. Michael Gaffey for the information in this table.
*See notes to Table 8.1.

Figure 8.9
A Comet Wasting Precious Ices
The head and tail of a comet are made up of vapors from evaporating ices, heated strongly by the Sun, and tiny grains of dust blown free of the comet's feeble gravity by the gas stream while the comet is close enough to the Sun to be strongly heated. The gas tail streams radially outward from the Sun, driven by the solar wind.

heads (comas) and tails show a complex mixture containing fragments of molecules, called radicals. Among the most common radicals are OH, CO, NH, CN, HCO, etc. It is hard to reconstruct what molecules were present in the ices which, after being evaporated by the Sun's heat and torn apart by the Sun's ultraviolet radiation, provided the coma and tail gases we have observed. Surely water vapor is very important, and both carbon monoxide and carbon dioxide are present as well. Ammonia and possibly methane are present in smaller amounts, as are hydrogen cyanide and organic compounds such as acetonitrile and methyl alcohol.

Our information about the dust component of comets is even less direct, since we are limited to observing the spectrum of the flash of light made when cometary meteors burn up in our atmosphere. Nonetheless, the most abundant of the elements in the dust are the same ones that dominate meteorites: silicon, magnesium, iron, calcium, aluminum, etc. Their relative abundances vary greatly from one dust speck (meteor) to another, so we cannot tell which meteorite class, if any, they most resemble.

Some asteroidal and cometary dust particles enter the atmosphere at low enough speeds to survive entry. They can be collected in the high stratosphere by balloons or high-altitude aircraft, brought into our laboratories, and analyzed. Many evidently cometary particles have been analyzed by Donald Brownlee, who finds that they are extremely complex carbonaceous aggregates of even smaller, heterogeneous particles generally reminiscent of carbonaceous chondrites. It is likely that an old short-period comet, if outgassed gently enough so that it did not blow its dust particles away, would end up with a deep, fluffy surface layer of carbonaceous dust.

Studies of the thermal evolution of such comet nuclei were begun in the late 1970s by David Brin and Asoka Mendis of the University of California at San Diego, before Brin became established as a Nebula- and Hugo-award winning science fiction writer* (fig. 8.10). These and later studies show that, after some tens of millions of years, the insulation provided by the dust layer would essentially prevent heating of the ice-rich interior of the nucleus, and evaporation would slow dramatically. The resulting comet would, to an outside observer, look like a C asteroid. Nonetheless, all of the body except the outermost few meters might be composed of cometary ices. The time required for the rest of the ice to evaporate could easily be billions of years in a typical short-period comet orbit. However, Earth-crossing comets are subject to the same hazards that Earth-crossing asteroids are: they tend to be destroyed by collisions with Earth after about a hundred million years.

Since the study of the Earth, the Moon, and meteorites has given us three closely similar estimates for the age of the Solar System, about 4.6 billion

*_The Practice Effect, Star Tide Rising, Sundiver, The Postman_ (New York: Bantam Spectra); also, _The Heart of the Comet,_ with G. Benford (New York: Bantam, 1986).

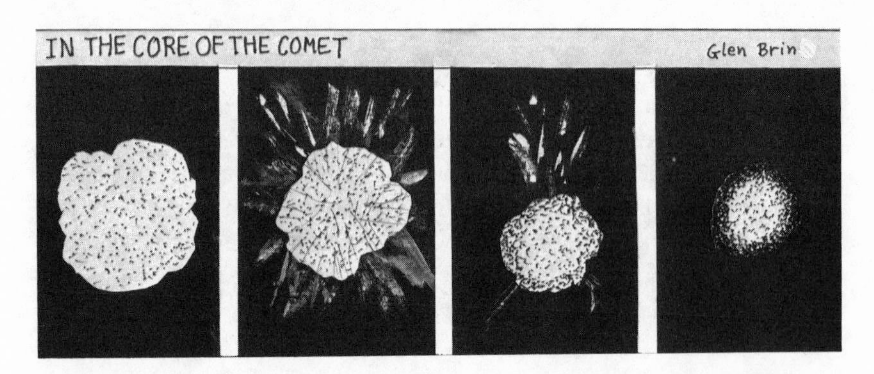

Figure 8.10
In the Core of the Comet
This strip shows the strange and interesting evolution of the dust and ice components of the nuclei of short-period comets. It is based upon the work of G. D. Brin, a confessed (but reformed) astrophysicist, witty raconteur, award-winning science fiction author, amateur genetic engineer, and manic four-wheel-drive operator. A short-period comet nucleus, a huge dirty snowball diverted into an orbit like that of a near-Earth asteroid by Jupiter's gravity, loses vapors and dust strongly each time it passes perihelion. But the gentler heating it experiences farther from the Sun produces less vapor, and drives it off less violently: dust cannot be blown away, and accumulates. The growing dust layer protects the icy cometary core against strong and rapid solar heating at the next perihelion passage, and the dust layer continues to grow slowly, preserving the ice beneath a thick insulating layer of dust. Such a "crypto-comet" has no readily visible coma or tail, and looks to an outside observer like a carbonaceous asteroid. The "asteroid" 1983 TB actually has a meteor swarm associated with its orbit, and may be a recently retired comet.

years, the presence of two families of objects with orbital lifetimes of only about 100 million years is rather startling. We do know of some mechanisms for replenishing the supply of short-period comets and Earth-crossing asteroids, but the number of bodies that can be provided by these mechanisms is poorly known. Orbital perturbations by the giant planet Jupiter can "kick" asteroids out of certain very narrow and specific locations in the belt into highly eccentric, Earth-crossing orbits. Further, close encounters of long-period comets with Jupiter can also kick them into low-inclination, short-period orbits. It is a well-known fact that many of the short-period comets and near-Earth asteroids are in orbits that take them out almost exactly to Jupiter's orbit. It has been suggested several times that the asteroids with aphelia near Jupiter's orbit are of cometary origin. These bodies often cross the orbits of two or more terrestrial planets, and hence are vulnerable to repeated perturbations that will alter their orbits, including erasing the "memory" of their past interaction with Jupiter.

So far as we can tell, it is reasonable to conclude that the present population of Earth-crossing bodies is extremely diverse in composition and extremely heterogeneous in origin. The overall balance between cometary and asteroidal matter is very poorly understood: the fraction of the Earth-crossing bodies that are of cometary origin could be as low as 10 and as high as 90 percent. Gene Shoemaker's best estimate is about 60 percent.

Little is known about the physical properties of the near-Earth bodies. Large airless bodies, such as the Moon and Mercury, are covered with deep layers of regolith produced by countless past impact events. Smaller bodies, such as Ceres, will lose regolith much more readily than the Moon will because the escape velocity of Ceres is so much smaller. Much smaller bodies, such as kilometer-sized near-Earth asteroids, will have sparse and probably very incomplete coverings of regolith (but may be shattered throughout). Mapping of temperature changes after sunset on an asteroid can permit mapping of the physical state of the surface, distinguishing clearly between bedrock (which stays warm at its surface by conducting heat out from its interior) and regolith (which cools rapidly because the poorly conducting dust stores and transports very little heat). This can be done from a flyby spacecraft, but can be done much more throughly with a rendezvous mission.

Fragments of asteroids and comets frequently enter Earth's atmosphere and break up because of the aerodynamic pressure of their very rapid flight through the atmosphere. Many such breakup events have been observed, and a wide variety of strengths have been seen. Some bright fireballs of cometary origin have been seen to disintegrate completely at altitudes where the ram pressure is only about 0.1 atmospheres. A "rock" with this strength could be easily crushed in your hand; it has about the strength of a moist clod of earth. Many other fireballs, whose vapor gives the spectrum of vaporized rock, have crushing strengths that are 200 to 2000 times as high. The famous Tunguska fireball of 1908, which exploded in the atmosphere and blasted and incinerated hundreds of square kilometers of Siberian forest, was about as strong as a sedimentary rock such as a sandstone.

CONCLUSIONS

We have followed the trail of the near-Earth asteroids from the dawn of the age of space flight to the present. The role of catastrophic impacts of asteroids and comets in the geological evolution of the Earth is now becoming clear. The classical uniformitarian picture, which attributes all the work of geology to the slow, steady operation of familiar processes, and which denies the significance of catastrophes, has broken down in the face of overwhelming evidence that catastrophic impacts do in fact occur. They are as natural a process as chemical erosion, sedimentation, or wind transport, but they most definitely are not slow and steady in their operation.

The philosophical basis of attributing almost all geological change to mundane processes is sound, but the claims of exclusivity made by uniformitarians sound hollow indeed. As for the great support lent uniformitarian geology by Darwinian biology, with its glacially slow processes of rare random mutations and gradual diffusion of successful genes throughout the gene pool, that too has changed.

It is ironic that many prominent biologists today, among them the eloquent Stephen Jay Gould of Harvard University, see most of biological history as bland and almost featureless, with most changes in species and genera occurring during brief and extraordinary periods of extremely high rates of change. This theory of "punctuated equilibria" is being increasingly adopted by paleontologists to explain the very erratic rates of change seen in the fossil record. Both the appearance of new species and the disappearance of old species seems to be very abrupt. Also, the species that are abruptly extinguished by major extinction events seem to have nothing in common with those that slowly fade away under more normal conditions. The role of local and global catastrophes in forcing (and permitting) biological evolutionary change is now a hot topic for research. The focal point for many studies is the Cretaceous-Tertiary boundary, where some 90 percent of the known species on Earth disappeared simultaneously, and after which a very impoverished cast of survivors proliferated into countless new species in a very brief time.

But the impact of the near-Earth asteroids (pun intended) extends far beyond simple historical issues in geology, biology, ecology, and climatology. There are asteroids in our future. Major impact events will continue to occur in a sporadic and as yet unpredictable manner. Multi-megaton explosions occur in every century (*vide* Tunguska, 1908), and megaton-sized impacts occur every few years. In a hair-trigger world of poised arsenals of weapons of mass destruction, the chance obliteration of Amman or Tel Aviv, of Brasilia or Jakarta, of Petropavlovsk or Phoenix by a megaton explosion of unknown origin could easily spell global disaster. Yet it is easily within the grasp of late-twentieth-century science to find these bodies, map their orbits, and establish an early warning system. With sufficient advance notice of an impending collision, it would be possible to destroy or deflect the threatening object with resources far smaller than those now poised in the silos and missile-firing submarines of the Great Powers.

Yet another option is open to us: the near-Earth bodies are among the most accessible in the Solar System. Many are easier to reach and far easier to return from than the Moon. They are astonishingly diverse in origin and composition. Among the materials abundant in them, life-support fluids (water, nitrogen, organic matter, oxygen, nitrogen), propellants (hydrogen and oxygen as water; organic matter) and free metals (iron, nickel, cobalt, platinum-group elements, and solutes such as sulfur, carbon, gallium, germanium, arsenic, indium, etc.) are most obvious. We should consider seriously the possibility that space-based industrialization will soon give us the ability to have our way with these asteroids. We shall accordingly devote the next chapter to an assessment of the exploitation of asteroid resources.

9

ASTEROID RESOURCE EXPLOITATION

Someday we will be able to bring an asteroid containing billions of dollars worth of critically needed metals close to Earth to provide a vast source of mineral wealth for our factories.—Lyndon B. Johnson

While we doubt that it will ever make any sense to risk moving a kilometer-sized asteroid into Earth orbit, the principle of using the great mineral wealth and diversity of asteroids for economic purposes is valid. To do so we need not embark on lengthy and demanding expeditions out to the distant asteroid belt to find rocks to tow home. Rather, we merely need to wait for them to come by.

The near-Earth asteroids are, at times, the nearest bodies to Earth. Many of them literally can pass between the Earth and the Moon, and, as we saw in chapter 8, about half of them are fated to collide with Earth eventually. This does not by any means guaranteee that it easy to travel from Earth to the surface of the asteroid: we need to consider specific examples, and take into account all the relevant factors.

One crucial question regarding the economic usefulness of asteroids is the cost of getting to them from Earth. The second factor that we must take into account is how well the accessible resources of asteroids match the needs of a spacefaring civilization. Of special interest are space sources of propellants, life-support fluids, and metals for construction and shielding.

The third major factor is the cost of returning materials from their sources, whether asteroids, the Moon, Phobos, or Deimos, to the places where they will be used. The most obvious point of reference, and the likely place for future processing, construction, and refueling operations, is low Earth orbit.

All spacecraft en route to deep space pass through low Earth orbit (LEO). Most projected future space activity, such as that involved in the Space Shuttle, Salyut, and Space Station programs, will occur in LEO. Most of the rest will involve the transfer of payloads to a 24-hour period geosynchronous orbit (GEO) above the equator, and most of those will be fired from LEO after having been carried that far by the Ariane or by large Soviet rockets such as the D1e "Proton" booster. Most materials returned from extraterrestrial bodies for use as propellants, life support fluids, or structures will almost certainly be destined for LEO. For our purposes, we shall concentrate on making resources available in LEO.

GETTING THERE

There are several distinct stages to the round trip from LEO to an asteroid and back to LEO. Each step corresponds to an engine burn with a particular associated velocity change, delta V. Delta Vs produced by firing engines cost fuel. For large delta Vs, a substantial amount of energy (and therefore fuel) must be expended not on lifting the payload, but on lifting the fuel for the remainder of the burn. The principle of rocket propulsion, which we explained in chapter 5, shows the enormous importance of selecting missions that have minimal delta V requirements (and also drives us in the direction of using propulsion systems that have the highest possible exhaust velocity, efficiency, and specific impulse).

The initial step is departure from LEO. If the spacecraft is equipped with a high-thrust chemical propulsion system, the burn is carried out in a time much shorter than the orbital period. It is almost as if the velocity change were made by a single impulse applied at LEO. The spacecraft is injected into an elliptical orbit that will carry it out to the distance of its destination, such as GEO or the orbit of the Moon. For escape missions, enough fuel is burned to put the spacecraft on a hyperbolic escape trajectory. For an asteroid mission, a small optional second engine burn, called a midcourse correction maneuver, may occur at a point remote from both Earth and the asteroid. A third burn will occur at the time of arrival at the asteroid to match velocity with it. Actual landing on and takeoff from the asteroid will use rather small amounts of fuel because of the minuscule gravity of the asteroid. Departure from the vicinity of the asteroid may involve a much larger delta V to secure a trajectory that intersects Earth. A minor mid-course correction burn may be needed during the return to Earth, and some fuel will also be needed to adjust the point of closest approach to assure that it is neither too deep in Earth's atmosphere (which might cause the vehicle to burn up) nor too shallow (which may permit it to skip out and escape permanently from Earth). Finally, after aerobraking has lowered the apogee of the orbit to the altitude of the Space Station, a small burn is required to lift the perigee out of the atmosphere and circularize the orbit near the Space Station, finally back in LEO.

Every one of these steps applies equally for returns from the Moon, asteroids, or other nearby bodies. What differs between these bodies is the amount of propulsion energy needed to execute these steps.

For many of the most accesssible known asteroids, the sum of the outbound delta Vs is between 4.5 and 5.5 kilometers per second. The lowest reasonable outbound delta V for any asteroid in a plausible orbit is about 3.4 km/s. The easiest known asteroid to get to, 1982 DB, requires about 4.4 km/s outbound. We may recall from the discussion in chapter 7 that the outbound delta V from LEO to the lunar orbit is 3.1 km/s, the delta V to match speeds with the Moon is 1.0 km/s, and the delta V to land on the Moon (through the inner Lagrange point, the gravitational saddle point between the Earth and the Moon), is another 1.9 km/s. Thus the total outbound delta V from LEO to the lunar surface is 6.0 km/s. Thus, as far as outbound propulsion requirements are concerned, the nearby asteroids are clearly superior to the Moon.

For many near-Earth asteroids, the total inbound delta V for return from the surface of the asteroid to LEO is under 0.4 km/s, with the very best candidates close to 0.1 km/s. For return from the Moon to LEO, 1.9 km/s is needed to depart through the inner Lagrange point, 1 km/s is needed to kill enough of the Moon's orbital speed to let the payload drop its orbital perigee into the Earth's atmosphere for aerobraking, and about 0.05 km/s should be budgeted for circularizing the orbit and rendezvousing with the Space Station after aerobraking. Thus the return delta V from the Moon to LEO is 3 km/s. The comparable figure for the near-Earth asteroid 1982 DB is 0.1 km/s. Thus, the best asteroids are *spectacularly* superior to the Moon for the return leg of the journey: the propulsion energy required per ton of payload is 900 times larger for return from the Moon than from 1982 DB!

It is important to realize that we are not talking about two or three asteroids with fortuitously good orbits. Even if only 20 percent of the Earth-crossing asteroids have round-trip delta Vs smaller than that for the Moon, then some 60,000 asteroids larger than 100 meters in diameter would be easier to get to than the Moon (see table 9.1).

There is another major difference between lunar and asteroidal propulsion requirements: for the trip to the Moon and back only about 4 km/s of the total delta V could possibly be done with an efficient, low-thrust, high-specific-impulse propulsion system. This means that specific impulses for most of the burns would be near 400 seconds (a hydrogen-oxygen chemical rocket) rather than near 4000 seconds (for a low-thrust, high-specific-impulse mercury ion engine). This combination of much higher delta Vs and much lower specific impulse has a devastating effect on the payload mass that can be returned from the Moon.

Finally, but not the least in importance, we must recall the evidence presented in chapter 8 regarding the great richness and diversity of asteroid compositions. Many valuable substances, such as water and native metals, are abundant in meteoritic and asteroidal material but rare or very difficult to extract on the Moon.

Two other features of asteroid resource use deserve mention. First, launch windows to a particular asteroid are typically two or three years apart, compared with lunar launch windows, which are only one month apart. This is, to our minds, more than offset by the fact that any three-year period should contain roughly 300 launch opportunities to kilometer-sized asteroids that are energetically easier to reach than the Moon and richly varied in composition. By comparison, there are 39 lunar launch windows in the same period of time. Counting launch opportunities to 100-meter bodies, the figure is close to 90,000 launch windows in three years (one new opportunity opening every 18 minutes). Second, there is some concern that the typical asteroidal mission is three years long, and therefore cannot compete with one-week round trips to the Moon. We see no problem with such mission durations: planetary spacecraft routinely operate much longer than three years, and the missions we envision are unmanned, so that the issue of life support never arises. With asteroidal missions that are capable of paying back 100 to 1000 tons for every ton invested, one can easily afford to wait three years for a return on the original investment.

Table 9.1
Accessibility of Near-Earth Asteroids

The search for near-Earth asteroids has barely begun, and is still terribly incomplete. These projections of the actual number of asteroids in near-Earth orbits were made by Gene Shoemaker of the U. S. Geological Survey, based on the statistics of asteroid search efforts to date and the cratering histories of the Moon, Mercury, Mars and Earth. The projected number of asteroids greater than 100 meters in diameter that are easier to reach than the Moon can be estimated as about 76,000. The numbers in parentheses are "educated guesses."

				Number Easier to Reach than Moon ($\Delta V \leqslant 6$ km/sec from LEO)		
Asteroid Class	*Number Known (10/84)*	*Projected Number* ⩾ 1 km*	*Projected Number* ⩾ 0.1 km*	*Known*	*Proj. ⩾ 1 km*	*Proj. ⩾ 0.1 km*
Aten (a < 1.000)	5	100±40	30,000	0	(20)	(6,000)
Apollo (q < 1.017)	37	700±300	200,000	7	140	40,000
Amor Earth-crossers (q < 1.3)	22	500±200	150,000	5	100	30,000
TOTAL	64	1300±40%	380,000±40%	12	260±50%	76,000±50%

*See E. M. Shoemaker, *Ann. Rev. Earth Planet. Sci.* (1983) 11:461.

WATER TO DRINK, WATER TO BURN

The first asteroidal resource of interest is water. Water is the central life-support material, both for use as liquid water and as a source of abundant oxygen: 89 percent of the mass of water is oxygen, with hydrogen supplying the remaining 11 percent. Closed ecological systems require a convenient source of both water and oxygen. Food production by hydroponic techniques also places a premium upon having a ready supply of water.

The water in carbonaceous asteroids is loosely bound in clay minerals, and can be released by gentle heating. A small solar furnace that can produce temperatures of 250 to 300°C suffices to extract almost all of the water in a typical carbonaceous meteorite. Liquid water can readily be condensed from the emitted gases. Extracting a ton of oxygen (1.125 tons of water) from 5 tons of carbonaceous asteroid uses about one tenth as much energy as extracting a ton of oxygen from ten tons of ilmenite. But there is another difference: the ilmenite process *requires an external source* of hydrogen, whereas the asteroidal process *provides* 0.125 tons of hydrogen per ton of oxygen extracted.

Manufacturing oxygen from water is quite a simple process. First, solar cell arrays make electricity from sunlight with an efficiency of about 20 percent. Electricity can readily be used to separate the oxygen from the hydrogen in water: this is the process of *electrolysis*. The hydrogen and oxygen gases thus liberated can be used for a variety of purposes. Together they make a superb chemical rocket propellant combination. Hydrogen is also an extremely useful chemical agent for processing a variety of ores.

We should bear in mind the likelihood that the cheapest source of water, hydrogen, and even oxygen for use on the Moon is probably near-Earth carbonaceous asteroids.

FRINGE BENEFITS OF CARBONACEOUS ASTEROIDS

We already have spectral evidence that a significant fraction of the near-Earth asteroids are carbonaceous. The most volatile-rich meteorites are the C2M and C1I classes of carbonaceous chondrites, with 5–10 percent and 10–20 percent water contents respectively. If extinct short-period comets are as common among the near-Earth asteroids as Shoemaker believes, then deep, fluffy regoliths of carbonaceous chondrite dust may be prevalent. Further, under the dust mantle there may lie a frozen and largely unaltered cometary core rich in water ice, other ices containing carbon monoxide, carbon dioxide, methane, ammonia, hydrogen cyanide, and nitrogen, and a wide range of volatile organic compounds.

The C1I chondrites (the I stands for the type example, the Ivuna chondrite), in addition to 20 percent water, contain 6 percent organic matter. This organic matter is chiefly in the form of an involatile polymer that is mostly carbon, but also carries much nitrogen, hydrogen, and oxygen. These meteorites also contain about 40 percent magnetite, a familiar magnetic mineral with the formula Fe_3O_4. This mineral is 27 percent oxygen and 73 percent iron. Using hydrogen from the electrolysis of water, it is easy to reduce magnetite to very pure metallic iron. The only byproduct of this reaction is water, whose usefulness we already know. The C1I chondrites also contain carbonate and sulfate minerals, some of which are soluble in water and may readily be extracted. Carbon dioxide can be released from the carbonates by heating or by treatment with acids. The total of all the readily extractable volatile components in the C1I chondrites ranges from about 28 to 40 percent by weight. Another 28 percent is iron metal (made from magnetite). Most of the remainder is a complex mixture of silicate minerals and rather insoluble sulfates, a combination sufficiently reminiscent of the recipe for concrete to suggest some research into the topic (see table 9.2).

Table 9.2
Volatiles in C-type Material

Hydrogen	CI chondrites contain about 20% water (2% H)
	CM chondrites contain about half as much
Carbon	CI chondrites contain about 6% organic matter
	CM chondrites contain about half as much
Nitrogen	CIs contain about 0.1% nitrogen, mostly in the form of organic nitrogen in heteropolymers
Sulfur	CIs contain about 6% sulfur, mostly sulfates, with some organic S and elemental sulfur
	CMs contain about the same amount of total S, but most of it is found as elemental S, with some sulfate and sulfide, and some organic S
	Chlorine and fluorine are abundant in both classes
Rare gases	CIs contain about 0.1 parts per million of rare gases

Cometary and asteroidal C material is present in the lunar regolith in tiny traces. Most volatiles are about 200 times as abundant in C asteroids as in the lunar regolith.

Cometary dust left behind by the evaporation of ices would closely approximate the C1I chondrites, so far as we can tell from the limited information we have on cometary dust. By remote spectroscopic observations alone, we may not be able to distinguish a loose, fluffy carbonaceous powder from a compacted carbonaceous "rock." But even a carbonaceous meteorite is easy to crush into a powder for processing: a pressure of a few atmospheres will suffice, compared to over 2000 atmospheres (30,000 pounds per square inch; 2000 kg per square cm) required to crush hard rocks.

Nitrogen is an important constituent of the organic gunk in carbonaceous meteorites. The organic polymer must be destroyed in order to liberate the nitrogen; however, oxygen from the electrolysis of water can accomplish this task easily: the polymer can be oxidized to a mixture of carbon monoxide, nitrogen, water vapor and a trace of sulfur dioxide. These gases can then be dried by freezing out water, and then treated with hot hydrogen gas (also derived from electrolysis of water) to make a mixture of methane, nitrogen, and more water. This second crop of water, made by the reaction of hydrogen with carbon monoxide, can again be removed by freezing it out. The nitrogen-methane mixture can then be liquefied and separated into its components by distillation.

Methane is a very promising fuel, since it is much more dense and much less volatile than liquid hydrogen, and hence is far easier to condense, store, and transport. Its performance as a propellant is not as good as hydrogen's, but its other virtues may compensate for that shortcoming in many kinds of use.

Nitrogen is of great interest as a life-support fluid. Curiously, one of its greatest roles is as a fire suppressant. An atmosphere containing adequate oxygen for human respiration will contain from 0.15 to 0.21 atmospheres pressure of oxygen, corresponding to conditions ranging from mountain tops to sea level on Earth. Earth's atmosphere dilutes this oxygen down with 0.78 atmospheres of nitrogen (and 0.01 atmospheres of argon, which is completely inert to life processes). Although nitrogen is an absolutely essential component of life, animals cannot utilize nitrogen gas directly. Indeed, "nitrogen-fixing" plants, so called because of their ability to consume nitrogen gas and synthesize nitrogen-rich proteins, do so not by any virtue of their own, but because their roots are colonized by astronomical numbers of specialized bacteria that do the job for them. Thus the legumes (members of the pea and bean family) produce highly nutritious seeds because of a benign "infection" of their roots! If their seeds are carefully cleaned of bacteria, and if they are planted in a soil devoid of these varieties of bacteria, they will germinate into weak, sickly seedlings that will often fail to grow to maturity. Hydroponic farms in space would benefit greatly from such an infection—it enables the production of large amounts of plant protein for human consumption.

If Earth's atmosphere were kept the same except for a threefold reduction of nitrogen content, nitrogen-fixing bacteria would still be able to do their work, and the food chain would survive intact. However, forest fires would burn with astonishing vigor. Any lightning strike could trigger a fire of such insatiable power that it could sweep a continent free of vegetation. The soot and air pollution from such fires would have a devastating effect on the global climate. The issue of fire retardation is equally important in a space station.

In the early days of the manned space program, Soviet and American spacecraft technology diverged on the issue of what the cabin atmosphere in spacecraft should be like. The Soviet approach conservatively required all manned spacecraft (and a wide variety of unmanned spacecraft as well!) to have internal nitrogen-rich atmospheres with total pressures of one standard atmosphere. This made it easier to design the systems in a spacecraft, since phenomena such as electrical breakdown and arcing in very low-pressure gases could be avoided. The Soviet manned spacecraft essentially carried small samples of the Earth-surface environment with them into space. The American space program was concerned with two much more sophisticated environmental problems. The first of these was the fact that large volumes of inert (and therefore evidently useless) gas weighed a significant amount: more useful payload could be carried if the nitrogen were omitted.

Second, there was concern about the consequences of explosive depressurization of a spacecraft cabin by, for example, a micrometeoroid puncture of the wall of its pressure shell. A person breathing normal air always carries a significant amount of dissolved nitrogen in his blood. A large pressure decrease permits the dissolved gas to come out of solution and form a very fine "fizz" of bubbles in the bloodstream. These bubbles can seriously impede the flow of blood into the brain, and cause excruciating pain in muscle tissue. The victim of rapid decompression may be immobilized for hours, preventing him from dealing with the emergency situation that threatens his life until it is much too late to do so. This affliction was first observed on Earth among deep-sea divers who were brought up to the surface too rapidly: it is commonly called "the bends."

The divers' cure for the bends was initially the adoption of very slow decompression. A diver would be brought up to the surface in many stages, with the pressure never changed very abruptly or by a large amount. This approach is, for deep dives, extremely time-consuming. The more modern approach has been for the diver to breathe a mixture of oxygen and helium, not nitrogen. Helium is much less soluble than nitrogen, and the diver may move upward far more rapidly without injury or pain. These considerations influenced the decision by American spacecraft designers to purge nitrogen from the atmosphere in our manned spacecraft and use essentially pure oxygen. This practice led to a disastrous result: a flash fire occurred in an Apollo command module while undergoing ground tests at Cape Canaveral. Astronauts Grissom, Chaffey, and White were killed by an intensely hot and very fast-moving fire that apparently started in a minor electrical short. Insulation and even wires burned in the pure oxygen atmosphere, leading to an intense pulse of heat, pressure, and toxic smoke and fumes. The astronauts did not live long enough to open the hatch. If this fire had occurred in space during an Apollo mission, there would have been a loss of communications without any warning or telemetered indication of trouble. Diagnosing the cause of the disaster would have been nearly impossible, and the entire Apollo program would have been placed in serious jeapordy.

During the Apollo-Soyuz Test Project one of the pervasive problems was that of an incompatibility between the oxygen-nitrogen atmosphere in *Soyuz* and the pure-oxygen atmosphere in *Apollo 18.* It is reasonable to anticipate further joint manned activities in space, especially Space Shuttle visits to future Soviet Salyut or Mir space stations. At some time in the not too distant future, a standard spacecraft atmosphere should be adopted by both parties. Nitrogen is likely to figure prominently in such an atmosphere. The cheapest source of vast quantities of nitrogen may well be carbonaceous asteroids.

Two other elements essential to life, sulfur and phosphorus, are also abundant in carbonaceous meteorites as sulfate and phosphate minerals. Other biologically essential elements, such as magnesium, iron, calcium, sodium, potassium, etc. are all present in quantity. When life-support technology advances to the point where space bases and colonies are striving to achieve self-sufficiency, virtually every component of carbonaceous meteorites will be useful in a life-support role.

FROZEN ORES FROM COMET CORES

As we discussed in chapter 8, many carbonaceous asteroids may turn out to be former cometary cores with deep, fluffy surface layers of carbonaceous dust covering a pristine icy interior. Even more so than the carbonaceous chondrites, these bodies exude promise. Water ice would be available *without significant processing.* Concentrations of carbon and nitrogen compounds are even higher than in a C1I chondrite. Overall, the composition of a comet core is remarkably similar to the composition of the human body. It is reasonable to expect that extinct comets are common among the near-Earth asteroids, and it is quite inconceivable that they would not be regarded as among the very choicest of resources.

As with the carbonaceous chondrites, the principal foreseeable uses of cometary ices are as propellants and life support fluids. Some chemial by-products, such as metallic iron from magnetite, may also achieve importance. However, our best estimates of the composition of a comet nucleus suggest that no more than about 10 percent of its mass should be iron. It seems more likely that the precious volatiles will be the target of early exploitation efforts.

NAKED TO OUR ENEMIES

There is another very important kind of material that will be much in demand in space: radiation shielding. Life on Earth has evolved under the protective cover of a massive atmosphere. There is about one kilogram of atmosphere above each square centimeter of the surface of the Earth (roughly 15 pounds per square inch). A blizzard of cosmic rays strikes the Earth at all times and from all directions, but the large majority of them strike the nuclei of atoms in the atmosphere and break them up into a shower of much less energetic particles which are unable to reach the ground. The magnetic field of the Earth helps somewhat to shield the surface from very low energy cosmic rays; this shielding works best near the equator, where incoming cosmic rays must cut across many magnetic lines of force, and works hardly at all near the magnetic poles, where the cosmic rays may slide unimpeded down the magnetic lines of force into the upper atmosphere.

On the surface of Mars there is no detectable magnetic field, and the atmosphere (the last line of defense) has a mass that ranges from about 3 percent of Earth's (over the deepest basins on Mars) down to about 1 percent on the upper slopes of the great volcanoes, high above most of the atmosphere. On the Moon there is no magnetic field, and also no protecting atmosphere. Even worse, at high Earth orbits, such as near GEO, there is the possibility that energetic radiation will become trapped in radiation belts embedded in the magnetic field, and may build up to high levels of concentration.

During times of violent flare activity on the Sun, great fluxes of energetic solar cosmic rays are sprayed out toward the planets. An astronaut in high Earth orbit or on the Moon at the time of a major flare could be made ill or even killed by the radiation dose he would receive before he could possibly arrive back to Earth. (Recall James Michener's *Space,* in which a fictional *Apollo 18* lunar mission is struck by a solar flare. The real *Apollo 18* participated in the Apollo-Soyuz mission in LEO.) For men and women stationed in a permanent lunar base, there is a simple and elegant solution: bury the base under a meter or two of lunar dirt. Natural lunar caverns extending much deeper under the surface of the Moon, such as ancient lava tubes, would be a valuable natural resource. For the crew of space stations or space settlements far from the refuges offered by Earth's atmosphere or lunar burrows, cosmic rays offer a continuing mild hazard, and solar flares threaten their lives.

It is easy to imagine a solution to this problem: simply place radiation shielding about the space station in sufficient quantity to absorb the cosmic rays. Consider a 30-ton cylindrical space station module that is three meters in diameter and 15 meters long. If we shield it surface with one kilogram of material per square centimeter, this will require 1500 tons of shielding. The bookkeeping is simple: one shuttle flight to lift the space station module into orbit, followed by 50 shuttle flights to lift the shielding. Allowing 15 Shuttle flights per year, it will take the full capacity of the Shuttle working for over three years to install each radiation-safe module. With present Shuttle costs of about $250 million per launch, the shipping bill alone would run to $12.5 billion.

It is simply not economically feasible to provide radiation shielding for structures in Earth orbit by lifting it from the ground. We have been extremely conservative even in our example above: we have assumed that only one module need be shielded (that is, we have one designated "storm cellar" module for a much larger space station), that there is only one space station, and that the assumed shielding thickness is adequate.

But what kind of shielding is needed to protect the crew from radiation? Do we need lead bricks, armor plate, or exotic materials like boron? No, not at all: all we need is mass. A space factory in LEO engaged in processing extraterrestrial materials would always have at hand a large inventory of unprocessed dirt, separated products, tanks of water, and byproducts from earlier processing awaiting future reworking in other processes. Even if we treat all these materials as just bags and boxes of dirt, they can do the job. A single asteroid return mission could bring back 100 to 1000 tons of asteroidal regolith to LEO! If the dirt were simply brought back to be used as shielding, then the cost of the shielding would be reduced by a factor of ten below the cost of lifting it from the ground, making it a minor element of the total space station program cost. If, however, the material is destined for other, money-making uses, and is simply used as shielding while it is sitting around for a few months being processed, the effective cost of the shielding will be zero: it is just a fringe benefit of having a factory in LEO.

THE NEW IRON AGE OF SPACE

Only one other space resource seems as attractive as the volatiles in carbonaceous meteorites. That resource is native metals. Most classes of meteorites, most asteroids, and all classes of chondrites that are not carbonaceous contain large quantities of native metals. Figure 8.4 summarized the major types of meteorites known on Earth. Of the seven subtypes listed, only the achondrites are essentially free of metal (although many of them contain more free metal than the lunar surface). Fully 87 percent of the known meteorites contain native metal. The meteorite types that fall on Earth are all necessarily derived from Earth-crossing orbits. They are thus an excellent guide to the materials available on Earth-crossing asteroids.

The statistics we have for meteorites recovered on Earth are, however, surely not representative of the bodies in Earth-crossing orbits. First, rather extraordinary conditions are required for a carbonaceous chondrite to survive entry and landing. Surely almost all of the ones that hit the atmosphere are destroyed as fireballs, and do not produce meteorite falls. Second, extremely tough customers like iron meteorites have an excellent chance not only of surviving atmospheric entry, but also of persisting on Earth's surface for times so long that a stony meteorite would either be utterly destroyed or at least rendered unrecognizable by weathering. Thus irons are surely over-represented in our meteorite collections compared to stones.

Some 75 percent of the recovered meteorites are chondrites, of which about 1 percent are carbonaceous (C), with very high oxidation states, no free metals, and abundant volatiles, all indicative of formation at low temperatures. Another 1 percent of the chondrites are extremely highly reduced, have much lower volatile-element abundances (and no detectable water), are rich in metallic iron-nickel alloy, and contain some exotic minerals, such as calcium sulfide and a nitride of silicon, that are completely unknown on Earth. Their dominant mineral is enstatite, a colorless silicate of magnesium, and these meteorites are therefore termed the enstatite (E) chondrites. The remaining 98 percent of the chondrites contain basically the same minerals, but differ significantly in their overall abundance of iron. These are the *ordinary* chondrites, which consist of three groups, the high-iron (H), low-iron (L), and very-low-iron (LL) chondrites (see table 9.3).

The carbonaceous chondrites are, as we discussed earlier, very weak "mudballs" dominated by water-bearing clay minerals, magnetite, and organic gunk. The ordinary and E chondrites, in comparison, are reasonably strong rocks. All contain metal, ranging from a low of about 3 percent in the most extreme LL chondrite up to 31 percent in the most extreme E chondrite; 96 percent of all chondrites contain between 7 and 19 percent metal (the L and H groups only). Spectral data on asteroids can detect and crudely measure the abundance of free metal, so that these various classes can be identified by spectra taken from Earth or from a distant space platform.

The metal in chondrites is found as small metal grains, usually less than a millimeter in size (see table 9.4). The largest metal grains are about the size of the head of a straight pin. The metal found in chondrites is very similar in composition to the metal in the familiar iron meteorites that form so prominent a part of museum meteorite displays. These huge chunks of natural stainless steel are for the most part products of melting of large bodies of chondrite-like composition. The various components are separated according to density once melting occurs, with the least dense silicates and volatiles rising to the top, the very dense metal and sulfide melts sinking, and the "mafic" silicates (named after magnesium and iron, their characteristic metal ions) left sandwiched between these layers. On a chondritic

Table 9.3
Compositions of Chondrites

The eight major classes of chondrites have, when viewed from the economic perspective, rather similar compositions with respect to the rock-forming (involatile and abundant) elements. This table shows, for example, that the total iron content of all the classes of chondrites spans the range from only 18 to 33%. The chemical nature of the iron is vastly different in these diverse classes, however; in the CI chondrites it is present mostly in magnetite and in clay minerals, while in the E4 (EH) chondrites it is found mostly as metallic iron-nickel alloy plus the sulfide mineral troilite, FeS. FeO-bearing minerals are abundant in the ordinary (LL, L, and H) chondrites.

Compositions of Chondrites (wt.%)

	C1(CI)	*C2(CM)*	*C3 (CV/CO)*	*LL*	*L*	*H*	*E4*	*E6*
Mg	9.6±0.2	11.8±1.0	14.3±1.0	15.3±0.8	15.2±1.0	14.2±1.0	10.7±0.7	13.3±1.0
Al	0.85±0.05	1.1±0.1	1.1±0.2	1.1±0.1	1.1±0.1	1.1±0.1	0.79±0.08	0.85±0.08
Si	10.3±0.4	13.1±0.5	15.5±0.5	18.8±0.7	18.7±0.8	17.1±0.9	17.0±0.6	19.5±1.5
S	5.9±0.9	3.4±0.6	2.2±0.5	2.3±0.4	2.1±0.8	2.0±1.0	5.8±0.2	3.3±1.0
Ca	1.1±0.2	1.3±0.2	1.7±0.3	1.2±0.2	1.3±0.1	1.2±0.1	0.84±0.03	0.90±0.08
Fe	18.4±0.6	21.9±1.0	25.2±1.0	20.0±1.5	21.8±1.7	27.6±3.0	33.0±3.0	24.5±3.5
C	3.6±0.4	2.4±0.5	0.7±0.3	0.1	0.1	0.1	0.5	0.4
H_2O	10±3?	5±2?	0.1?	<0.1?	<0.1?	<0.1?	<0.1?	<0.1?

Data from B. Mason, ed., *Handbook of Elemental Abundances in Meteorites* (New York: Gordon and Breech, 1971). The ± values indicate the approximate range of compositional variation.

asteroid that has a regolith, metal grains of high purity should be easily collected with a simple electromagnet "rake."

The metal in "iron" meteorites is indeed an iron-rich alloy, but the other components of the alloy are at least as interesting. Nickel is very abundant, ranging from low concentrations of 6 or 7 percent in the E chondrites, where iron is fully reduced to the metallic state (and hence dilutes down the other metals), to as high as 30 percent in the fairly oxidized LL chondrites, where most of the iron has been oxidized and removed from the metal, leaving the other components in high concentrations. The cobalt abundance is almost always 5 percent of the nickel abundance, or usually 0.5 to 1 percent of the mass of the metal grains. Two other important families of elements are present in much smaller amounts, dissolved in the metal grains. These

Table 9.4
Concentration of Components of Chondrite Metal Phases
Note the high concentrations of nickel and cobalt in chondritic metal grains. The platinum-group metals in the metal phases in chondrites have higher concentrations than in the richest known ore bodies on Earth.

Composites of Chondrites

	Class		
	LL	*L*	*H*
% total metal in meteorite	4±1	9±2	16±3
Ni concentration in metal	25±5%	15±3%	10±2
Co concentration in metal	1.2±0.2%	0.7±0.1%	0.5±0.1%
concentration in metal:			
nonmetals (ppm)			
Ga	1 to 15	6 to 30	?
Ge	200±30	110±30	?
As	1.2±0.2	1.7±0.2	2.1±0.2
Pt-group metals			
Pt	21±5	13±1	11±2
Ru	12±1	8±1	5.7±0.6
Os	10±2	6±1	4.7±0.4
Ir	10±2	5±1	4.8±1.2
Re	1.0±0.2	0.6±0.1	0.5±0.1

are, first, the precious and platinum-group metals gold, silver, platinum, rhenium, osmium, iridium, and ruthenium, and second, such elements as sulfur, phosphorus, carbon, gallium, germanium, arsenic, indium, antimony, etc.

The concentrations of the platinum-group metals are extraordinarily high. We, as natives of Earth, are accustomed to living on the crust of a planet that has been thoroughly melted and differentiated. The formation of the Earth's metallic and sulfide core has efficiently extracted all of the elements that dissolve readily in metal or sulfide melts. Thus the concentration of platinum-group metals is extremely low. Even worse, the distribution of the small amount that remains is very uneven: some 96 percent of the world's commercial production of platinum metals comes from the Republic of South Africa, most of it from a handful of very deep mines (see fig. 9.1). Threats by the South African government to retaliate against

```
                      0%        25%       50%       75%       100%

Rhodium               AAAAAAAAAAAAAAAAAAAAAAAAASSSSSSSSSSSSSSSS

Ruthenium             AAAAAAAAAAAAAAAAAAASSSSSSSSSSSSSSSSSSSSSS

Palladium             AAAAAAAAAAAAASSSSSSSSSSSSSSSSSSSSSSSSSSS

Gold                  AAAAAAAAAAAAAAAAAAAAAAAAASSSSSSSSSSSSSS

Platinum              AAAAAAAAAAAAAAAAAAASSSSSSSSSSSSSSSSSSSS

Diamond               AAAAAAAAAAAAAAAAAAAAAAAAASSSSSSSSSS

Vanadium              AAAAAAAAAAAAAAAAAAAAAAAAAAASSSSSSSS

Chromium              AAAAAAAAAAAAAAAAAAASSSSSSSSSSSSSSS

Germanium             AAAAAAAAAAAAAAAAAAAAAAAAAAASSS

Cobalt                AAAAAAAAAAAAAAAAAAAAAAASSSSSS

Manganese             AAAAAAAAAAAASSSSSSSSSSSSSSS

Beryllium             AAAAASSSSSSSSSSSSSSSSSSSSS

Uranium               AAAAAAAAAAAAAAAAAASSSSSSS

Antimony              AAAAAAAAAAAAAASSSSSSSS

Copper                AAAAAAAAAAAASSSSSS

Mercury               AAASSSSSSSSSSS

Iron                  AASSSSSSSSSSS

Lead                  ASSSSSSSSSSS                  A = Africa

Nickel                AAAASSSSSSSS                  S = Soviet Bloc

Tantalum              AAAAAASSSSSS                  0 = OPEC

Tungsten              AASSSSSSSSS


OPEC + Soviet Oil     00000000000000000000000SSSSSSSSS
```

Figure 9.1
Soviet-Bloc and African Production of Strategic Materials

Table 9.5
Strategic Materials Vulnerability Index
(U.S. Army War College)

Material	*Index*	*Major Suppliers*
Chromium	34	Soviet Union, South Africa
Platinum metals	32	Soviet Union, South Africa
Tungsten	27	Canada, Peru
Manganese	23	Brazil, Gabon
Aluminum	22	Jamaica, Canada
Titanium	20	Australia, Canada
Cobalt	20	Zaire, Canada
Tantalum	16	Zaire, Brazil, Canada
Nickel	14	Canada, Norway

American pressure on the apartheid issue have taken the form of interruptions of our supply of strategic metals (see table 9.5). Chromium, manganese, cobalt, and the platinum metals are at the top of the list of our strategic import needs. Of these, the threatened South African interruption of American supplies affects the platinum metals and chromium most severely.

In view of President Botha's threats of October 21, 1985, it is at least interesting to know that every major class of meteorite material except the achondrites is a richer source of platinum metals than the richest known ore body on Earth. Further, a magnetic rake sample from an LL asteroid would contain more than 50 parts per million (ppm) of platinum metals, more than ten times as rich as the best known terrestrial ore.

Any metal-bearing asteroid with a regolith should readily yield metal grains to the magnetic rake. It should be possible to load metallic "dirt" with a purity of at least 95 percent, and possibly over 99 percent metal. A simple, brute-force approach, but one wasteful of energy, would be to melt down the asteroidal metal and pour castings of useful parts. The strength of these castings would be limited by the large amount of impurities, especially nonmetals, in the feedstock. Ironically, these impurities contain a wide variety of critical and strategic metals.

Fortunately, it is possible to define a process that is very well understood, has a century of industrial experience, can separate all of the element groups in the metal and hence provide very pure metals for specialized uses and for high-strength and corrosion-resistant applications, and which uses much less energy than casting. This is the gaseous carbonyl process (introduced in chapter 7), long used by International Nickel to extract and purify most of the free world's supply of nickel (see table 9.6). As an added bonus, it is a process that is optimal for processing metallic feedstocks. So efficient is the process that huge tonnages of nonmetallic ores were reduced to the metallic state in order to be fed into the carbonyl process. The most familiar example to Americans is INCO's giant Sudbury nickel plant in Ontario.

The principle of the carbonyl process is simple but, at first sight, a little hard to believe. When metallic iron-nickel alloy is treated with a few atmospheres pressure of carbon monoxide gas at a temperature near the normal boiling point of water, the metals react to form gasous carbonyl compounds. Nickel forms a gaseous compound with four carbon monoxide (CO) molecules, called nickel tetracarbonyl, $Ni(CO)_4$ and iron reacts with five CO molecules to make iron pentacarbonyl, $Fe(CO)_5$. These compounds are about as volatile as water. They condense at room temperature to make clear liquids. Both compounds can be decomposed by heating them to higher temperatures, or by keeping them near 100°C, and lowering the pressure. They decompose by depositing a mirror-bright film of pure metal, releasing carbon monoxide gas.

When a meteoritic or asteroidal metal alloy is treated by this process, all the iron and nickel volatilize and depart with the carbon monoxide gas stream. Left behind are cobalt and the platinum-group and other solutes, in the form of a fine magnetic dust. The concentration of the platinum-group metals in this cobalt dust is about 0.5%, and the concentration of the gallium, germanium, and arsenic group will be around 1.5 to 2.0 percent. This residue can in turn be extracted with moist carbon monoxide at a pressure near 100 atmospheres. Under these conditions, cobalt also reacts to form a gaseous carbonyl. Cobalt extraction leaves behind a platinum-group metal residue that is worth roughly $20,000 per kilogram (nearly $10,000 per pound) on the terrestrial metals market.

Imagine a near-Earth L chondrite asteroid with a diameter of one kilometer. It would have a mass of about two billion tons. Of this, 200 million tons would be metal. Extraction of the metal (one ton for each person in the United States; call it a new car for each family) will free 30 million tons of high-purity nickel, 1.5 million tons of the strategic metal cobalt, and about 7500 tons of the even more strategically critical platinum-group metals.

Table 9.6
Carbonyl Processing

CARBONYLS
$Ni(CO)_4$
$Fe(CO)_5$
$Os(CO)_5$; $Os_2(CO)_9$
$Ir_2(CO)_8$; $Ir_4(CO)_{12}$
$Ru(CO)_5$; $Ru_2(CO)_9$; $Ru_3(CO)_{12}$
$Rh_2(CO)_8$; $[Rh(CO)_3]_n$; $Rh_4(CO)_{11}$
$Co_2(CO)_8$; $Co_4(CO)_{12}$

CARBONYL HALIDES
$Pt(CO)_2Cl_2$, etc.
$Pd(CO)Cl_2$, etc.

BASIC FEATURES OF THE GASEOUS CARBONYL (MOND) PROCESS

Iron and nickel form gaseous carbonyls readily:
$Fe + 5CO = Fe(CO)_5$ (gas)
$Ni + 4CO = Ni(CO)_4$ (gas)

Optimum formation conditions are:
CO pressure of about 1–10 atmospheres
temperature of about 110°C
trace of sulfur or sulfur gases present

Iron or nickel carbonyl decomposes to 99.97% pure metal plus CO gas near 0.1 to 1 atm pressure and 200°C

90 years of industrial experience
International Nickel plant at Sudbury, Ontario

Many metals form gaseous carbonyls
Fe, Ni, Co, Os, Ir, Ru, Rh, W, Cr, etc.

Chemical Vapor Deposition of many metals from carbonyls
W (tungsten) coatings from tungsten carbonyl
Cr (chromium) coatings from Cr carbonyl
Ni (nickel) "castings" commercially made from nickel carbonyl- very strong, complex, precise

Laser Chemical Vapor Deposition
Strong, thin metal films of controllable thickness

Assuming that the iron and nickel, and probably the cobalt, could be profitably used in space, we still would have a byproduct worth exporting to Earth: 7500 tons (about 300 shuttle loads) of platinum metals, worth $150 billion at present market prices. Thus the platinum-group byproduct would by itself more than pay for all the shuttle flights that could be flown over a 25-year period! This is also enough platinum to meet Earth's total use rate for several decades. Clearly, the present high price of these metals, and their vulnerability to manipulation by ideologically hostile or resentful foreign governments, need not be permanent fixtures of the planetary economy.

Let us suppose that the carbonyl process was in use to process asteroidal native metal ores. What would the immediate products be, and how would they be made into useful (marketable) commodities? To answer this question, we turn first to a relatively new industrial process called chemical vapor deposition (CVD). In CVD, a compound, in this case nickel carbonyl (or iron carbonyl), is passed through a mold at a temperature of about 150°C and about one atmosphere pressure. The nickel carbonyl decomposes, plating the inner surface of the mold with a very pure, strong layer of nickel. Hollow nickel objects may be made in this way, or the gas flow may be continued until the mold is filled. Pure nickel deposited in this way can be annealed to a strength of 100,000 pounds per square inch (6600 kg/cm^2). If the carbonyl gas stream is doped with about 100 parts per million of a boron compound, strengths of 200,000 psi (13,000 kg/cm^2) are attainable. Deposition rates are fast enough so that most "castings" could be made in one or two days. This technique was pioneered by Vaporform Products of New Kensington, Pennsylvania. The technique can be adapted to use mixtures of iron and nickel carbonyls, but no serious research on this possibility has yet been done (see fig. 9.2).

One of the highly desirable traits of this process is its ability to produce finished products in one step. If the mold is of optical quality, then the finished product, fresh from the mold, will be just as good. No cutting, machining, milling, lathing, grinding, or polishing is needed. This makes the process exceptionally simple and clean, and very suitable for use in any location where labor-intensive, dirty processes are highly undesirable—such as on a space station.

Another exciting prospect for direct use of carbonyls of iron and nickel is the laser CVD process, in which iron (or nickel, chromium, or tungsten) carbonyl at low pressure is placed in a cold chamber. A low-power laser, operating at a wavelength at which the carbonyl gas is highly transparent,

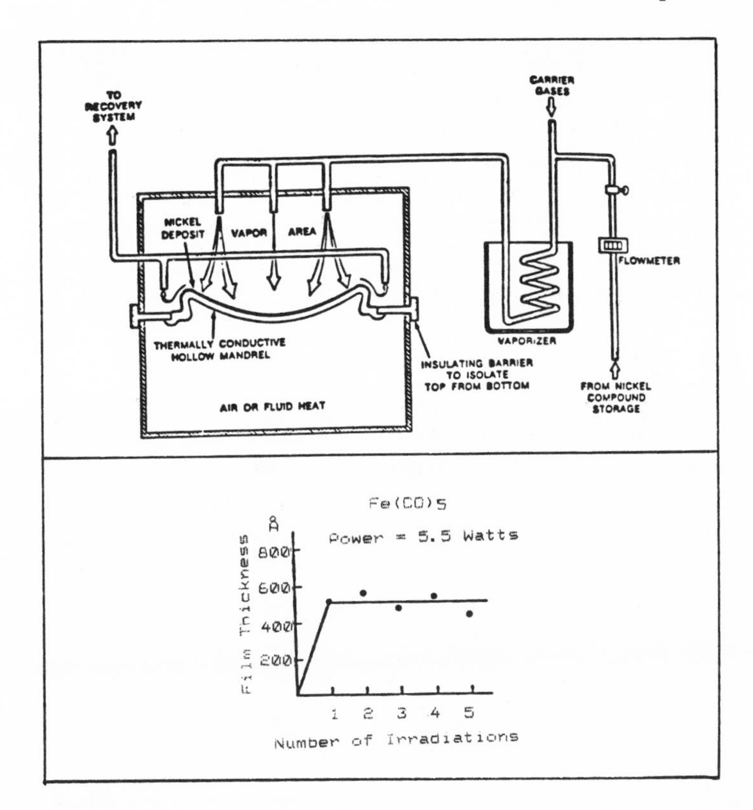

Figure 9.2
Vaporforming Nickel
Asteroidal or lunar nickel can be "cast" at temperatures of only about 200–250° C into extremely detailed, precise shapes by chemical vapor deposition of nickel metal from gaseous nickel carbonyls. Nickel chemical vapor deposition is practiced commercially by Vaporform Products of New Kensington, Pennsylvania.

Laser chemical vapor deposition has been shown by Susan D. Allen of USC to be capable of depositing very thin, strong films of metals such as iron or tungsten on a surface upon which a laser beam "writes" a moving hot spot. The film thickness can be controlled by regulating the power density in the laser-illuminated spot.

shines through the gas and "writes" on a target surface. The laser heats a very thin line on the target, and the gaseous carbonyl decomposes upon contact with the hot surface to deposite a thin, tough film of metal. In variants of this process, integrated circuit elements can be "written" on a chip by the laser. Experiments done by Susan Allen at USC have shown that it is possible to control the film thickness quite precisely by regulating the power density in the laser-heated spot. This may prove to be a superb technique for fabricating ultra-lightweight mirrors and solar sails in space.

Another interesting sidelight to the carbonyl process is afforded by the fact that the platinum-group elements (and many others as well) form volatile compounds in combination with carbon monoxide and halogens such as chlorine (see table 9.1). This makes accessible a wide range of simple, exotic techniques for fabricating unusual products. A good example is the solar thermal propulsion system discussed in chapter 5. That system depends upon a truly remarkable piece of hardware, a highly refractory thrust chamber made of rhenium metal. So refractory is rhenium that it is quite impossible to melt and cast it in the usual manner. What mold or crucible material could hold it? Instead, the thrust chamber is made at quite ordinary temperatures by the vapor deposition of rhenium from a gaseous compound. Several rhenium compounds, such as rhenium carbonyl chloride, can be used in this way.

What, then, are the likely products from an "iron age" metals plant in space? They must, first of all, be products that are made of readily available space materials that can readily be fabricated in space. Further, it is not worthwhile to develop processing facilities in space unless the products are made in considerable mass, so that their production significantly reduces the mass that must be lifted from the ground. Among the most likely metal products are beams, plates, fixtures, wires, cables, filament-wound containers, and thin films.

A deep-space mission departing from LEO may well ride on a solar thermal rocket stage with structure made of asteroidal metals, insulated with silicate glass fiber and thin-film reflective metal layers, powered by acres of LCVD-deposited thin metal film mirrors, engines made of asteroidal refractory metals, and pumps brought up from the ground, fueled by oxygen or hydrogen made from asteroidal water in LEO. Permanent space structures, such as solar power satellites, would be made almost entirely of asteroidal materials processed in space by astronauts brought up from Earth.

Only the high-tech components, those that are most complex and labor-intensive, would be made on Earth. After a decent interval of space industrialization, even the astronauts could be made in space.

BEYOND ICE AND IRON

We have so far emphasized the two most important early asteroidal resources, water and native metals. Although these two materials provide a very large proportion of the materials needed in space with relatively simple technology, there is a great amount more to a modern industrial civilization than can be fashioned from them alone. Are there any other space materials that might be of comparable value?

For structures, very high strength-to-weight ratios are desirable. Unfortunately, the technology for producing known supermaterials is complex, fussy, and difficult to automate or export to space. An important goal for near-term research is to identify likely materials and processes for use in space. For example, recent work strongly suggests that gallium arsenide devices may be at the heart of the next generation of high-speed computers. Integrated GaAs circuits run faster and use less power than silicon-based chips. Further, it appears that the best and cheapest way to make the extremely pure, clean GaAs crystals needed for these chips is to grow these crystals in micro-gravity; i.e., on the space station. If it should, for example, prove possible to develop an efficient new process for making superstrength silicon carbide fibers in space, the structural use of SiC composite cables would become very attractive. Further, a large number of very interesting activities in space become possible only if tethers and cables of such extraordinarily high strength become available. The payoff for learning how to make only one or two new supermaterials may well pay for the entire research program many times over. Also, such space-fabricated supermaterials, like the platinum-group byproducts mentioned earlier, could be bartered in exchange for high-tech items made on Earth.

In general, we should constantly be on the lookout for space products that might profitably be exported to Earth. Gallium arsenide crystals, platinum-group metals, zero-gravity *objets d'art,* and purified pharmaceuticals all come to mind, but this list is certain to be incomplete. What about the other strategic metals not already considered, such as cobalt, chromium, and manganese? There is only a small chance that space sources will ever be able to compete with the prices of these metals from Earthside mines (unless, of course, a major producer embargoes its products). However, in our politically unstable world there is always the possibility that we will be forced to find other sources almost irrespective of cost. Of these elements, cobalt is the most obvious choice for import because it will already be available in large quantities as a result of iron processing.

Both chromium and manganese, however, are generally *not* found concentrated in distinct and easily separable minerals. In ordinary chondrites, achondrites, and the lunar crust both elements are widely dispersed. The only classes of meteorite that look promising as a source of these elements are the enstatite meteorites, in which manganese is found as a sulfide mineral, alabandite (MnS), and certain rare iron meteorites, in which chromium forms a very unusual nitride, called carlsbergite (CrN). Zirconium, vanadium, and hafnium are three other potential targets. Also, one rare class of meteorites, the ureilite achondrites, contains about 1 percent by weight of black, industrial-quality diamonds. If the name South Africa suggests chromium and platinum, it surely even more forcibly conjures up diamonds.

We are aware of only one other space resource that might be suitable for economically profitable export to Earth. That is the lighter isotope of helium, ^{3}He (read as "helium-three"). This isotope is extremely rare on Earth. Very little of Earth's originally meager supply is left, and it escapes from Earth's upper atmosphere just as fast as it trickles out of the interior in volcanic gases. But this isotope is a *superb* fuel for fusion reactors. The Moon of course long ago lost its even more meager supply of primordial gases, but airless bodies are directly exposed to the solar wind: the lunar regolith contains a small trace (0.1 grams per ton at most) of ^{3}He that has been implanted in surface mineral grains by the solar wind. Asteroids, too, have been irradiated by the solar wind and contain the same implanted gases, as we know from our studies of meteorites. The economics of helium extraction have yet to be explored well enough to compare the attractiveness of the lunar and asteroidal sources, but we can surely expect this to be done in the near future.

HOW TO GET THERE FROM HERE

The first prerequisite for the exploitation of asteroidal resources is the discovery, orbit determination, and compositional identification of a large number of near-Earth asteroids. This can be carried out by the selective expenditures of a few million dollars over the next decade, coupled with a dedicated infrared satellite similar in sensitivity to IRAS, but programmed to emphasize asteroids. It will be necessary to carry out the Earth-based optical search in coordination with observers who can measure the reflection spectrum of the newly discovered asteroid while it is still close to Earth and readily visible. This probably requires the existence of at least one small dedicated observatory with instrumentation of the sort that has been used so successfully in recent years to determine the compositions of asteroids.

Some direct, *in situ* studies of a few carefully selected asteroids would also be extremely valuable. A spacecraft with virtually all of the capabilities needed is under development for the *Mars Observer* mission in 1990. This spacecraft is planned as the first of a series of closely similar missions, with such high commonality of parts that the cost per flight can be dramatically reduced relative to recent mission costs. This is basically the same spacecraft that is planned for use as a Lunar Geosciences Observer.

Simple flybys of several near-Earth asteroids would be valuable as a means of testing the validity of our inferences from Earth-based multicolor photometry. However, within a few years we should have orbital data on so many near-earth asteroids that it will be easy to design spacecraft missions that can rendezvous with—and orbit around—several different asteroids during the spacecraft's useful lifetime. These missions could provide us with very detailed information on the chemical composition and physical state of the surfaces of these selected bodies.

These and a variety of other asteroidal, cometary, and deep-space missions would profit immensely from use of a low-thrust propulsion system with high specific impulse, such as electrical (ion) propulsion. We already have a well-demonstrated ability to operate spacecraft for many years. With the potential of reaching and orbiting six to ten different asteroids within the lifetime of each multiple-asteroid rendezvous spacecraft, there is a clear promise that a few such missions could, within a decade, characterize dozens of near-Earth asteroids well enough to attempt their exploitation.

The asteroids chosen for such close looks would be those whose compositions appear promising from Earth-based studies, and whose orbits are highly accessible from Earth. This means that sample return from a small number (a few thousand) such asteroids might be extremely easy. Each multiple-asteroid rendezvous spacecraft might be equipped with one or two capsules capable of returning kilogram-sized regolith samples from apparently promising asteroids.

Simple approaches are worth exploring, since return is so easy: the spacecraft needed for return may be very small and simply constructed. The heat shield needed to return 10 kilograms (22 pounds) from a near-Earth asteroid to the surface of the Earth would weigh between 1 and 2 kilograms. The propulsion system needed to fire this 11-12 kg package on a trajectory taking it back to Earth would weigh only 0.2 to 1 kg (for any of about 30,000 of the closest asteroids). The velocity change needed for return to Earth might most readily be provided by the spacecraft propulsion system: after picking up a dirt sample in the same way that the automated Soviet Luna sample-return vehicles (*Lunas 16, 20,* and *24*) did, the spacecraft could simply carry the sample capsules along until, in the course of its maneuvering between asteroids, the spacecraft aim point should happen to pass near Earth. The spacecraft could then be tweaked into a collision path with Earth, the capsules released, and the spacecraft then diverted to its next destination, leaving the sample return capsules to "fall home." Note that every one of the target asteroids would be chosen on the grounds that its path already takes it near Earth. Thus this return technique should be convenient to use.

Also, if the return propulsion is provided by a highly efficient ion engine with a very high exhaust velocity (a specific impulse near 4000 seconds) instead of by a small space-storable chemical rocket (with a specific impulse of 300), the fuel requirements could be reduced by at least a factor of four. There is a serious possibility that return of near-Earth asteroidal samples will prove so easy that many more returns could be carried out for negligible additional cost. Our experience with the Moon has been that, even on a body that has been visited by six manned expeditions and orbited by two dozen satellites, the little 0.1-kilogram samples returned from the *Luna 16, 20,* and *24* landing sites have still had a major scientific impact on our understanding of the Moon.

It is reassuring to recall that, after all, spacecraft have been hitting, orbiting, photographing, mapping, landing upon, and sampling the Moon since 1959. *Tens of thousands of the near-Earth asteroids are much easier to reach than the lunar surface*. The automated return of samples from the Moon was carried out by the Soviet Union as early as 1970. Return from an asteroid is a much less demanding propulsion problem, but requires better navigation. However, with current American computer and tracking expertise, the navigation problem is insignificant.

If 1982 DB had been discovered in 1962 instead of 1982, we probably would have many pounds of it in our laboratories and museums today. Unfortunately, it was discovered after the cessation of both the Soviet and American lunar programs in the mid 1970s. It has therefore remained wholly undisturbed, protected from intrusions by the erosion of our willingness to exercise our abilities in space.

10

HOMESTEADING MARS
The Dream and the Difficulties

If we are interested in Mars at all, it is only because we wonder over our past and worry terribly about our possible future.—Ray Bradbury

BASIC FACTS

The greatest advantage of Mars as a future home is its similarity to Earth. The greatest disadvantage of Mars as a source of material resources for development in space is its similarity to Earth. Gravity is at the heart of both judgments.

Mars has a surface gravity about 40 percent of Earth's, and an escape velocity (about 5 kilometers per second) slightly less than half of Earth's. In order of difficulty getting off the surface and into space, Mars ranks third in the inner Solar System, less challenging than Earth and Venus, but harder than Mercury, the Moon, and the asteroids.

Mars is about half the diameter of the Earth, and has about one tenth its mass. Earth is nearly 75 percent covered by water, but oceans are absent on Mars. The area of exposed land on Mars is thus about equal to the total area of all the continents of Earth.

The planet rotates every 24 hours and 35 minutes about an axis that is inclined 25 degrees to the pole of its orbit. Both of these features are uncannily similar to Earth.

Mars today has no liquid water on its surface; no oceans, lakes, or rivers. It has a thin, cold atmosphere made mostly of carbon dioxide, with about 2.5 percent nitrogen and small but highly variable amounts of water vapor. There also are tiny traces of the rare gases and about 0.01 percent carbon monoxide and oxygen, made by the destruction of carbon dioxide molecules by sunlight. That there is any atmosphere at all is directly attributable to gravity: a planet only slightly smaller than Mars (such as Mercury) would not be able to prevent its atmosphere from escaping. The only smaller bodies that do have atmospheres are much farther from the Sun and hence much colder than Mars—so cold that their molecules are traveling much slower than the escape velocity.

OUT OF THIN AIR

The average atmospheric pressure on the surface of Mars is only 0.006 atmospheres, lower than the vapor pressure of liquid water at its freezing temperature; if liquid water were exposed to conditions on the surface of Mars it would rapidly boil away. Even ice water would boil!

The temperature near the equator in early afternoon in midsummer can reach a comfortable level, roughly normal room temperature. Every night, however, the temperature drops at a fantastic rate. The very thin atmosphere has little ability to hold heat, and is so transparent that it scarcely hinders the surface from radiating off its warmth into black space. Nighttime temperatures, even in midsummer, plummet to under −80° C (−112° F) just before dawn (see fig. 10.1). Every night the temperature falls to near the lowest temperature ever recorded on Earth, as cold as the deepest freeze of Siberia or Antarctica.

If you were to stick your bare hand out into the bracing night air of Mars, you would *not* instantly turn into a pillar of ice. Because the air is so cold and thin, and because the molecules in it are therefore traveling so slowly and striking your skin so infrequently, the Martian air cannot remove heat from your hand very rapidly. Your warm blood is 100,000 times as dense as the atmosphere of Mars: one cubic centimeter (about 20 drops) of blood cooling from your body temperature to near the freezing point releases enough heat to warm 10,000 liters (350 cubic feet) of Martian atmosphere up to room temperature. Still, a plant immersed in such cold air for several hours would be thoroughly frozen (plants are not warm-blooded).

Another consequence of the thinness of the atmosphere is that killing ultraviolet sunlight penetrates all the way to the ground. On Earth, ultraviolet light with wavelengths less than 0.3 micrometers (3000 Angstroms) is absorbed by atmospheric gases, mostly ozone. Ozone in Earth's atmosphere is made by the effects of ultraviolet sunlight on the abundant

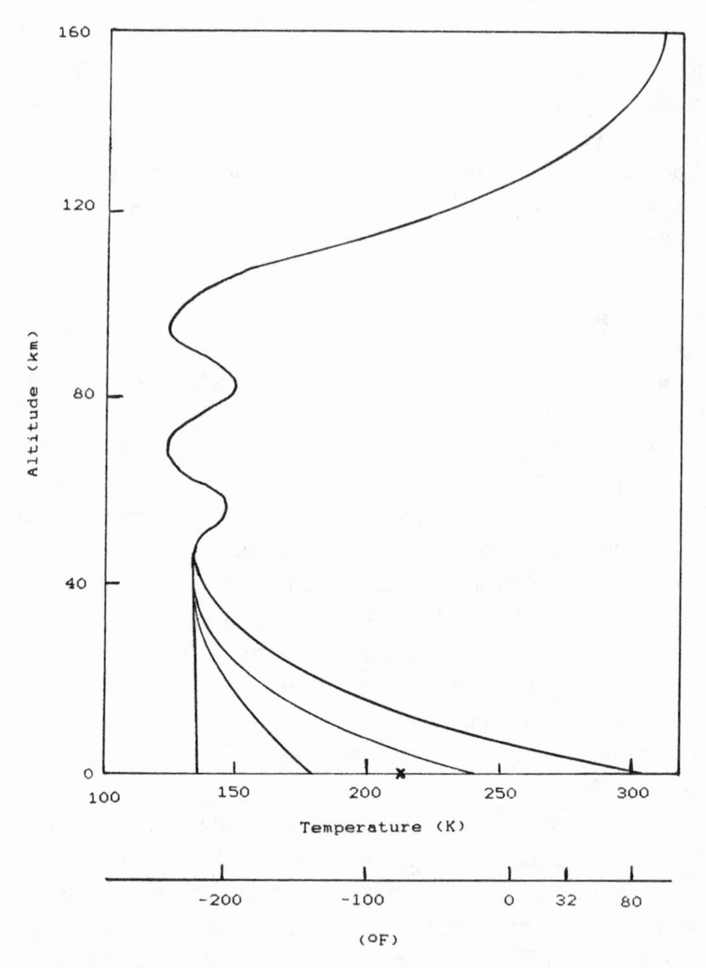

Figure 10.1
Temperatures in the Atmosphere of Mars
Temperatures at the surface of Mars range from a high near normal room temperature at low latitudes, shortly after noon in mid-summer (S) to a low cold enough to condense carbon dioxide as dry-ice snow near the winter pole (P). At middle latitudes the daily range of temperatures is quite large, from about −20°C in the afternoon down to −140°C shortly before dawn. In the stratosphere the temperatures are generally very low, often with a distinct wave-like structure. Temperatures are shown here in absolute (Kelvin) degrees, which count from absolute zero. Zero°C is 273.15 K.

gas oxygen: oxygen molecules, O_2, consisting of two oxygen atoms, are torn apart into separate atoms by the ultraviolet light. Atoms of oxygen are very reactive, and can even react with (oxidize) oxygen molecules to make new molecules containing three oxygen atoms, O_3, which we call ozone. Earth's atmosphere is so dense that the solar radiation capable of making ozone cannot penetrate to the surface. The ozone is made in a layer high above the tops of the tallest of Earth's mountains.

But the atmosphere of Mars is so thin that the solar ultraviolet radiation can penetrate all the way to the surface. At the surface of Mars, this energetic solar UV light breaks apart carbon dioxide and water molecules. The main fragments from these photolysis reactions are atoms of H and O, hydroxyl radicals (OH), and carbon monoxide (CO). After reaction of these gases with each other, the final products from carbon dioxide are carbon monoxide and atomic oxygen, while destruction of water makes hydrogen, oxygen, and hydrogen peroxide (H_2O_2). Attack of oxygen molecules by atomic oxygen also makes a small amount of ozone. Thus solar UV light shining on the surface of Mars provides a continuous supply of atomic oxygen, hydrogen peroxide (H_2O_2) and ozone. All of these are used commercially on Earth to destroy living cells and organic matter.

The Martian atmosphere is not only too thin to shield its surface against UV radiation from the Sun, but it also admits a large proportion of the cosmic rays striking the planet. To provide protection equivalent to that afforded by Earth's atmosphere, about one kilogram of shielding is needed per square centimeter (about fifteen pounds per square inch, or one ton per square foot). Without such shielding, radiation damage to living tissue becomes a serious problem, especially at times of severe solar flare activity.

From time to time it snows lightly. Winter at high latitudes brings a snow blanket, only centimeters thick, that persists for months. At dawn, hoar frost or light snow are commonly present, only to evaporate quickly when the rays of the rising Sun strike the ground (see fig. 10.2). In the winter, near the poles, solid carbon dioxide (dry ice) may fall as snow. At all latitudes above about 30 degrees the subsurface temperatures are low enough so that there should be permafrost present year round (see fig. 10.3).

Seasonal variations are especially striking on Mars, since two separate "seasonal" effects are at work. The first of these, as on Earth, is the cycle of seasons produced by the axial tilt of the planet. When one polar region is

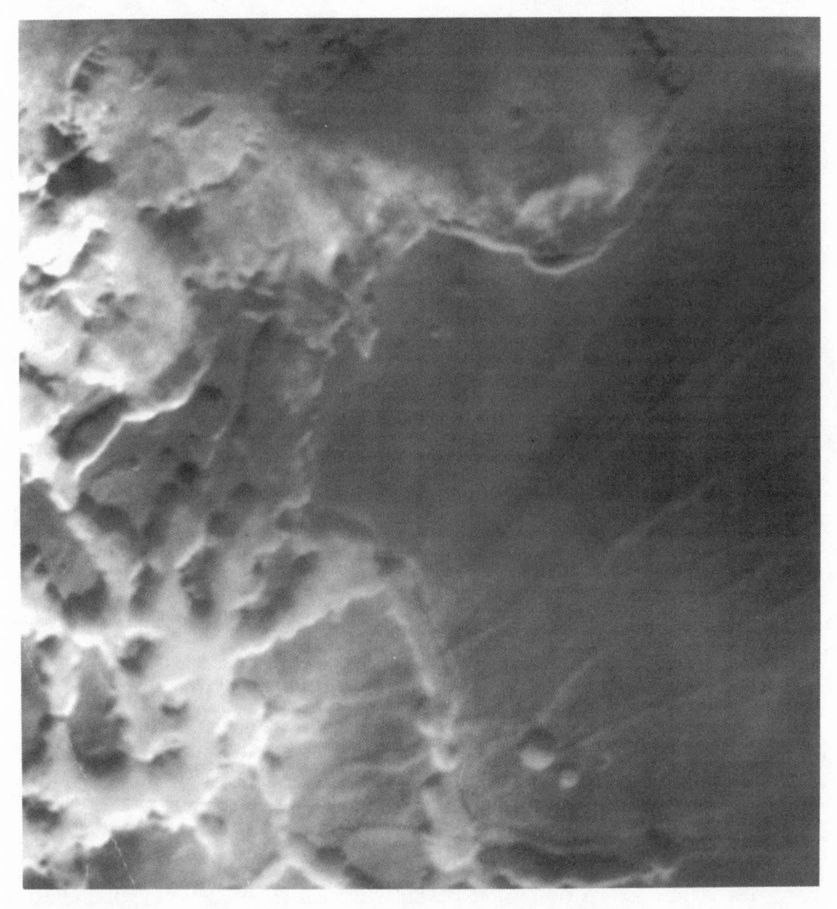

Figure 10.2
Morning Frost and Haze in Martian Valleys
This *Viking 1 Orbiter* picture of the Noctis Labyrinthus canyon system shows dense clouds of water ice particles trapped within the canyons shortly after local dawn.

tilted toward the Sun it experiences permanent daylight for hundreds of days, causing a local summer. The opposite pole is at the same time pointed away from the Sun, thus causing equally prolonged periods of perfect darkness. This corresponds with winter in that hemisphere.

The second annual cycle contributing to the seasons is due to the eccen-

Figure 10.3
The North Polar Cap of Mars
This *Viking 2 Orbiter* picture shows the residual ice cap surrounding the Martian North Pole near the time of midsummer. The dark serpentine bands are ice-free valleys too warm for persistence of ice. The higher ridges remain covered with a layer of water ice. In winter the water ice cap is far more extensive, and solid carbon dioxide snow (dry ice crystals) also falls onto the ice cap.

tricity of Mars' orbit about the Sun. With a semimajor axis (mean distance from the Sun) of 1.5237 AU and a present eccentricity of 0.0934, the actual distance of Mars from the Sun can range from 1.393 to 1.681 AU. The intensity of sunlight on Mars when it is at perihelion is 1.456 times as large as at aphelion, a truly impressive difference. (Earth's orbital eccentricity is only 0.0167.) Thus the average temperature of the whole planet is much lower near aphelion than it is near perihelion.

As on Earth, the direction of the polar axis of Mars precesses in response to tidal forces exerted by the Sun: the size of the axial tilt stays about the same, but the poles of rotation circle slowly about the pole of Mars' orbit. Earth's axial precession takes about 26,000 years, and Mars' takes about twice as long.

When the spin axis is oriented so that aphelion occurs during midwinter in one hemisphere, then that pole will experience extraordinarily severe winters while the other pole is experiencing a very cool summer. Half a

Martian year later, the pole that was so severely frozen is pointing toward the Sun during perihelion passage, leading to an extremely warm summer. At the same time, the other pole has cycled from a cool summer to a very mild winter. Thus, for thousands of years at a time, one hemisphere may experience enormously exaggerated seasonal variations, while the other hemisphere has very muted ones. The changing orientation of the spin axis relative to the locations of perihelion and aphelion of the orbit causes severe global climate fluctuations.

When solar heating on Mars is near its peak (near perihelion), wind speeds become high enough to lift dust from the ground. Solar heating of the dark red dust clouds causes them to accelerate upward, which drives even faster surface winds, which in turn lift more dust (see fig. 10.4). The result, in the most extreme cases, is a severe global dust storm near the time of perihelion passage. Wind speeds can reach one third of the speed of sound!

There are also longer-term changes that affect the climate. The eccentricity of the orbit is pumped up and down, from nearly zero to over 0.2, by gravitational interactions with the other planets working over hundreds of thousands of years. The laminated polar terrain apparently preserves a record of cycles of sediment deposition driven by these long-term orbital variations.

These climate cycles surely have profound effects on the polar ice caps and on the global distribution of permafrost. We are still unable to make a good estimate of the total amount of ice and other volatiles accommodated in the crust and regolith, and hence we cannot tell whether a modest global warming of Mars would permit liquid water to flow on its surface (fig. 10.5). Recent estimates of the present amount of water locked up in the regolith and ice caps of Mars suggest that, if it were all released onto the surface, it might form a layer of liquid water covering the planet to an average depth of 10 to 100 meters, with over 1000 meters present in the past.

Even though pure liquid water cannot exist at the surface, very concentrated salt solutions (brines) could be present under ideal conditions. Indeed, Robert L. Huguenin of the University of Massachusetts has assembled evidence that a region on the western rim of the huge Hellas impact basin may have a number of brine "vents" very close to the surface, releasing water vapor into the atmosphere. These brines, because of their high electrical conductivity, would be almost as good at reflecting radar waves as metals. Interestingly, the radar reflectivity of the Martian surface

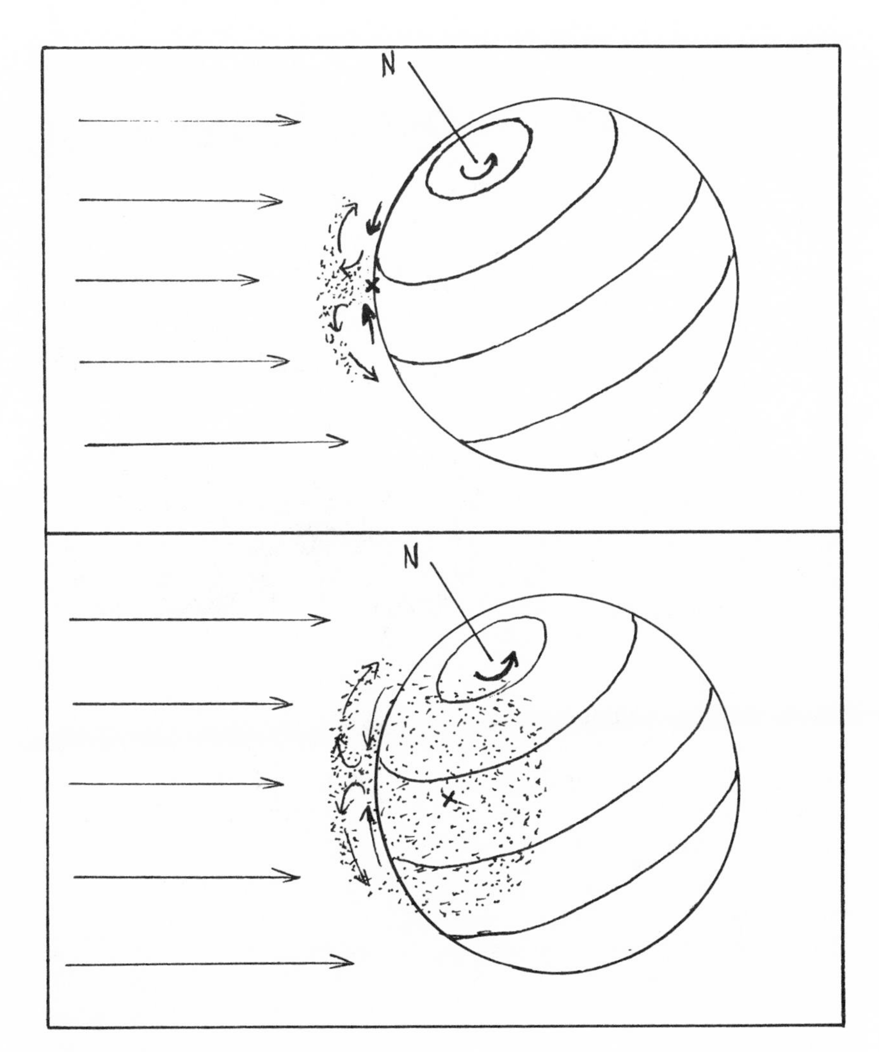

Figure 10.4
The Birth of a Dust Storm
When Mars is close to perihelion its tropical and temperate regions experience the greatest possible solar heating and the highest possible ground temperatures. The strong heating drives hot air columns, which rise rapidly. The flow of cooler air along the ground to replace the rising warm air travels with unusual speed, and lifts vast quantities of dust. The dust becomes entrained in the rising columns of air and spreads widely over the planet. The rotation of the planet (bottom frame) assists in spreading the dust around the planet, and the dust itself makes the atmosphere more opaque to heat radiation, preventing the surface from cooling off dramatically at night. A global dust storm may then develop.

Figure 10.5
Ancient Stream Beds on Mars
This stream bed, several billion years old, was photographed by *Mariner 9.* Vast areas of Mars contain features carved out by brief, catastrophic floods of water. Much of the water that carved these channels is probably still present, frozen into permafrost and tied up in clay minerals.

near Huguenin's "oases" is dramatically higher than elsewhere on the planet. A Mars expedition might be richly rewarded for the effort spent in visiting and exploring these oases.

THE SANDS OF MARS

The surface of Mars is covered with a fine red dust. Even the naked eye shows Mars to be red; indeed, this is the origin of the sanguinary reputation and martial name of the planet. This red dust is essentially rust. It is a mixture of iron oxide compounds, similar to rouge, made by the oxidation of iron-bearing minerals in the surface rocks. It is the dried blood of a lifeless planet.

The surface minerals on Mars are mostly similar to those found in abundant rock types on Earth and in sedimentary deposits. Areas with the spectral signature of basalts are common. The soil at the Viking landing sites appears to contain feldspars, clays, iron oxides, sulfates, and other minerals of the sort familiar in terrestrial deserts.

The geology of Mars has many features familiar to residents of Earth, but is utterly lacking in many other traits that we take for granted. The absence of open water is of course a fundamental difference. But even more fundamental is the absence of continental drift (see fig. 10.6). Earth is covered by about two dozen plates of crustal material that move about, collide, and slide under and over each other. Most of the general features of the geology of Earth are due to these crustal motions. The wide, deep, and mobile ocean basins, midocean ridge systems, mountain-building arcs where crustal blocks collide (the Andes, Rockies, Alps, Pyrenees, Himalayas, Appalachians), volcanic chains and deep oceanic trenches along the margins of ocean basins, and the young average age of the crust are all results of plate mobility. But the surface of Mars seems to be all one thick plate.

Without plate tectonics and continental drift, sediments containing weathering products such as carbonates, sulfates, sulfides, chlorides, and clays are not subducted into the hot mantle, and the volatiles in them are not boiled out and recycled. Crustal rocks are not reworked, and hence the range of surface rock types must be much narrower than on the busy, complex surface of the Earth.

But why isn't there continental drift on Mars? Basically, because Mars is so small and far from the Sun. Because of its small size (10 percent of Earth's mass; 25 percent of its surface area), Mars produces only 10 percent as much heat from radioactive decay in its interior. Radioactivity is the main source of energy that drives the geological activity of Earth. The heat is conducted very slowly out through the crust to the surface, whence it is radiated off into space. This heat flow on Mars is like putting 0.1 of the Earth's heat flow through 0.25 of the Earth's surface area: the heat *flux,* which is the amount of heat carried out per unit area per unit time (calories per square centimeter per second, for example), is then only about 40 percent as large as on Earth. Crustal rocks on Earth and Mars are about equally poor as conductors of heat.

Because of the insulating effect of the rocks, heat can flow outward only if the temperature in the crust increases steadily with depth. The familiar fact that the deep mines of South Africa have temperatures of about 130°F (55°C) at their bottoms is directly due to the insulating effect of the Earth's crust, which holds in the heat coming from the interior of the Earth.

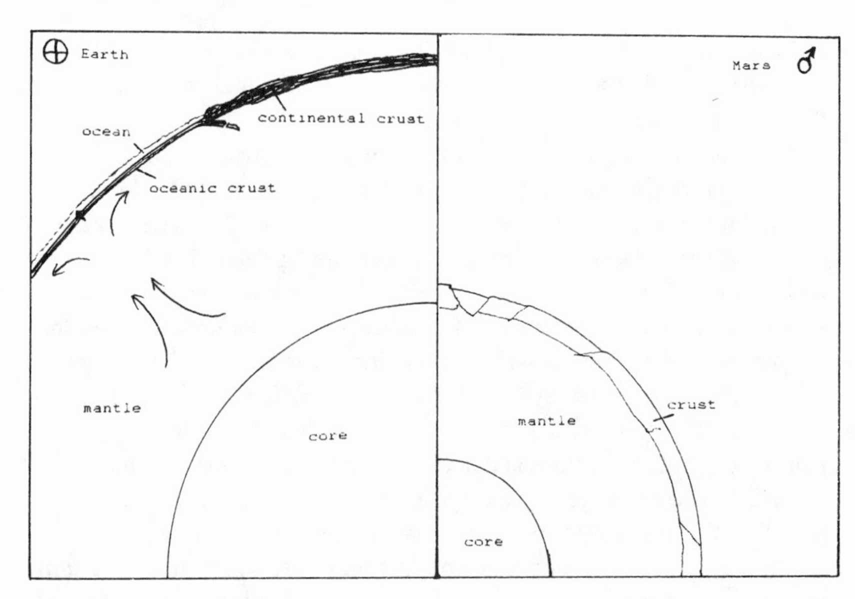

Figure 10.6
Continental Drift on Earth and Mars
The section of Earth, on the left, shows a thin, dense layer of oceanic basalt crust, carrying its burden of oceanic sediments, being showed against a thick, less dense continental block by currents in the mantle. The ocean floor cracks open where it is under tension, allowing hot magma from the mantle to erupt through the ocean floor to form mid-ocean volcanic ridges. The oceanic crust is shoved beneath the continental crust at the edges of the continents, breaking up (to produce earthquakes) and melting (to produce volcanic eruptions) as it is shoved down into the hot mantle. The much smaller Mars, on the right, is cooler inside and has a smaller rate of increase of temperature with depth: the "lithosphere" of cold, hard surface rocks is much thicker on Mars than on Earth. This inhibits continental drifting, and leaves Mars with essentially a single "frozen" plate.

The heat flux passing through the crust of a planet is proportional to the conductivity of the rocks (which varies very little with rock type) and to the rate at which temperature increases with depth. On large solid bodies in the Solar System, this *temperature gradient* is typically a few degrees per kilometer of depth. The larger the body, the larger the heat flux must be, and (since the conductivities are all about the same) the larger the temperature gradient must be. On Mars, therefore, the temperature gradient must be only about 40 percent of what it is on Earth. Thus temperatures in Earth's crust increase with depth 2.5 times as rapidly as on Mars. Also, of course, the temperatures start from a much lower surface temperature on Mars.

The outer, cold, rigid shell of a planet, which acts like a solid, is called the *lithosphere*. At the bottom of the lithosphere, temperatures are just getting high enough, and close enough to the melting point of the rocks, so that they can begin to flow like a very viscous, sluggish fluid. The entire crust and the upper portion of the mantle make up the lithosphere. On a very hot planet, like Venus, the surface temperature is so high that all temperatures in the crust are much higher than on Earth. Thus the lithosphere is thinner than on Earth. On Mars, the surface temperature is much lower than on Earth and the temperature gradient is much less. Thus the lithosphere on Mars is nearly three times as thick as on Earth. The thickness of the rigid lithosphere is so great that crustal plates cannot be bent and subducted as on Earth.

On Mars, it is very hard for volcanic activity (which originates at the base of the lithosphere) to penetrate the crust. Also, unlike Earth, Mars lacks a very widespread pattern of volcanic vents covering the planet (fig. 10.7). Where Earth's crust is thin and brittle, along midocean ridges and subduction zones, hot gases can readily escape from the interior.

But holes through the crust are much harder to make on Mars, and are much rarer. Once a volcano becomes well established on Mars, it is the only show in town. Magmas flock there from great distances to participate in the action. Also, because the magmas rise from immense depths, and do so in a gravity field much weaker than Earth's, they can be lifted to immense heights. This gives rise to one of Mars' most striking features, a handful of immense volcanic cones that tower to heights in excess of 23 kilometers. For comparison, Mount Everest, the highest mountain on Earth, rises about 9 km above sea level, but only 5 km above the surrounding Tibetan plateau. Terrestrial volcanoes, such as Kilimanjaro in Kenya, rarely rise more than 4 km. (Mount Everest is not volcanic; it is being squeezed upward by a violent collision between two formerly separate continents, Asia and India. The Tibetan plateau, where the two continents mix and mingle, is twice as thick as normal continental crust.)

More than half of the surface of Mars is covered by an ancient, heavily cratered terrain. Craters of all sizes are everywhere. The first spacecraft to fly by Mars and photograph its surface, *Mariners 4, 6* and *7,* had the ill for-

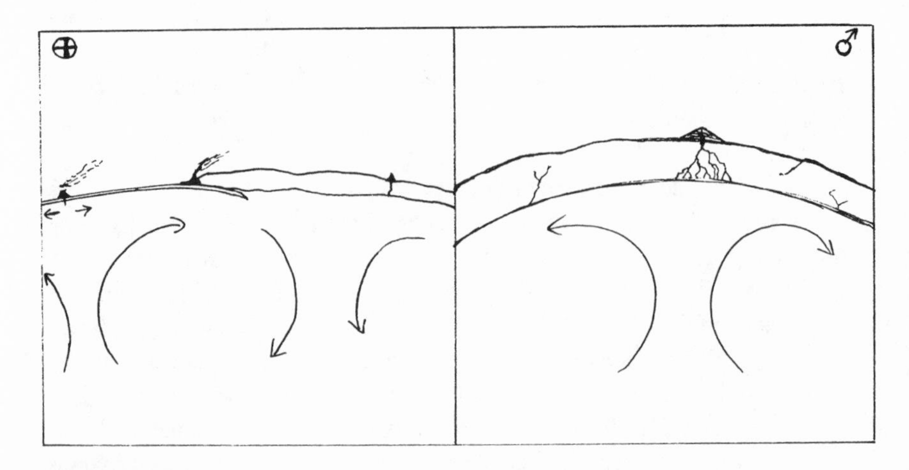

Figure 10.7
Volcanic Styles on Earth and Mars
On Earth (left) most volcanoes are associated with spreading centers in the middle of ocean basins or with subduction zones near the edges of the continents. Occasional volcanoes (mostly extinct) are found in the interiors of continental blocks, commonly associated with hot spots in the mantle. Volcanoes are very widespread on Earth, and magma has a hard time rising to great heights in Earth's strong gravity. On Mars (right) there are very few cracks that extend all the way through the very thick lithosphere, so all the magma generated in the planet is apportioned among just a few volcanoes. These volcanoes can rise to un-Earthly heights because of the great pressure exerted by the very thick lithosphere, and because of the lower gravity on Mars.

tune to make their closest passes to the planet over this cratered terrain, thus establishing the image of Mars as a desolate, dull, lunar-like landscape reflecting only external influences on the planet, devoid of any clear evidence for past or present internal geological activity. Only upon closer inspection by the *Mariner 9, Viking 1,* and *Viking 2* orbiters did other geological features become apparent in the cratered terrain. There is pervasive evidence of wind erosion and landslides, neither of which is particularly surprising. But there are also many short streamlike features that appear to be ancient watercourses produced by brief, catastrophic floods.

Mariner 9 presented us with a "new" Mars, far more complex and interesting that the "lunar" images from earlier flyby missions suggested.

Consider: Olympus Mons, the largest volcano in the Solar System, rearing its head almost out of the Martian atmosphere (figs. 10.8 and 10.9); the Hellas basin, a huge, deep circular impact structure with atmosphere filling it like a lake, its dense air uncommonly dusty, obscuring our view of the ground most of the time; the Valles Marineris, the Grand Canyon of Mars,

Figure 10.8
The Largest Volcano in the Solar System
The immense Olympus Mons volcanic pile towers some 26 kilometers above the lowland plains of Mars. The craters in its summit are calderas, caused by collapse of magma pockets within the volcano. This morning *Viking Orbiter* view shows the surrounding lowlands blanketed by clouds. This volcano alone has dozens of times the mass of any volcano on Earth.

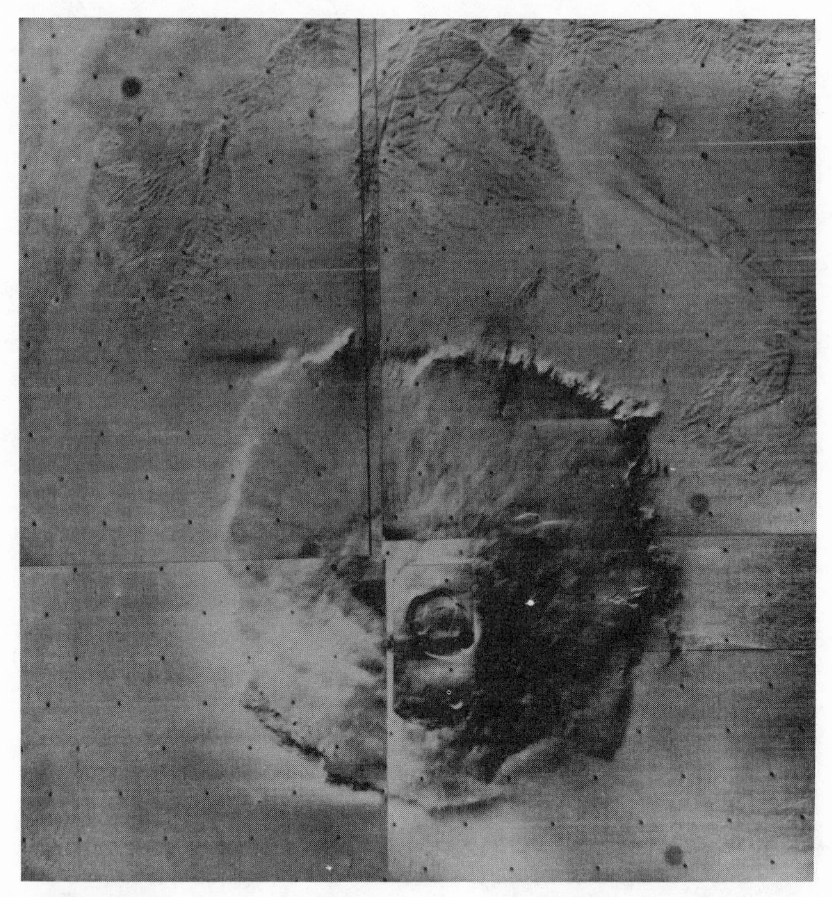

Figure 10.9
Mount Olympus
This afternoon view of Olympus Mons shows vast ancient lava flows originating from the volcano and reaching out to distances of hundreds of kilometers. For comparison, Mount Fuji, Kilimanjaro, or Mt. St. Helens would easily fit inside one of the summit calderas.

stretching a distance as far as from Los Angeles to New York, with many minor branches that are several times as wide, deep, and long as Arizona's Grand Canyon (Fig. 10.10); beds of streams that must have rivaled the Amazon in volume of flow; great sheets of layered (*laminated*) terrain circling the poles, with alternating layers rich in dust and ice, exposed and

Figure 10.10
A Small Portion of the Valles Marineris
The enormous Valles Marineris complex is a deep crack in the surface of Mars, stretching a distance as far as from New York to Los Angeles, with many minor side canyons larger than Arizona's Grand Canyon. In this scene, several huge landslides have swept across the canyon. The main landslide on the far side of the canyon is about 50 km (30 miles) wide. This is a mosaic of pictures taken by the *Viking 1* orbiter.

grooved by the winds; polar caps so cold each winter that enough carbon dioxide condenses to lower the total atmospheric pressure of the planet; huge expanses of jumbled, chaotic terrain, looking as though thousands of cubic kilometers of ice or rock had simply been pulled out from under the regolith.

The chaotic, collapsed structures, including the immense Valles Marineris, may be due to the extraction of underlying carbonate rocks by ground water that has been acidified by dissolved carbon dioxide. Features of this sort on Earth are called *karst.* Alternatively, some of these structures may be due to the melting and draining away of vast deposits of buried ice.

Since liquid water cannot exist on the surface of Mars today, the existence of large river beds cut by running water is most interesting. It has often been suggested that conditions on Mars in the distant past were more benign than at present, and that liquid water was then able to exist on the surface because of a higher surface pressure. Recent work by Hampton Watkins of the University of Arizona suggests very strongly that the major impact events that shaped the cratered terrain must have had an even more devastating effect on the atmosphere. He finds that these giant impacts "splashed" most of the early atmosphere of Mars off into space, leaving it today with an atmospheric pressure that may be 100 to 1000 times smaller than the original pressure. Such events also appear to be responsible for ejecting rocks from the sites of major Martian impacts. The rocks fall on Earth as the rare shergottite, nakhlite, and chassignite (SNC) classes of achondrites. (Such ejection is virtually impossible from Earth because of its much larger mass and escape velocity.)

HOW HOSTILE IS MARS?

Mars is indeed a harsh and alien planet on which to be stranded. But then, Roanoke Island was so harsh and alien that the first settlement of Europeans died out without a trace. The Viking settlements in Vinland and Greenland were utterly severed from Europe and died out when the climate over the North Atlantic deteriorated somewhat in the late fourteenth century. The Labrador coast, the Siberian Arctic islands, the Kalihari desert of Namibia, the Great Erg, the Rub al Khali, the Outback, Sonora's Gran Desierto, the Altiplano, and the Tibetan Plateau are legendary for their harshness. Yet human societies survived, adapted, and flourished in these locations. The depths of the Yucatecan jungle and the peaks of the Andes saw great cities of high cultures. Hundreds even spend years at a time in Antarctica. Several men have spent months in zero gravity aboard Earth-orbiting space stations. Is Mars harsher than these environments? Is it utterly hopeless to think of people adapting to live on the surface of Mars—carving habitats out of the Martian environment?

Figure 10.11
A Martian Landscape
The *Viking 1* Lander returned this panoramic view of the lifeless Martian surface. The scene is reminiscent of the most desolate of terrestrial deserts, with marching dunes and wind-eroded rocks filling the scene.

If we are thinking of humans as we know them and Mars as we see it, the answer is unequivocal: we cannot live on Mars. But if we fold in human technology and expertise in controlling environments, and if we address the particular hazards present on Mars in a logical way, several fascinating possibilities become apparent.

First of all, consider the list of environmental factors that affect habitability:

1. temperature
2. pressure
3. atmospheric composition
4. gravity
5. radiation level
6. achievability of self-sufficiency

Some human habitats (Antarctica, space stations) are inhabited despite their lack of self-sufficiency because the cost of importing necessities such as air and water (for space stations) and food (Antarctica and space) is deemed reasonable in relation to the benefits gained. Environmental temperature extremes are commonly dealt with by technological manipulation (clothing, housing, active and passive heating and cooling).

Pressure extremes can be dealt with to a limited extent by acclimation (as Andean and Himalayan natives attest), but for total pressures less than about 0.15 atmospheres even supplementation with pure oxygen cannot sustain mammalian life and activity. Encapsulation and pressurization are then required.

Atmospheric composition must be maintained within safe limits, either by natural biological homeostatic mechanisms (on a world ship like Earth) or by mechanical, chemical, or other artificial means.

Absence of gravity is a severe problem after several months, but artificial gravity can easily be provided by centrifugal force, as in the "von Braun wheel" space station concept. Excessive gravity will not be a problem—no body on which humans are likely to land in the next century would confront us with gravity beyond safe human tolerance levels.

The radiation hazard has been mentioned previously in chapters 7 and 9. The solution is to maintain "storm cellars" shielded by thick layers of dirt.

Finally, there is the issue of self-sufficiency. In point of fact, there is scarcely a nation on Earth that is self-sufficient *in sensu stricto.* The resources of our planet are so widely and inequitably distributed that virtually every country has shortages of some necessities and surpluses of others. In the ideal state of affairs, nations would simply trade with each other.

In practice, however, the total per capita endowment of different nations is far from equal. Humanitarian considerations dictate the dispensation of food for famine relief, of medication and clothing, and of knowledge regarding the means of achieving self-sufficiency from the richer nations to the poorer. The Green Revolution has for now removed food production from the list of limiting factors for life: Earth is now producing enough food for its entire population. There is not one single credible scientific or technical reason why anyone on Earth should starve. (Then why do people still die of starvation in Ethiopia and Kampuchea, in barrios, favelas, and urban ghettoes? Because of ignorance, inefficiency, bureaucratic bungling, personal greed, and utterly inhuman and anti-human political and ideological considerations.)

But what of self-sufficiency on Mars? For decades to come, Mars will be an exporter of ideas and scientific data in the same way that the Antarctic is today. Thus Mars will be worthy of subsidy for many years. The question of self-sufficiency may be solved adaptively over a matter of decades, rather than faced all at once before we are familiar with the lessons that can be gained only from actual experience on Mars. Further, once we know more of Mars, we may find it easy to identify items and materials that can be exported to Earth in trade for the high-technology assistance that Earth alone can render.

We may then face a much less demanding problem: to identify what can be done to assure that environmental conditions suitable for life can be maintained in a Martian base, and to sketch out the initial steps than can be made toward self-sufficiency of a Martian colony.

TOWARD MARSBASE

The problem of establishing a small base on Mars is in many respects similar to that of maintaining a base on the Moon or at the South Pole. Minimizing the mass that needs to be transported is crucially important, but being quite certain that all the necessary supplies and spare parts are included is even more important, since the lives of the residents depend on having them when they are needed. Even more than on a family trip from New York to San Francisco, it is important to identify and use as many local resources as possible, since use of them might greatly reduce the mass that needs to be lifted from Earth.

The first problem to be faced by the builders of Marsbase is transporting the necessary components from LEO to the Martian surface. This is at first sight a daunting task: the least-energy flight path to Mars from Earth follows an elliptical orbit that departs from Earth on one side of the Sun and swings out to Mars on the opposite side of the Sun. The trip is about 600 million kilometers long, equivalent in distance to traveling from Earth to the Sun and back—twice! The trip to Mars would last between eight and nine months. Three days into the trip, and still well inside the realm of Earth's gravitational influence, the Mars vehicle will pass the moon. Mars lies 260 days ahead. What is the penalty that we pay for being so ambitious as to go to Mars rather than just shipping the components of our base to the Moon? The answer is fascinating.

First, we shall compare what could be accomplished with a heavy-lift launch vehicle that can lift a 200-ton expedition off the Earth. Let us first imagine a 200-ton lunar vehicle in Low Earth Orbit. It has a high-specific-impulse stage to lift it from LEO and hurl it out to the orbit of the Moon, plus a second stage with space-storable propellants to ease it down onto the surface of the Moon, The payload delivered to the Moon is 15 tons.

Now let us imagine a 200-ton Mars vehicle in LEO. It needs only one major propellant burn, to kick it from LEO into a trajectory that takes it out to Mars. The necessary delta V is larger than that required for the first leg of the lunar journey, *but there is no second burn to land.* Instead, the landing is carried out by dissipating the spacecraft's excess energy in the thin Martian atmosphere using an aerobrake heat shield. About eight percent of the payload mass must be devoted to the heat shield to dissipate the energy of the arriving expedition during a single aerocapture pass through the upper Martian atmosphere. The total mass that can be sent to Mars is 29.4 tons, of which 25 tons is landable payload. Thus the "penalty" for sending the payload to Mars instead of to the Moon is that you get to install 25 tons of equipment on Mars instead of 15 tons on the Moon!

For expeditions intended to stay on the Moon or Mars for an extended period of time, the life-support requirements are comparable—except that, on Mars, there is the real prospect of being able to extract water, oxygen, nitrogen, and so on from the atmosphere and surface. A small mass of extraction equipment can be translated into a very large mass of life-support fluids and propellants which, because they are made on Mars, need not be carried along from Earth.

The difference between the Earth-Moon trip time of three days and the Earth-Mars trip time of eight or nine months does indeed make the Mars trip (the actual time spent in travel) much more demanding. However, nine months on the Moon is just as hard on life-support systems as nine months in the interplanetary void. But nine months on the surface of Mars affords ample time to begin living off local resources. Also, don't forget that Soviet missions aboard the *Salyut 6* and *7* space stations have shown that eight or nine months is readily achievable. They are on the verge of being able to do it *without resupply!*

DOMES AND HOMES

The initial Marsbase is certain to use Martian soil for radiation shielding, in the same manner that we discussed for a lunar base. Beyond that point, however, the divergence is wide indeed. The Martian base is surrounded by volatiles in easily handled form. Water may be condensed directly from the atmosphere or extracted from the regolith. Fully half of Mars is covered by permafrost, so water ice may be "mined" in many locations (see fig. 10.12). The water is directly useful, and of course electrical power may be used to make hydrogen and oxygen by electrolysis of the water.

Figure 10.12
A Martian "Mud" Crater
Thousands of large Martian craters are surrounded by thick blobs of what appears to be frozen mud. These craters are found at latitudes were permafrost is expected, and are virtually absent in the warmer regions near the equator.

Electric power for the base can be provided by a solar cell array, and stored in banks of batteries or as hydrogen and oxygen gas made by electrolysis of water. During the twelve-hour Martian night, the hydrogen and oxygen can be used to generate electricity in fuel cells.

The lower intensity of sunlight on Mars means that solar cell arrays will pick up only about 500 watts of sunlight per square meter, less than half the amount at Earth (and the Moon), and generate about 100 watts of useable electric power per square meter. A base with an average energy use rate of

100 kilowatts needs to produce 200 kilowatts during the time that the Sun is up. This requires about 2000 square meters of solar cells. The amount of energy that must be stored for nighttime use is about 1200 kilowatt-hours.

A lunar base with similar size and power demands would present far greater problems. Although it also sees sunlight half the time, and gets sunlight with twice the intensity of that reaching Mars, it has great difficulty storing enough energy to get through the night. It can supply all its needs with only 1000 square meters of solar cells, a single collector just under 6 meters (19 feet) in radius—but it needs to store enough energy to get through the lunar night, which is 14 days long! The amount of stored energy must be over 33,000 kilowatt hours, 28 times the amount that must be stored on Mars.

Next comes the problem of supplying a suitable atmospheric composition and pressure for the base. An inflatable dome bag capable of bearing a pressure difference of several millibars between inside and outside can be brought from Earth. The dome is inflated with Martian air to its maximum pressure, and then sprinkled or sprayed over with dirt-based concrete until it begins to sag. The internal pressure can then be increased, more "dirtcrete" added, and so on until the internal pressure reaches a few hundred millibars. The outside is then stabilized with a final spray-on layer to protect against wind erosion. The air in the dome is of course mostly carbon dioxide and nitrogen, with traces of carbon monoxide and oxygen. Mixing in more oxygen (extracted from water) and passing the gas mixture through a catalytic converter bed at moderate temperatures will thoroughly oxidize the poisonous carbon monoxide. The thermal inertia and insulating properties of the dust blanket will stabilize internal temperatures near 0° C without any day-to-night variation and without any heat being applied.

Now let us set up a few light, deployable solar reflectors outside the dome. Each reflector will collect sunlight from an area of tens of square meters and focus it crudely into a "blob" of light about 30 centimeters across on the outside of the dome. We may now install windows in the dome at each of these illuminated points. Simple sun-tracking systems of the sort used in rooftop solar water heaters on Earth will keep the collected beams of light shining through the windows so long as the Sun is above the horizon. The amount of heat deposited in the dome during the daytime will greatly exceed the amount that will be lost by conduction at night. And of course, the windows can be covered with insulation at night to prevent heat loss.

The inside of the dome can now be made as bright as desired (since the area over which sunlight is collected may be larger than the area of the dome), and temperatures sufficient to keep Gila monsters happy can be built up quickly.

The next step is to provide a temperature-moderation system (presumably a tank of water) to smooth out the daily temperature fluctuations caused by solar heating through the windows. While all this is going on, the air is slowly purged: carbon dioxide is taken from the dome, passed through an extremely hot solar-powered "cracking tower" and partially separated into carbon monoxide and oxygen. These gases can then be separated by their vapor pressure differences. The carbon monoxide fraction can be "baked" to cause it to decompose into carbon black and carbon dioxide, or simply discarded. The oxygen is returned to the dome, and the carbon dioxide is fed back through the "cracking tower." After the dome atmosphere has been converted into a nitrogen-oxygen mixture, the dome may be occupied as living quarters or workspace. The crew will surely be happy to get out of the cramped quarters of the ship that they have been living in for the last nine months!

The nitrogen necessary to provide a measure of security from fire is available in the atmosphere: if each dome is 5 to 10 meters high and takes in all the atmospheric nitrogen from the volume of Martian atmosphere directly above the dome, then nitrogen pressures of 150 to 300 millibars can be provided inside the dome. Although this is less than the 780 millibars at Earth's surface, it is quite sufficient from practical use.

The hydrogen liberated as a byproduct from the decomposition of water to make oxygen has a number of exciting uses. It may be used to reduce the ubiquitous iron oxide powders to metallic iron for use in construction, or it may be reacted with carbon monoxide to make methane or methanol for use as a rocket propellant to get off Mars again.

The second dome can be built in exactly the same way except that the floor of the dome bag will be lined with about one meter of dirt prior to pressurization. This is the "soil" of the first Martian farm (see fig. 10.13). This dome will be pressurized with Martian air, the solar heating system will be installed, and oxygen will be injected. It will then be ready to receive all the organic wastes of the base. Aeroic composting of the refuse will provide rich

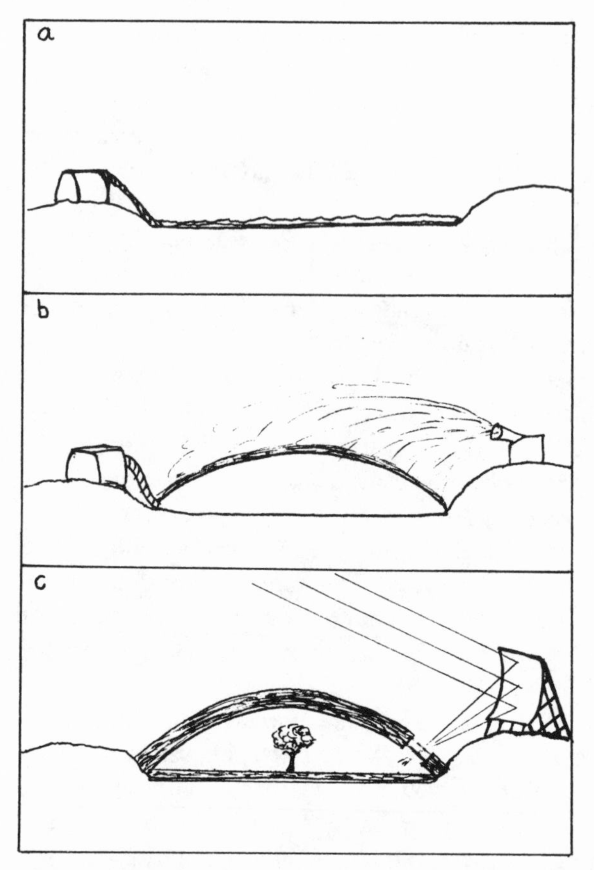

Figure 10.13
Making a Dome on Mars
A circular area is leveled and an inflatable dome is laid out in it (a). A compressor inflates the dome to its full capacity, and alternating layers of dirt and "Marscrete" are sprayed on to give the dome rigidity. The air pressure is increased to bear the weight of the added dome material. Once the dome has set, small windows may be cut in it to admit sunlight collected by steerrable mirrors. A layer of dirt may be added onto the dome floor, and agriculture may then begin.

humus to build the soil, which can be planted with food crops within a few weeks.

Once vegetable production has become established, Marsbase may well choose to work in the direction of a wider diet. The most useful and efficient farm animals for use in space are probably domestic ducks, chickens, and goats. The main products will be milk, eggs, cheese, butter, ice cream, yogurt, pate de foie gras, and roast duckling. (Well, no, I'm not a *strict* vegetarian—why do you ask?)

Anyone familiar with goat husbandry will attest to the insanity of letting a goat run loose in the garden or living quarters of a space settlement. Goats are likely to be hyperkinetic wizards of navigation in zero gravity, as well as voracious consumers of vegetable garden wastes, insulation, notepaper, grey tape, and floppy disk envelopes (when the opportunity arises). We pity the poor astronaut who has the responsibility of ferrying the first doe kids to Mars: the cute and cuddly bottle-feeding kids of launch day will be rambunctious yearlings by the time they arrive. But once they are there, their milk and milk-derived dairy products will provide the highest possible return of animal protein per pound of feed. The secret of living with goats is the same as achieving successful nuclear fusion reactors: perfect containment.

TERRAFORMING MARS

The approach outlined above for adapting stepwise to the Martian environment can produce an "underground society" that is largely, although not completely, self-sufficient. The residents of Marsbase can use local resources to increase their level of self-sufficiency to a remarkable degree, but the development of a sufficiently diversified industrial society to meet all their needs would require a population of at least tens of thousands of people. Further, there is the question of whether there will be ore deposits suitable for the extraction of all the elements needed for such an industrial structure. These issues cannot be answered at present. Nonetheless, Martians will probably need lots of cash to pay for essential imports.

What could Martians export to Earth that would have a sufficiently high unit value to pay for its transportation costs? Possibly rare minerals or precious metals, but more likely not. We suggest transplanting a few struggling young artists to the colony. (Established governments often feel the urge to export writers, scientists, dissidents, religious minorities, artists, adventurers, innovators, conscientious objectors, and other misfits . . . . the supply should be ample.) Their sculpture and painting, stimulated by the novel environment, may well be treasured imports on Earth. Also, do not forget the value of Martian mineral samples, museum specimens, and the like.

Notice that we are now discussing "Martians" and a "colony." Where did they come from? Think back on the Dutch investors who first contracted workers to come over to Nieuw Amsterdam to cut timber for the European market. The Dutch West Indies Company (Geochtroyde Westindische Companie) forbade its employees to manufacture *anything* in the New World: they wanted to be certain that they controlled every market and made a profit on every item sold. They wanted to have the undivided, full-time loyalty of all of their employees. And they most definitely did not want the growth of a colony that would have its own interests, distinct from those of the board of the DWI Company.

Some of the more enterprising workers declined their free passage back to the Netherlands at the end of their hitch. Instead, they used their hard-earned money to bring their families over to join them in the New World. Jacques Cortelyou, a French Huguenot desirous of escaping religious persecution in his native France, fled for his life to the Netherlands. As an unlanded resident of a densely populated country, he was a natural candidate for an adventure in the New World, and so he took ship to Nieuw Amsterdam. He became the first surveyor in North America, the first European to live on Long Island—and the first person to commute to a job in New York!

Resolved Waldron, a Puritan cousin of the Waldrons who came over on the *Mayflower,* stayed in the Netherlands for a while, where he joined the Reformed Dutch Church. He took ship to Brazil, where he found the Dutch influence waning and the religiously intolerant Portuguese in the ascendant, so he again pulled up roots, this time moving to Nieuw Amsterdam. He soon had a position of responsibility, serving as the chief magistrate and law-enforcement officer ("schout") of the colony as second in command to Pieter Stuyvesant.

Then there was James Clark, who, driven from his native Ireland by the great famine, elected to face the vast promise and equally vast uncertainty of the New World. In the free frontier environment of the new land, where individuals could act decisively to affect their own destinies, these men and their descendants flourished.

We can better understand the hearts of the pioneers by having read their histories, and we think we can better grasp the excitement of New Worlds because we are direct descendants of all three of these pioneers. So now you know why we changed "Marsbase" to "colony," and now you can guess who the "Martians" are. They are our descendants.

Now, what are the chances that these Martians can remake their planetary environment so that human beings could live "outside" on Mars? Rather than encapsulating themselves and their society in dirt-banked domes, is there any way that they could make the surface environment hospitable to terrestrial life forms?

The first problem with living outside is the lethal nighttime cold. This problem could be greatly alleviated if some means could be found to make the atmosphere more opaque in the infrared. I (JSL) recall discussing this issue with Ronald Prinn of MIT in about 1978, while he was working on the effects of Freon-like compounds on Earth's atmosphere. Freons, used as refrigerants and as propellants in aerosol spray cans, are compounds of carbon and the halogen gases chlorine and fluorine, called *halocarbons* as a class. Prinn did some calculations on the mass of halocarbons that would be needed to warm the Martian nights (and days!), and concluded that the mass was much too large to be transported from Earth. We simply forgot about the matter at that point.

Recently James Lovelock has attacked the problem from a different perspective. He likewise concluded that, of all the things that could be added to the Martian atmosphere to improve the climate, the halocarbons were indeed the best. However, he also pointed out that the logical solution to the transportation dilemma was to make them on Mars out of local resources.

There are surely abundant chlorine sources on Mars. The crust anywhere on the planet should yield ordinary salt (among other compounds) when washed with water. Salt, sodium chloride, is a direct source of both sodium metal and chlorine gas. The chlorine gas can be reacted with methane or simple hydrocarbons made from hydrogen and carbon monoxide to yield freons. Whether fluorine is an essential ingredient of the freon mixture depends on several considerations that we cannot now foresee. First, we do not know what, if any, fluorine ores are present on Mars. Second, we do not know what the optimum gas composition is to cause the greatest warming per ton of freons. And finally, we have not yet studied all the processes that may destroy or remove freon gases from the atmosphere. These questions can be answered by geochemical exploration of Mars and a few man-years of computer modeling of the atmosphere.

Warming of Mars inevitably will cause an increase in atmospheric pressure as adsorbed carbon dioxide gas is released into the atmosphere from the regolith, and as the carbon dioxide snow in the polar caps evaporates. It is possible that the total amount of carbon dioxide released could amount to several hundred millibars. If so, it would provide a very large portion of the cosmic ray shielding and atmospheric pressure needed by the Martians, and hence would make the job of building adequate domes much easier.

It would also mean that anyone equipped with an oxygen tank and mask could walk the surface of Mars in the daytime in normal clothing, without any special suit. At night it would still get chilly, perhaps dropping below freezing in the open air every night. Fortunately many plants, including a number of food crops, can survive such temperature conditions. The atmospheric composition, however, would still be nearly pure carbon dioxide, with traces of nitrogen and oxygen. We know that, to a point, plants thrive on high carbon dioxide levels. What we don'd know is how well a wide variety of food plants thrive on *enormous* carbon dioxide levels. We thus cannot yet asses the likelihood of raising food plants "outside" on Mars. Of course, plants growing in the open would convert carbon dioxide into organic matter and release oxygen gas into the atmosphere . . . .

A warmer surface means the melting and evaporation of a huge mass of permafrost at middle latitudes. Some of this will find its way to the polar regions; indeed, there is a risk that most of it will unless precautionary measures, such as darkening the polar regions with carbon soot to make them warmer, are taken. Irrigation of plants on the surface will be an expensive proposition. It would be much more sensible to enclose the crops both above (to recapture evaporated water) and below (to prevent water from sinking out of sight into the regolith). The techniques of interest are very similar to those under consideration for use in arid lands on Earth, where water is too valuable to simply be thrown away.

There is a second, profound hazard of increasing the atmospheric water content. This is the resulting great increase in the rate of destruction of water vapor by solar UV light. A moist atmosphere will lose water by photolysis, making mostly oxygen and hydrogen. Most of the hydrogen will find its way to the top of the atmosphere and rapidly escape from the planet.

Since water is in such short supply on Mars, such a loss would be insufferable.

Whether this loss can be slowed to an acceptable level is unclear at present. The carbon dioxide itself competes with water for UV light with wavelengths less than about 0.195 micrometers, and hence water vapor under a thick blanket of carbon dioxide would be reasonably well shielded from the Sun. The problem gets much worse when warm, moist air rises to high altitudes, above much of the carbon dioxide shield. This may occur if Mars develops a thick, warm, moist atmosphere.

Another promising but complicating factor is the consumption of carbon dioxide by plants. Photosynthesis turns carbon dioxide and water into carbohydrates plus free oxygen. Thus, once green plants can be established on the surface, the atmospheric content of oxygen rises steadily as the carbon dioxide pressure declines. The total pressure would remain unaffected. Where there is oxygen, an ozone layer forms (Mars already has a tiny trace of ozone). The ozone layer helps shield the surface from dangerous UV in the 0.2 to 0.3 micrometer wavelength range. What effect this would have on the rate of hydrogen loss is a complex issue, one that deserves careful study.

Nitrogen is in very short supply in the atmosphere. Much of the success of plant communities depends on the supply of fixed nitrogen. There is serious reason to doubt whether nitrogen-fixing bacteria will be able to work on nitrogen pressures as low as the 0.00015 atmospheres of nitrogen now present on Mars (recall that Earth, where these bugs learned their trade, has 0.78 atmospheres of nitrogen).

It is possible that there are nitrogen compounds in the soil, polar caps, and permafrost. UV-driven chemical reactions in the atmosphere can make small amounts of nitrogen oxides. Most of these are destroyed and made back into atmospheric oxygen and nitrogen, but it is likely that some small proportion of them gets frozen out in the polar caps or adsorbed in the regolith. In the presence of water, nitric acid will form. The nitric acid reacts readily with many minerals to make water-soluble nitrates. These nitrates are superb fertilizers. If they are present in useful amounts on Mars, plants may assist in liberating some of that nitrogen indirectly: the plants make proteins, animals (Martians) eat the plant proteins, their excreta undergo aerobic decay (composting), and some nitrogen gas is released. At present, however, the nitrogen content of the surface of Mars is unknown. It is actually easy to make fixed nitrogen in Martian chemical factories, using nitrogen and oxygen gas as the reactants. Such a fertilizer plant might quickly pay for itself on Mars.

Suggestions have been made that the water budget of Mars may be augmented by diverting comets so as to to make them collide with Mars. The wisdom of this venture is not evident. Comets are extremely fast-moving projectiles carrying huge amounts of energy. It is very likely that the average comet impact causes a *decrease* of the mass of the atmosphere because the violent impact explosion blows off a significant mass of atmosphere. There is mounting evidence that Mars formed as a low-temperature planet, with a higher content of volatile elements than Earth. The present paucity of atmospheric gases is understandable as the result of a long history of major impacts, which caused explosive loss of at least 99 percent of the original atmosphere. Causing more big impacts may simply make the problem worse.

THE FIRST STEPS

A great deal may be accomplished by paper studies of the effects of altering the composition of the Martian atmosphere, the design of chemical schemes for making useful materials from atmospheric gases, and the extraction of useful materials such as metals and gases from surface rocks and minerals. However, no definite conclusions are likely to be possible without further knowledge of Mars. An absolute requirement for progress is a highly capable geochemical mapping mission that will survey the entire planet from polar orbit.

A second requirement is a series of Mars landers with highly sophisticated instrumentation for determining the chemical and physical properties of the surface over a very wide area. These landers might be similar in concept to the Lunokhod lunar rovers of the early 1970s. They could be launched and emplaced on Mars by either the Space Shuttle (with an appropriate upper stage carried in the payload bay) or by the Soviet D1e heavy-lift booster. We should recall that the weight that a given booster can deliver to Mars is *larger* than the mass it can deliver to the Moon. A significantly upgraded Lunokhod is possible. A valuable part of the same project would be the deployment of a network of seismometers on the Martian surface, either from the rovers or via penetrators dropped from orbit. These seismometers could pinpoint the location of current volcanic activity or crustal fracturing, where highly concentrated volatiles may be venting out of the planet's interior, and possibly even, as on Earth, producing rich ore deposits.

A third possible mission is the automated return of surface samples from Mars. While there are clear advantages to having even one kilogram of Martian material in our terrestrial laboratories, it is not obvious that this mission need take precedence over a first manned visit to Mars. A manned expedition would be materially assisted by the $2 billion or so needed for the sample return mission, and there are plausible schemes for a first manned mission that would cost under $10 billion. The manned mission would be more than five times as valuable in every respect: samples chosen for return could be much more carefully screened, much larger in mass, and sampled from a very wide area.

Finally, there is an important niche for automated (unmanned) materials-processing experiments, designed to test promising techniques for processing Martian atmosphere and crust under the most realistic possible conditions. Such missions have so far not been discussed by NASA. It would be reasonable to combine the planning and development of such processing experiments with the design of similar packages to be sent to the front porch of Mars—its two natural satellites Phobos and Deimos.

11

PHOBOS AND DEIMOS
Special Cases with Special Benefits

The potato is my favorite example.—Jane Brody

BACKGROUND

Phobos and Deimos, the two small natural satellites of Mars, were named after the horses that drew the war chariot of the Roman god Mars. The Greek meanings of the two names are "fear" and "dread; terror." They were both discovered by the American astronomer Asaph Hall in 1877.

Curiously, descriptions of the two small satellites of Mars had already appeared in the fictional writings of Voltaire and Jonathan Swift, in both cases complete with rather circumstantial (but inaccurate) descriptions of their sizes and orbits. (Voltaire's inspiration may have come in 1726 during his exile in England: he was there, and was in contact with Swift, at the time of the publication of *Gulliver's Travels,* in which the Laputan astronomers describe their discovery.) The reasoning behind these literary predictions was simple in the extreme: as we move out from the Sun, we find first Mercury and Venus with no moons, then Earth with one, then Mars, then Jupiter with the four Galilean satellites then known, then Saturn with a flock (eight were known at Hall's time, but only five in Swift's). What could be more logical than that Mars should have two?

Phobos, the inner of the two moons (Figs. 11.1, 11.2), orbits 9378 kilometers (5814 miles) from the center of Mars in an almost perfectly circular orbit inclined less than one degree to the equator. It circles Mars about 2.5 times in every Martian day. Since it is traveling around Mars fast-

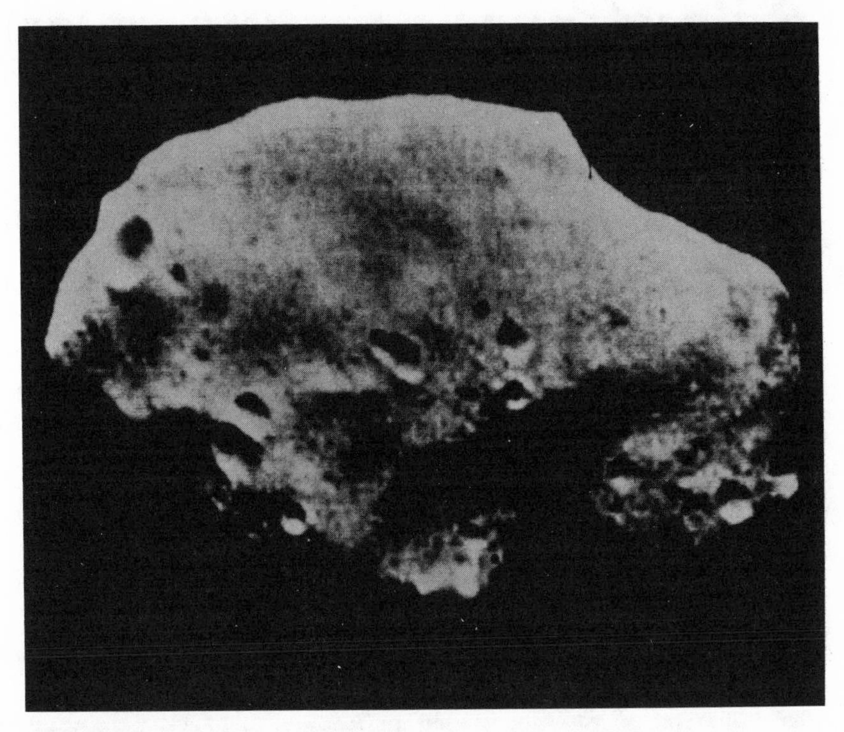

Figure 11.1
Phobos: a Solanaceous Satellite
This *Mariner 9* view of Phobos showed a battered, fragmented and irregular chunk of rock, deeply mantled with debris from impact events. The largest crater seen is approximately the largest crater that could be recorded by a body this size: any more energetic impact would have fragmented Phobos. Debris ejected from Phobos by impacts will go into a nearby orbit about Mars and will soon be reaccreted by Phobos.

er than the planet spins on its axis, Phobos appears from the surface of Mars to rise in the west and set in the east. In two Mars days Phobos makes almost exactly five orbits, which means that, as seen from the surface, Phobos passes by three times every two days. Its altitude over the equator (its mean orbital radius minus the mean radius of Mars) is 6939 km (4302 mi) or 2.845 Mars radii.

Deimos orbits 23,459 km (14,545 mi) from the center of Mars, also on a path that is nearly perfectly circular and coplanar with the equator. It com-

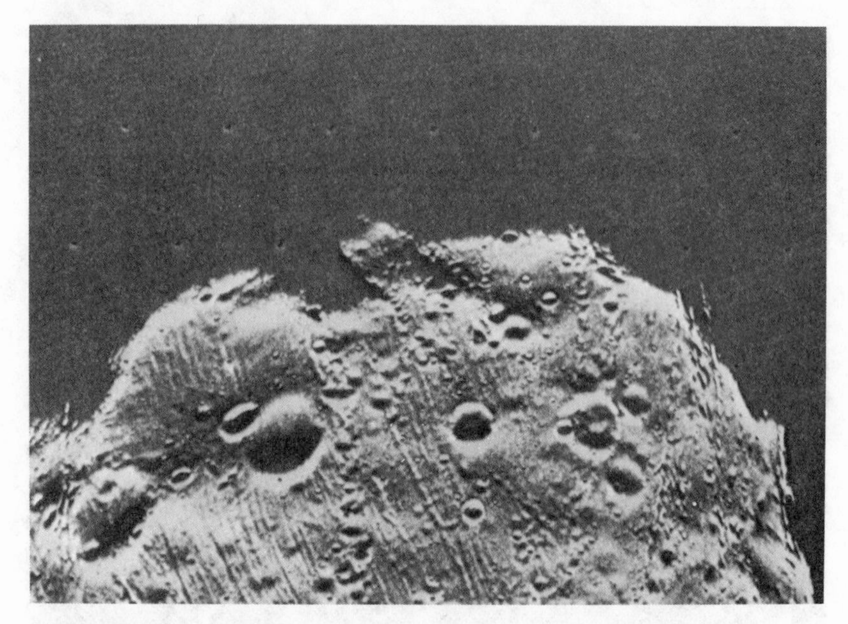

Figure 11.2
Phobos: A Closer Look
The Viking Orbiters improved considerably upon *Mariner 9*'s coverage of Phobos. The best pictures show vast numbers of roughly parallel grooves in the regolith, apparently due to dirt falling into deep cracks that pervade the interior of this shattered worldlet. The smallest features visible in this picture are about the size of an average building.

pletes an orbit in 1.23 Martian days. Since it orbits more slowly than the planet revolves, an observer on Mars would see it move across the sky from east to west. In five Martian days, Mars spins five times and Deimos completes just over four orbits.

Both satellites are very small and very irregular in shape, more similar to Earth-crossing asteroids (or potatoes) than to Earth's massive, nearly spherical Moon. The average equatorial radius of Phobos is about 12 km, while that of Deimos is only 7 km. Phobos has about 0.01 percent of the mass of Mars, and Deimos has about 10 percent of Phobos' mass. The masses, sizes, and shapes of both satellites have been studied by the *Mariner 9* and *Viking 1* and *2* Mars orbiters, and are well enough known so that their densities can be estimated crudely. Incredibly, the densities are close to 2 grams per cubic centimeter, similar to the very volatile-rich carbonaceous (C) chondrite meteorites, which contain 10 to 20 percent water and about 6 percent organic matter. Photographs of these moons taken by the Viking orbiters through colored filters show that they are also very dark in color, like C asteroids.

It seems likely that both satellites are very ancient, largely unprocessed remnants of the population of small planetesimals out of which the planets formed. As such, their scientific value is immense: they may be windows giving a view of processes at the time of planetary formation, more than 4.5 billion years ago.

The gravitational acceleration on the surface of Phobos is less than a thousandth of that on Earth, and its escape velocity is about 15 meters per second, compared to Earth's 11,200 meters per second. The surface gravity and escape velocity are both about 40 percent smaller on Deimos. This means that even rather gentle impacts can knock debris off of these satellites.

Impact cratering on Phobos and Deimos will generally place debris in orbits about Mars that will continue to intersect the orbits of the satellites. This means that there is an excellent chance that material ejected from these satellites will later be reaccreted by them.

Asteroids of the same approximate size are not so lucky: they orbit the Sun, and the volume of space into which their ejecta can expand is far larger. Also, these ejecta are subject to other disturbing forces, such as solar radiation pressure and perturbations by the other planets. These forces change the orbits of the debris more quickly than the debris can possibly find its way home again. Thus Phobos and Deimos, unlike asteroids, are extremely successful at recapturing their own collisional debris. They therefore are covered with a deep regolith. Indeed, photographs taken by the Viking orbiters show that their surfaces are heavily mantled with regolith.

We therefore expect that both Phobos and Deimos will present resources closely similar to C chondrites, the same as expected for C asteroids and for the surface layers of extinct comet cores (see table 11.1). Since we have already explored the extraction and uses of water, hydrogen, oxygen, carbon, methane, methyl alcohol, nitrogen, etc. in chapter 9, we will not take the time to repeat ourselves: it must suffice to say that such bodies are very superior sources of a wide range of resources, especially volatiles for use as life-support fluids or propellants.

We may propose, following S. Fred Singer of George Mason University,

Table 11.1
Extraterrestrial Resources

Moon	*Asteroids*	*Phobos & Deimos*
differentiated	both differentiated and undifferentiated	primitive
volatile-poor	very low gravity	very dark
moderate gravity	many volatile-rich	very low gravity
metal-poor	many metal-rich	low density
basalts (some Ti-rich)	sources of meteorites	carbonaceous?
anorthosite	(See chapter 8)	(See chapter 11)
KREEP		
regolith with asteroid component		
solar wind trace gases		
polar volatiles???		
(See chapter 6)		

to send a simple landing vehicle from Low Earth Orbit into a trajectory that grazes Mars. After a pass through the Martian upper atmosphere, the payload is decelerated by aerocapture, braking into an elliptical orbit that, at periapsis, grazes the atmosphere, and at apoapsis, reaches out to the orbit of either Phobos or Deimos. Up to this point, the velocity change requirements for spacecraft destined for the surfaces of Mars, Phobos, or Deimos would all be the same, about 4.8 kilometers per second. If the spacecraft is left alone, drag forces will quickly drop it out of orbit and allow it to "land" on Mars without further fuel expenditure. With the use of a parachute, the landing could be made survivable.

However, if the spacecraft is to land on a satellite, it must fire its engines at apoapsis to give it the additional velocity it needs to stay in circular orbit at the height of the satellite, rather than simply falling back to graze the atmosphere again. This velocity change, including a small allowance for rendezvous and landing, is about 700 (600) meters per second for Phobos (Deimos). Thus a Phobos (Deimos) landing requires a total delta V from LEO of about 5.5 (5.4) kilometers per second, compared with 4.8 for a Mars landing, 6.2 for landing on the Moon, or 4.5 for landing on the asteroid 1982 DB. This Phobos/Deimos landing scheme is termed the "PhD mission" by Singer.

It should be obvious from these numbers that it does not make sense for an expedition from LEO to Mars to land on Phobos or Deimos to take on fuel for the descent to Mars: no fuel is needed to make a direct descent to

Table 11.2
Velocity Changes for Missions in the Mars System

Mission	*delta V (km/s)*
Mars to Low Mars Orbit (LMO)	4.4
LMO to Phobos	0.54
LMO to Deimos	0.87
LMO to Mars	0.05
LMO to escape	1.43
LMO to Earth return	3.42
Phobos to LMO or Mars	0.56
Phobos to Deimos	0.74
Phobos to escape	0.89
Phobos to Earth return	2.88
Deimos to LMO or Mars	0.67
Deimos to Phobos	0.74
Deimos to escape	0.56
Deimos to Earth return	2.55

Mars, but a significant amount *is* needed to stop off at a satellite. Furthermore, even more fuel is needed to get from the satellite down to Mars! The delta Vs required to travel from various important locations in the Mars system are given in table 11.2.

Fuel delivered from Phobos or Deimos to LMO or to the surface of Mars could provide the propulsion energy required for a Mars expedition to return to Earth. We saw in chapter 10 that the availability of potential propellants (water) on the surface of Mars is excellent at latitudes greater than 30 degrees, where permafrost abounds at shallow depths. Unfortunately, we cannot be sure that the equatorial regions will be as convenient a source of water as the permafrost. If fuel must be brought down to the surface of Mars from Earth or from one of the Martian satellites for use in the liftoff of the expedition, then there are clear advantages to using Phobos or Deimos as the source of propellant.

Consider a tank of space-storable propellant, with a specific impulse of 300 seconds, that must be present on Mars to return a 20-ton payload to Earth. If we start with a two-stage rocket on the surface of Mars, each stage providing half of the 7000 meters per second required, the takeoff weight must be 270 tons. Of this, 236 tons is fuel, 20 tons is the return payload, and the remaining 14 tons is tankage, engines, and structures, all of which must in any event be shipped to Mars.

The first and most conventional assumption is that the Mars mission would be carried out in a manner similar to the Apollo landings on the Moon: there would be no attempt to use local resources, and all propellant needed for the mission would have to be carried from Earth. The total mass that must be landed on Mars is therefore 270 tons. The mass that must arrive at Mars includes the aerobrake, which would be near 19 tons. Thus 289 tons must be sent from Earth. The hydrogen-oxygen engine required in Low Earth Orbit to send this weight to Mars (a delta V of 4.8 km/s) would require a mass of 1086 tons in LEO. Above 70 percent of this mass is hydrogen and oxygen.

At present shuttle launch costs, $4000 per pound, just placing this mass in LEO would cost $9000 million (and 36 shuttle flights!). This is unfortunately still an impossible scenario, since the present three shuttle orbiters could hardly be expected to average more than 12 flights per year: liquid hydrogen is not a space-storable propellant, and could not be accumulated for 36 months or more; it would boil away in a few days. Thus this scenario would also require the construction in LEO of a very large cryogenics handling facility with active refrigeration. This is expensive.

The second possibility is that fuel will be made on Phobos and brought down to the surface of Mars. A processor, emplaced on Phobos long before the arrival of the manned expedition, produces space-storable propellants such as methanol or methane for the exclusive use of the returning expedition. The mass of propellant that must be landed on Mars is 236 tons, plus 14 tons of tankage. The mass of fuel burned to drop the payload into the atmosphere is 76 tons. Thus 312 tons of fuel must be made on Phobos. The mass landed on Phobos must be at least 14 tons for tankage and engines, 14 tons for an aerobrake, and two tons of extraction equipment. The mass sent directly to Mars is the 20 ton return vehicle with its 1.5 ton aerobrake. Thus the total mass arriving at Mars from Earth must be about 52 tons, and launch costs are therefore approximately $1540 million.

In the third case, if fuel is made on the surface of Mars, the only penalty incurred by the lander is the weight of the processing equipment and power supply to make the fuel. That seems unlikely to be more than one or two tons. The weight delivered to the surface of Mars must then be 34 tons. Counting the weight of the aerocapture heat shield, the mass sent to Mars must be about 37 tons. If we then consider the size of the hydrogen-oxygen stage that must be used to kick this weight from Low Earth Orbit to Mars (a delta V of 4.8 km/s), a weight of 139 tons must be placed in LEO initially.

Of this, 96 tons is liquid hydrogen and liquid oxygen. The transportation costs for lifting this weight from the surface of the Earth to LEO are, at $4000 per pound, just under $800 million for the fuel, or $1100 million for the entire expedition. Three shuttle flights would be required to deliver the cryogenics to LEO, and all the liquid hydrogen could be delivered in the external tank on a single flight.

The fourth case is most striking: to use asteroidal water to make hydrogen and oxygen in LEO at the space station. In this case, the cost of the propellant would be much less than the cost of lifting it from Earth. The weight that must be lifted from Earth then declines to roughly a single shuttle payload, for a transportation cost of only $250 million.

The choice between these possibilities is not difficult: staging a manned Mars expedition without using space resources is extremely expensive and barely feasible. A conventional "conservative" approach would guarantee that flights to Mars would be so expensive that they would not be attempted more than once. In short, the Mars flight program would be a dinosaur doomed to die of its own crushing weight, another enormously expensive dead end like the Apollo program.

If space resources are used, repeated flights become feasible. The differences in cost between making propellants on Mars and making them on Phobos are not large; such factors as the choice of landing sites on Mars (permafrost or no permafrost) and the technical details of operating processors on Phobos or Deimos instead of Mars could be decisive. If both types of propellant production become feasible, then the satellites may be of more value as staging and refueling points for missions to the inner part of the asteroid belt than for Mars missions. It is possible that, several decades from now, most of the people passing through the Mars system will never even set foot on Mars.

But Fred Singer has also pointed out that Phobos and Deimos have another, perhaps even more important, role to play in opening Mars to exploration. The first manned flights to the Mars system could be targeted at the satellites, not the planet. Indeed, many scientists would place the geological exploration of the satellites at a level of importance comparable to that of a manned Mars landing. Such expeditions could set up or expand a propellant-producing facility on one of the satellites, use the propellant to carry out maneuvers such as visiting and exploring both of the satellites, and even skim through the upper atmosphere of Mars itself. Satellite-derived propellant would then be used for the return to Earth. Return to

Earth from Mars orbit, or from the surface of Phobos or Deimos, is far easier than return from the surface of Mars itself. Such a mission would be very rewarding in its own right, would be much easier than a return from the surface of Mars, and would provide both facilities and experience in the Mars system upon which future Mars surface missions could be based.

12

WHAT SPACE RESOURCES WILL LET US DO

Earth is the cradle of mankind; but one cannot live in the cradle forever.—Konstantin Tsiolkovsky

CUTTING COSTS TO LEO

Almost all of the potential large-scale activites that are now being contemplated by the American and Soviet space programs would profit enormously from the use of space resources. However, there is one area that must be mentioned first because it is of extreme importance, because it is one of the very few areas that would *not* immediately benefit from space resources, and because it is the major expense in many future programs. That is the cost of launching payloads from the surface of Earth into Low Earth Orbit. Indeed, even those programs which bring back space materials to LEO at much lower cost will for decades to come be dependent upon Earth launch of all of their complex and high-technology components. It is likely that the expense of extraterrestrial materials in LEO will still be dominated by costs of launch from Earth. Thus, the reduction of launch costs is fundamental and contributes to the economic feasibility of *all* competing sources of materials in comparable ways.

Having said this, it should now be clear that two separate tacks must be followed. First, the cost of launch from Earth must be brought down as far as possible. Second, the fraction of the components lifted from the surface of the Earth to LEO must be minimized, and the fraction derived from space resources must be maximized.

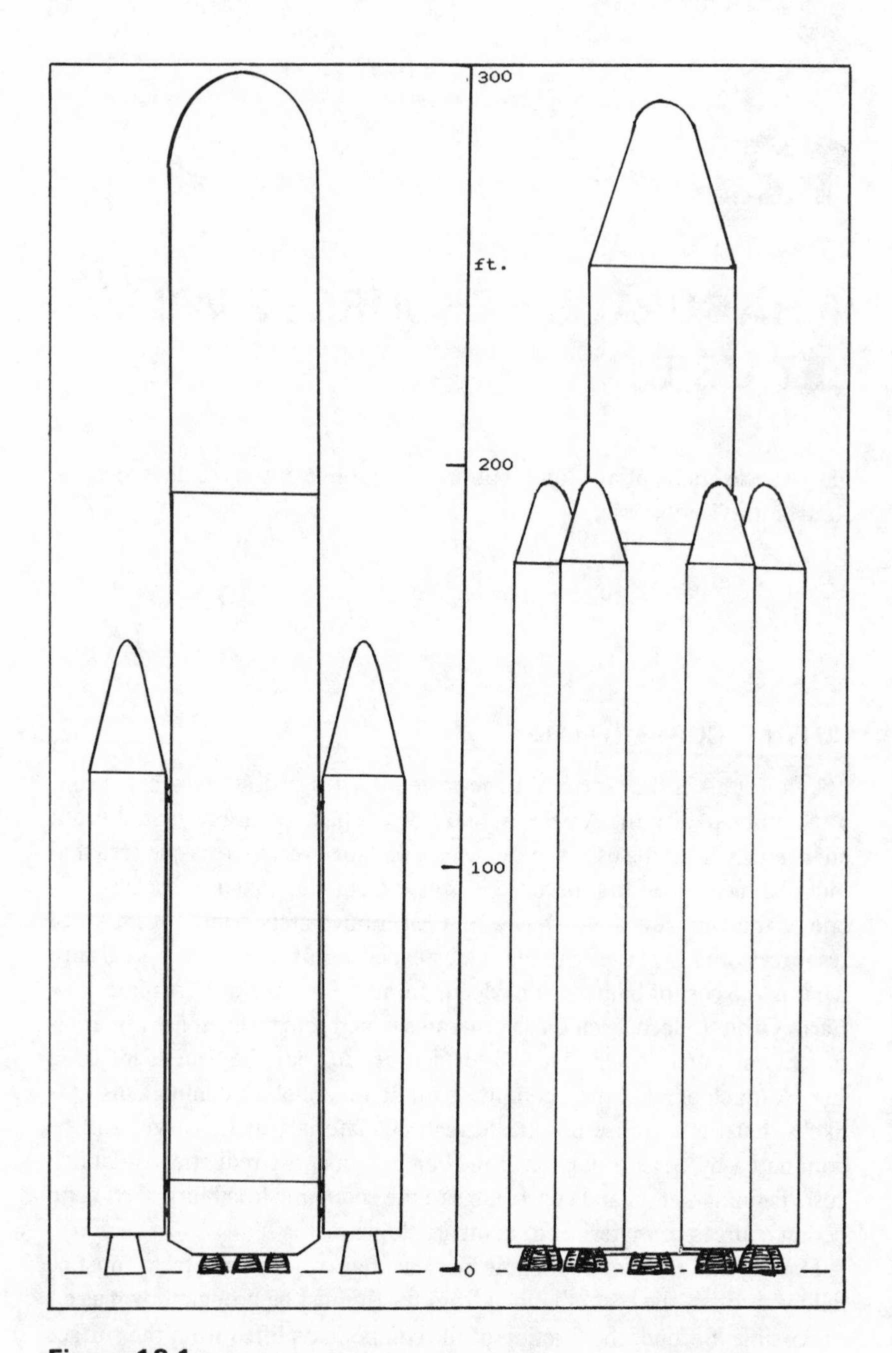

Figure 12.1
American and Soviet Shuttle-Derived Launch Vehicles
An unmanned cargo-carrying booster derived from Space Shuttle technology is shown on the left. The Solid Rocket Boosters (SRBs) are evolved

The prospects for achieving drastic reduction of launch costs in the near future are not bright. If, as we have done so far in this book, we restrict our attention to the years between now and 2010, there are only two possible developments that might bring down launch costs appreciably. They would be the development of an unmanned launch vehicle based on Space Shuttle technology, but not using the winged, reusable shuttle craft, and the orbital version of the Aerospace Plane discussed in chapter 5.

Many different conceptual designs have already been explored for a Shuttle-Derived Launch Vehicle (SDLV), but no firm decisions have been made concerning the design and capabilities of this booster (see fig. 12.1). Nonetheless, a typical design would use two or more huge strap-on boosters like the solid rocket boosters (SRBs) on the Shuttle. These solid-propellant boosters would be attached to a core stage based loosely on the present External Tank (ET), but, unlike the ET, equipped with its own engines. The orbiter would be replaced by an unmanned cargo cannister. The exact size, payload capacity, and other characteristics of the SDLV, and the degree of reusability expected of it, will probably remain undecided for years to come.

There is a small possibility that the loss of Challenger may stimulate a decision to accelerate the development of the SDLV. If the payload capability of the SDLV can be made large enough, the Department of Defense may support its prompt development. Their interest will be es-

---

from 1986 post-*Challenger* designs. The winged, manned Shuttle orbiter has been replaced by an unmanned, nonrecoverable cargo cannister, and the Space Shuttle main engines have been moved from the orbiter to a recoverable pod on the aft end of the External Tank (ET). Cargo may be carried either on the side of the ET in place of the orbiter, or on the front end of the ET. The Soviet Space Shuttle design uses four liquid-propellant strap-on boosters plus a very large core stage analogous to the Shuttle-Derived stage shown at left. The Soviet Shuttle will from the start operate with its main engines on the ET, not on the orbiter. Thus replacing the orbiter with a nonrecoverable unmanned stage would be simple. The cargo carrier would either be placed on the side of the ET (in this view, the far side of the vehicle) or atop the ET. In both cases, the Shuttle-Derived Launch Vehicle (SDLV) could be modified to use more strap-ons and carry heavier payloads.

pecially great if the Strategic Defense Initiative research program succeeds in developing a requirement for large-scale deployment of space systems, and if the SDLV can deliver payload for substantially lower costs than the Space Shuttle.

The hope of the designers of the SDLV is to achieve marginal costs (the cost of adding one more flight; development costs are neglected) as low as $250 per pound of payload delivered to LEO. Those who feel inclined to accept this figure at face value are strongly encouraged to return to chapter 5 and reread the sorry history of launch cost reduction from the 1950s to the present.

The development cost of the SDLV is likely to surpass $2 billion. The expected availability date is after 2010 (and as late as 2050) in all except the most ambitious projections, according to Jesco von Puttkamer, an advanced-program planner at NASA headquarters. The minimum development lead time for such a large, ambitious undertaking is about three years, and the program is so large that it could not expect to receive funding until the Space Station program has passed its funding peak. It is therefore not reasonable to expect the initiation of the SDLV program until at least 1994, with its operational date close to 1997 or later.

However, the loss of Challenger again enters in: with only the three surviving orbiters, it is very hard to see how NASA could support the Space Station flight schedule along with its many other committments. It is possible that the development of low-cost heavy-lift booster may be given higher priority than the Space Station in order to assure that, once the Space Station is operational, it will be possible to fly to and from it.

President Reagan's Science Advisor, George Keyworth, set in motion before his departure a serious effort to crystallize DoD and NASA support around the concept of an Aerospace Plane, perhaps to be operational in the mid 1990s, in time for the operational date of the Space Station. At present, there are several major obstacles in the way of such rapid development. First, the multiple-function engines (turbojet plus scramjet, or scramjet plus rocket) are a complex new technology that is not ready for even preliminary design of flight hardware. Second, the development costs of the Aerospace Plane can scarcely be expected to be less than $10 billion: there is no way that NASA could afford to pay for it while struggling to build the Space Station. This forces the ball into DoD's court—and if Defense pays for the vehicle, you can bet that it will be just what they want, not what NASA wants. This conflict between the different fundamental conceptions of the spaceplane is the third major obstacle to its early development: it is all things to all people.

THE LUNAR BASE

A lunar base has long been discussed as the next major program for NASA after the completion of the Space Shuttle. As such, the funding for the lunar base would be phased in as the Space Station passes its funding peak. The earliest feasible date for the establishment of a lunar base is therefore shortly after 2000. Note that the development of the lunar base would conflict directly with the SDLV program for funding. It is very unlikely that NASA could afford to do both at the same time.

But just exactly what is the lunar base? How expensive would it be? Could anything else be done at the same time, or is it another case of having to place all of NASA's eggs in a single basket?

The best available answer is that the scope of the lunar base program is still very poorly defined. Many options remain open for the simple reason that no one has a clear idea how much money NASA will have in its budget in the 1990s.

Rather than sketching out a number of different lunar bases of different levels of ambition and cost, let us pick just two distinct alternatives that lie at opposite ends of the cost spectrum.

First, there is the "cheap" version, with a pricetag in the vicinity of $8 billion. Let us call this the Lunar Research Shack (LRS). This would be a small, largely automated research facility, capable of accommodating two or three astronauts. The LRS would be manned from time to time to install and check out new experiments and repair old ones, to serve as a base camp for manned explorations using advanced roving vehicles based on those used in the Apollo program, and as the site of experiments on the use of lunar resources. The base would, from the start, use lunar soil for shielding, and as such demonstrate the simplest form of lunar resource use.

The Lunar Research Shack would be the ideal place to develop the ilmenite process for extracting oxygen from the lunar surface. It could perform a wide range of experiments, installed by astronauts and then left to run for long periods of time unattended. Monitoring the seismic activity of the Moon (to probe the lunar interior) requires very long observation times: native moonquakes are very rare because the Moon has essentially "died" in the geological sense. However, infrequent large meteorite impacts on the

Moon create seismic events that can be used for the same purpose. Also, roving vehicles from the LRS could easily deploy an extended seismic network during routine geological field work and geochemical prospecting.

Such a small station, dedicated to basic research and to the assessment and development of processes for the ultimate exploitation of lunar resources, would enjoy fairly wide support in the scientific community if certain conditions were met. First, the LRS must not be so expensive as to cause another round of "slaughter of the innocents," in which one expensive program squeezes out many very valuable but less expensive ones. Second, the return to the Moon must not be a monomaniacal manned *tour de force* designed to re-create the halcyon days of Apollo: it must present a balanced mix of manned and unmanned elements chosen to optimize cost-effectiveness and maximize the return of useful data.

The program should involve unmanned lunar mapping missions in high-inclination orbits, using philosophy and technology borrowed from the American Landsat and Soviet Priroda Earth-resources programs, the *Luna 19* and *22* and Lunar Orbiter photographic mappers, and the Galileo Near Infrared Mapping Spectrometer. It should draw heavily on the use of unmanned, automated surface sample return missions of the sort developed by the Soviet Union in the *Luna 15, 16, 18, 20, 23* and *24* mission series, which would sample remote or hazardous areas not readily accessible to manned expeditions. It would also use both manned roving vehicles and unmanned rovers similar to *Lunokhod 1* and *2*, which were flown on the *Luna 17* and *21* missions.

The second version is a permanently manned base with a staff of about 12 people, with crew rotation carried out continuously. This program would involve a commitment to maintain the base for at least a decade. The program, from new start to the end of its first decade of operation, would span about 18 years and consume about $80 to $100 billion in 1985 dollars. All the desirable unmanned missions listed above could be flown for 2 or 3 percent of this cost: however, if this version of the lunar base is funded, it would almost certainly spell the death of all such complementary programs, since they would find themselves competing against a monster for their sustenance.

Now, we would be the first to admit that $100 billion is a huge amount of money. In fact, it makes Apollo, with a runout cost of $24 billion, sound stingy. But in fact, $100 billion in 1985 dollars is almost the same as $24 billion in 1968 dollars; the difference in cost, after allowing for inflation, is actually smaller than the uncertainty in how much it would cost. So call it another Apollo. The budgetary impact of Apollo was concentrated in the years 1962–1972, a span of 11 years, while the lunar base would spend a closely similar amount of real money over 18 years. Thus the impact of the lunar base program on the taxpayer would be only about 60 percent of Apollo's.

Nonetheless, the real NASA budget is much tighter now than it was in the late 1960s. The scientific community is deeply concerned that the high-budget approach to a lunar base would leave us with far less knowledge of the Solar System, and with a far narrower range of experience, than a cheaper but broader program would offer. Further, very few obvious benefits, even that of "national prestige," would be conferred by electing to spend $100 billion rather than $8 billion on the Moon. It is fair to say that most space scientists could with good conscience endorse a small lunar station, but most would have great difficulty finding anything good to say about a ten-times-larger program that relives all of the most painful mistakes of the past. There is also a strong likelihood that the public perception will be "Go back to the Moon? Why? That's really boring! And the *cost!*"

But the lunar base is being strongly advocated inside NASA, especially at Johnson Space Center in Houston. Why? We can identify three major reasons.

First comes the Next Logical Step argument: the lunar base is pictured as picking up where Apollo left off—never mind if it's a dead end or very expensive, it's where we were when the Vietnam war killed the American space program: it's unfinished business.

Second is the Nostalgia argument: the lunar base does not require any truly innovative new technology. The new features, such as techniques for extracting lunar resources, would be largely Moon-specific, and would not rub off on other programs. This appeals to the vested interests of the Apollo-era managers in NASA, those who were educated in the 1960s, who joined NASA for the race to the Moon, and whose greatest career thrill was participation in Apollo: they can relive the old days without having to learn new tricks. Also, nowadays in Washington the argument that "no new technology is required" is considered a strong positive selling point. How the mighty have fallen!

Third, less important but much more legitimate to our minds, is the Lunar Resources argument. The advocates of this school *have* learned something since Apollo: they cite a vast array of studies done from the mid 1970s to the early 1980s that point out ways (some straightforward, others Byzantine) of exploiting lunar resources both for use on the Moon and in Earth orbit. Their rationale for return to the Moon is that it might, after all, *not* be a dead end!

Our sole argument with this school of thought is that the Near-Earth Asteroids are much more attractive than the Moon in several fundamental ways: hence this argument of the utility of space resources leads us not to the Moon, but to the asteroids. The Moon takes on significance as a source of resources principally for local use, not for the export trade.

The bottom line is simple: it makes no sense to commit to anything beyond a minimal lunar base for any known reason. A minimal Lunar Research Shack should have as its two main goals the pursuit of basic scientific research and the development of technology for processing lunar ores. If such a research program produces any economically viable scheme for export of any lunar commodity, the base should be expanded and adapted to that end. If pleasant surprises await us, then a full-scale manned lunar base program might arise out of the LRS in a natural and evolutionary manner, based on its demonstrated merits, not mandated by arbitrary imposition or political meddling.

MARS, PHOBOS, AND DEIMOS

A prominent alternative to the Shuttle-Derived Launch Vehicle and the Lunar Base is a manned expedition to Mars. Obviously, projections made in the late 1960s that placed the first manned expedition to Mars around 1980 are no longer taken seriously. Indeed, a feat that then seemed 12 years away, appears today, from NASA's perspective, to be 20 years away from now.

The reason for the increasing remoteness of a Mars landing is not that it is a technically staggering feat (there were credible schemes for developing the Mars mission directly out of Apollo hardware) but that NASA has become accustomed to doing one thing at a time. Many at NASA headquarters subscribe to the Central Dogma that the Shuttle naturally leads to a Space Station, the Space Station naturally leads to a permanently manned lunar base, and *voila,* suddenly it's 2010 and it's time to do the Next Logical Step. By then, with all the Apollo-era managers approaching 70 years of age, it may be possible to actually do something new, something outside the path foreordained by 1965 logic. Of course, we still might get more Business as Usual, with the construction of a Truly Huge, Absolutely Useless Lunar Orbiting Palace (THAULOP), or some such aberration. Anything is possible.

Well, if anything is possible, what about learning some new tricks? What about doing something more interesting than returning to the Moon, something that requires greater flexibility of thought and less cash than a permanently manned lunar base; indeed, something costing about the same as the Lunar Research Shack? That something is a manned mission to Mars.

The resources of the near-Earth asteroids and Phobos and Deimos are the key to Mars. The benefits conferred by space resources on even a single expedition are considerable. Asteroidal hydrogen and oxygen could provide the fuel to lift the expedition's vessels from Low Earth Orbit (see fig. 12.2) and hurl them on their crossing to Mars. Propellants manufactured out of the water and carbon dioxide of the Martian atmosphere could provide the energy needed to climb from the surface of Mars to Low Mars Orbit (LMO). Propellants made from the water and carbonaceous matter of Phobos or Deimos could be brought down to LMO (or even to the surface of Mars) to refuel the expedition for its return to Earth.

In addition to the enormous savings resulting from the *in situ* production of propellants, it is possible that a major part of the mass of the vehicle, especially tankage, insulation, and the aerobrake for entry into the atmosphere of Mars, might be made from asteroidal materials in LEO at the Space Station. Such a space factory could later be duplicated elsewhere, such as in LMO, within easy reach of the main asteorid Belt.

Further, the industrial base needed to carry out any of these functions would be of much broader utility than just this one mission. Permanent propellant-production facilities on Phobos and on Mars open the way for a prolonged series of missions, making a permanently manned Marsbase (or colony) an accessible option. The ability to fabricate structural members, tanks, and heat shields in LEO would be of great value to all missions going beyond LEO.

We have seen in chapters 9, 10 and 11 that three sources of space materials could be involved: Mars itself, Phobos or Deimos, and a carbonaceous near-Earth asteroid. The use of any one of these sources already would be of great help. Combine these considerations with the fact, brought out in chapter 10, that a given-size rocket in LEO can deliver far more mass

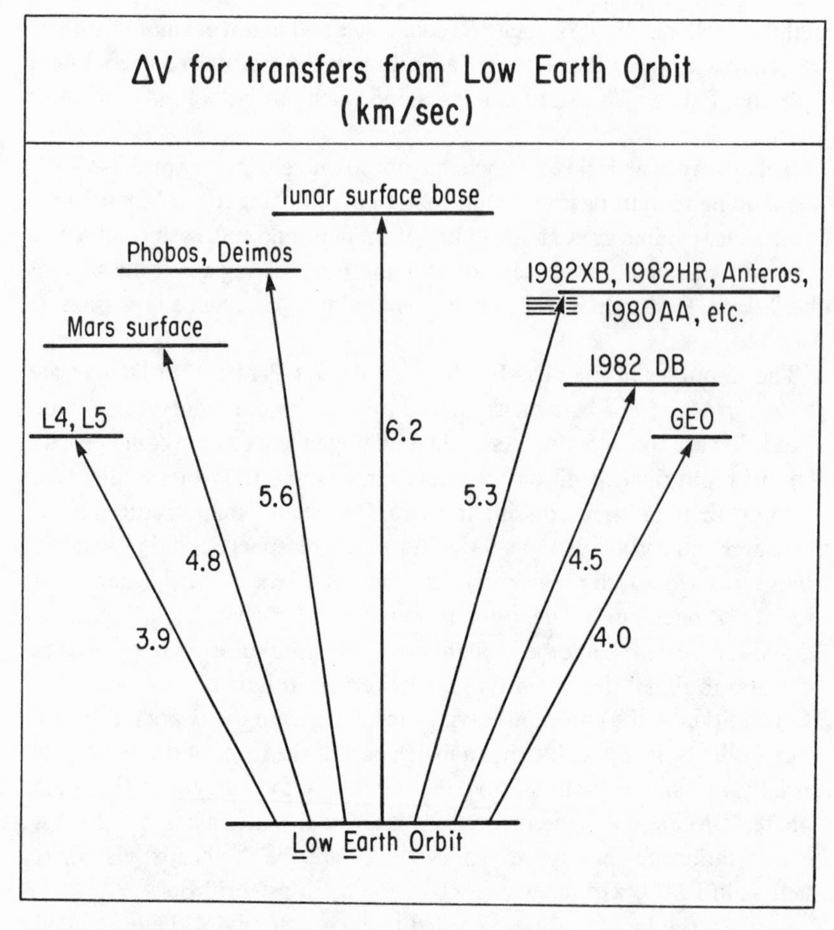

Figure 12.2
delta V for Transfers from Low Earth Orbit
The minimum velocity change (delta V, in kilometers per second) required to reach each of the specified destinations from LEO is displayed to permit comparison. The easiest destinations to reach, the L4 and L5 points on the Moon's orbit and GEO, are not sources of space resources. The easiest asteroid to reach is 1982 DB. About 76,000 other asteroids larger than 100 meters are more accessible than the Moon. Phobos and Deimos, although not as easily reached as the best asteroids, are still more accessible than the Moon.

to the surface of Mars than to the lunar surface, and the attractions of a stepwise technology-development program aimed at Mars become evident.

Another attraction is the low cost of the initial Mars visit, about $8 billion, or 8 to 10 percent of the runout cost of the large version of the lunar base, even without the cost-saving use of space resources. We could have *both* this expedition and a small Lunar Research Shack at much less cost than the large lunar base alone. If, for budgetary or political reasons, we can have only one or the other, then there is no contest: Mars and its satellites outweigh the Moon in interest, scientific importance, and resource promise by a wide margin.

LARGE-SCALE DEVELOPMENT OF LEO

The most conservative NASA visions of our future in space have us huddling in the Space Station in Low Earth Orbit for the next 30 years, growing crystals. If we seriously believed that this was the American future in space, we would immediately start studying Japanese and Chinese.

Growing crystals in LEO undoubtedly has a future. So does growing crystals on Earth. But to define our future in space in such incredibly narrow terms is just exactly as rational as defining the purpose of Earth over the next 30 years as "growing better crystals." About 0.01 percent of the manpower of this country is involved in growing crystals, and in the long run it will not be much different in space.

In the meantime, however, there are certain industrial processes that would surely benefit from the conditions in space. One example is the purification (and, yes, even the crystallization) of biochemicals and pharmaceuticals with very high unit costs. A second example seems to be the growth of gallium arsenide crystals to make extremely fast, low-power integrated circuit chips that, if they were readily available, would revolutionize the computer industry. This process, on Earth, gives a yield of about 2 percent useable chips and 98 percent junk—but GaAs crystals made in space are so good that yields of 98 percent may be attainable! A third example is the extraction of precious and strategic platinum-group metals from asteroidal metal alloys for return to Earth. In all of these cases, space processing of materials should easily pay for itself.

An industry will arise in which workers will at first be brought up from the ground to work in monastic isolation for a few weeks in zero gravity, after which they will be returned to Earth, and to their families, before any serious deterioration of their health can occur. A little later, someone will realize that it is easy to build a station in which the living quarters are spun (on the outside of a wheel) at high enough speeds so that centrifugal force provides an artificial acceleration of 0.2 to 1 normal Earth gravities. This is the famous Von Braun Wheel concept from the early 1950s, a great improvement over the American Space Station now being planned for 1996. Then workers may stay in orbit for years at a time without physical deterioration, at enormous saving in launch costs. Of course, this will create a demand for construction materials, life-support fluids, and radiation shielding in LEO, all demands that can be met at reasonable cost by the near-Earth asteroids, but not by delivery from Earth.

The only remaining step required is to make life on the space station complex psychologically tolerable by allowing the workers to bring along their families and set up housekeeping in space. This may be discouraged by policy, just as it was discouraged by the Dutch West Indies Company in New Amsterdam. And the policy will eventually fail for exactly the same reason. (Tom Jones has reminded us that neither large colonization efforts nor revolts have occurred in Antarctica—but neither has Antarctica been exploited economically to generate exports, income, and growing economic self-sufficiency. Without a local source of income, there is no incentive to stay and establish a colony.)

Just as in any centrally controlled economy, the caring efforts of a few motivated individuals will begin to provide commodities of quality superior to, or wholly different from, those brought up from Earth and dispensed through the company store. Someone will experiment with raising vegetables hydroponically using hoarded water and wastes. Someone will produce art work, some will provide counseling, some will perform music, some will develop educational programs geared to the need of a transcultural and highly technical space society, some will discover the incredible possibilities of low-gravity dance. More and more job opportunities will open up outside the range of interest of the crystal growers. Administrative positions will move from Earth to space as the complexity of the space culture increases. Some degree of autonomy will be afforded the on-site administrators. Residents of the space station not employed by the company will discover ingenious uses for the byproducts from the asteroidal processing operations and begin business on their own. Finally, businesses that are actively competitive with the parent company will arise, prosper, and buy out the parent.

Any economically viable business carried out in space will open the door to the large-scale industrialization and colonization of space. The key to this development is the availability of low-cost materials from sources in space and, at first, the preferred sources will be those that can most readily provide materials to operations in LEO (See fig. 12.3).

VERY LONG MANNED FLIGHTS

Flight times for minimum-energy missions to Mars are about eight to nine months, depending on the year of launch. The trip time for manned flyby of Venus, from launch to arrival at Venus, would be about five months. A manned flyby of Mercury would occupy about eight months each way. Round trip times are always more than twice as long as the one-way trip times because the spacecraft generally cannot depart on a minimum-energy return to Earth starting on the date of arrival from Earth. Such round-trip times for visits to the nearby planets are typically two or three years.

Manned flights beyond Mars, whether dispatched from Mars or from Earth, would presumably be targeted for main-belt asteroids or for the Jovian planets. Such trips would almost certainly not be made on minimum-energy trajectories because very substantial trip-time savings can be realised on the longest trips by modest increases in departure speed from LEO. Nonetheless, trip times will be at least three to ten years for all such missions.

The classical method of provisioning a spacecraft for prolonged trips is to include a "pantry" with the necessary supply of dried, preserved, or otherwise insulted food. Can you imagine spending a year or so eating C rations? If you think you can, chances are you never have. Their food value is seriously degraded by prolonged storage, and, perhaps equally important, the morale of a crew subjected to such a diet would suffer severely. The provision shipments arriving at the Salyut space stations on Progress resupply vehicles often contain gustatory goodies to sustain the crew, and of course Soyuz ferries often bring up two or three visitors for a few days at a time to provide a welcome diversion. Such diversions will not be available to astronauts sailing the interplanetary void. But what alternative is there?

A diet containing freeze-dried noodles, toothpaste-tubes of borscht, and Tang is not objectionable if these foods are interspersed with a wide variety of other foods. Even better, if fresh produce were available to the extent of 10 percent of the diet, such preserved foods would fit naturally and inoffen-

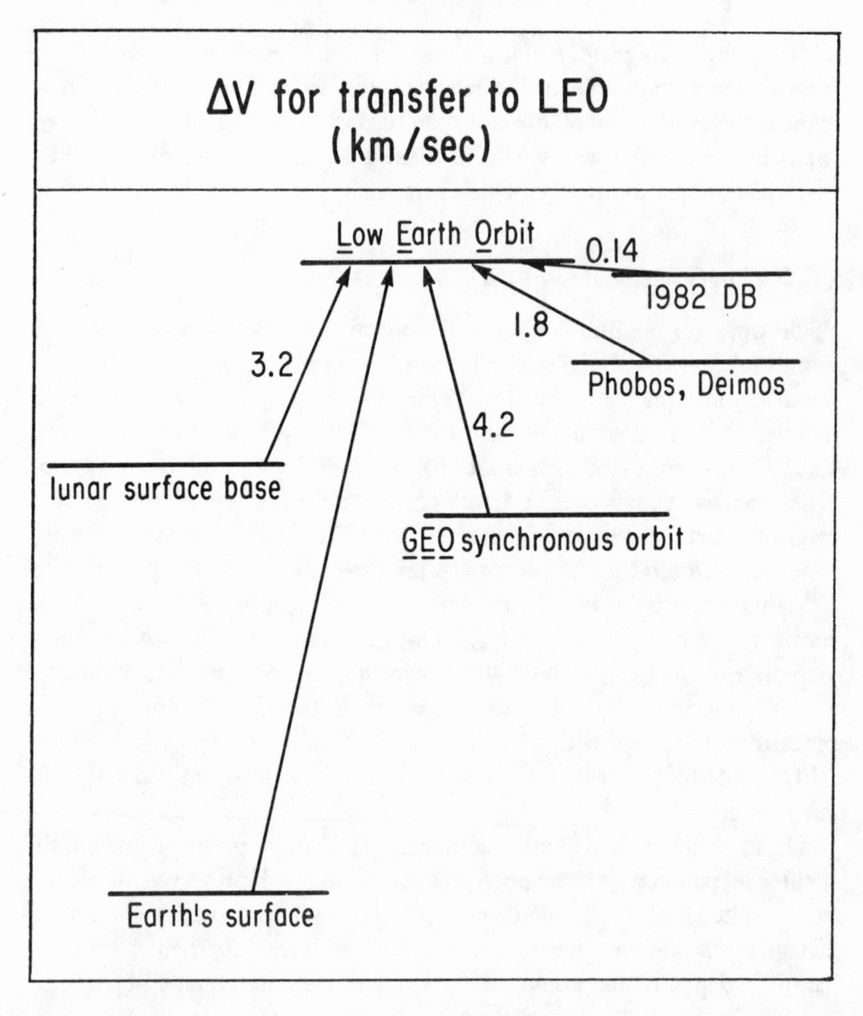

Figure 12.3
delta V for Transfers to Low Earth Orbit
The minimum propulsive velocity change (kilometers per second) required to return to LEO from each of several locations is given for purposes of comparison. Note the enormous superiority of the near-Earth asteroids (here represented by 1982 DB) and the very attractive position of Phobos and Deimos. The figure of 0.14 km/s given for 1982 DB allows 0.06 km/s for return to entry into Earth's atmosphere, plus 0.08 km/s allowance for orbit circularization to rendezvous with the Space Station. If the payload is instead captured by a tug and towed to the Space Station, the number will be close to 0.06.

sively into the diet. But when you are 100 million miles from the nearest planet, having fresh produce means growing your own.

To date, almost no attempt has been made to grow food in space: all manned spaceflights with durations greater than a few days have been in LEO, close to Earth-surface sources of food. "Seed-to-seed" experiments have been carried out on Salyut, demonstrating the feasibility of zero-gravity food production. Also, intensive research into hydroponic techniques on the ground has demonstrated incredible crop yields of high-quality produce. Productivities are so high that hydroponic culture in greenhouses can now compete profitably with dirt farming.

The time is ripe for serious experiments with space agriculture. The Soviets, looking ahead to a long and exciting future in space, began such experiments several years ago. Carbon on present-day space missions is carried up as prepared food, ingested by the astronauts, partly oxidized to carbon dioxide to liberate energy, and excreted as body wastes and exhaled as carbon dioxide gas. Chemical air purifiers such as lithium peroxide absorb this carbon dioxide and release oxygen. When the absorbent is used up, the old material (now lithium carbonate) must be discarded and a new purifier installed. Body wastes are collected, bagged, and discarded.

A planet designed to operate like this would be a total failure! Valuable materials should be recycled, not discarded. A carbon cycle should be established on a space station by growing a sufficient mass of plant material to provide useful quantities of food and to consume the carbon dioxide released by the respiration of the astronauts. It is not easy to design and build a highly complex, fully closed, completely self-sufficient environmental system in one step (Earth took billions of years to come up with the present system)—in fact it is currently impossible! This argument has been used by defenders of the status quo to argue that space agriculture should be ignored.

Our conclusion is exactly the opposite. For long-duration flights, the mass that must be devoted to growing a small part of the travelers' diet is at worst about the same as the mass of additional stores that would have to be carried. At best, the ability to grow food confers major advantages in saving weight, as well as the health and happiness factors discussed above. We strongly urge that "greenhouse modules" be developed for early use in space. They make little sense on flights with durations less than one or two months, but should generally be carried on all longer missions, *even if they confer no weight advantage,* as valuable technology-development experiments.

If space agriculture lives up to its apparent promise, the relationship of this technology to space resources is profound. First, the components of the hydroponic greenhouse are structure, windows, water, soluble nutrients, carbon dioxide, and human and animal wastes. Of these, the water and soluble nutrients, and even carbon dioxide, can readily be supplied from carbonaceous asteroidal material. The structure itself would be as simple and cheap as possible: one easy solution would be high-purity corrosion-resistant iron or nickel from asteroidal metal alloy. The windows may be any translucent silicate or oxide composite made from any of a wide range of iron-poor silicates, and certainly do not need to be clear window glass. The human and animal wastes and respiratory carbon dioxide consumed by the plants represent a bonus, since they save mass that would otherwise be discarded. Space agriculture not only uses available space resources extensively, thus minimizing the amount of mass that must be lifted from Earth, but it also uses waste material that otherwise would be discarded, and would have to be replaced at great expense with more provisions and air-purification equipment from Earth.

The limits to which space agriculture can be pushed are far from clear. It is reasonably obvious that a little experience would permit the design of a second-generation module that would better fit the input and output needs of the space community it serves. Whether total self-sufficiency in food can be achieved will not be known until years of experience have accumulated. But much of this experience can be gained at modest expense in closed-system studies right here on Earth.

The contribution of such experiments to the development of space settlements and Mars colonies is obvious. The needed experiments are simple and inexpensive, and could be begun at once on the ground, using exclusively terrestrial materials.

A complementary research program would be to search for efficient ways to make the necessary components of the space farm out of carbonaceous and ordinary chondrite material. Such knowledge might permit us to build entire space-farm modules out of readily available extraterrestrial materials, thus saving large amounts on launch costs.

SOLAR POWER SATELLITES

Our search for space resources to exploit has generally focused our attention outward, away from the Sun. The bodies that lie between Earth and the Sun are Venus, Mercury, and a small number of Aten asteroids. While some of the Atens may be made of useful material and may be energetically accessible if they approach Earth closely, Mercury and Venus are hopeless as sources of useful materials because of their deep gravity wells and energetic distance from Earth. Yet our fascination with more distant bodies should not obscure the fundamental importance of the Sun itself, a prodigal source of light and power.

There is no shortage of power in space. A single square meter of area facing the Sun near Earth's orbit receives more than 1000 watts of power continuously, most of it between the wavelength of 0.3 micrometers in the ultraviolet and 2.5 micrometers in the infrared. The peak emission from the Sun is in yellow light, near a wavelength of 0.65 micrometers.

High-efficiency solar cells can turn 1000 watts of incident sunlight into 200 watts of electricity (and 800 watts of waste heat). There is active research into ways to improve this efficiency, and it is likely that, within a decade, significantly higher efficiencies will be possible.

Let us imagine a space structure in geosynchronous orbit (GEO) nearly 40,000 kilometers above the equator, A square 10 km on each side is filled with a lightweight framework of beams and wires. Suspended from the frame is a complete covering of thin, high-technology solar cells. The structure rotates once relative to Earth on each orbit, so as to always keep the solar cells facing the Sun. A single such structure would intercept 100 billion watts of sunlight and could generate, *even with present technology,* an incredible 20 billion watts of electrical power. The average American home uses less than a kilowatt: this is enough energy to power 20 million homes housing 80 million people! Sixty such power satellites could produce enough power to provide everyone on Earth with the same amount of electrical energy now used by the average American or Western European.

The power from these satellites would be broadcast as microwave energy to huge "antenna farms" covering many square kilometers on Earth. The received microwave energy would be used to generate electricity, which would then be fed into the power grids. There have already been discussions of building a connection between the Siberian and North American power grids across the Bering Strait. Combined with Solar Power Satellite (SPS) energy, this would present global delivery of electrical power.

Building a network of Solar Power Satellites out of terrestrial materials, while hideously expensive, is not much worse than building an equivalent amount of power-generation equipment on the ground. Note that every generator ever built has a finite life: they all will have to be replaced sometime in the next 50 years in any event.

Studies done in 1977, however, show that use of lunar materials to build the SPS system would lower the cost enough to make them competitive with terrestrial power generation. That study, remarkably enough, was given the list of materials that arose out of studies of Earth-launched SPS systems, and was not permitted to optimize for the use of materials actually available on the Moon.

Since 1977, however, it has become abundantly clear that all the necessary resources are available from near-Earth asteroids at much less cost than from the Moon (see fig. 12.4). The time is ripe for a reassessment of SPS technology with asteroidal resources in mind. It seems obvious that the resulting SPS system would be economically superior to any other power source studied to date, and would eliminate coal strip-mining and Co_2 pollution on Earth.

Consider a requirement for 1 million tons of metal in GEO: the equivalent mass that must be launched from LEO is about 3 million tons, of which most is fuel (hydrogen and oxygen). If that mass were lifted from the ground, 100,000 shuttle launches would be required. Replacing the shuttle with a Shuttle-Derived Launch Vehicle with a 100-ton payload capacity reduces the number to a mere 30,000 launches. At one launch per day, it would take a century to place that much mass in orbit!

Alternatively, consider two factories located in LEO, one making hydrogen and oxygen fuel from asteroidal water, and the other making metal structural members, wires, cables, etc. from asteroidal metals. In addition, each asteroid on which mining operations are under way would have a small "factory" that makes up low-tech heat shields out of local dirt and, perhaps, an organic resin brought from Earth (microwave sintering of the material may be preferable).

For each ton of material launched from LEO to an asteroid, approximately 1000 tons of asteroidal material will be returned to LEO. The million tons of metal and 2 million tons of hydrogen and oxygen propellants needed to get to GEO could be returned by a mere 3000 tons of equipment sent from Earth to nearby asteroids. The total mass that would have to be launched to carry this cargo from Earth to LEO could be lifted by 100 shuttle launches or by 30 SDLVs.

Direct SDLV launch cost for the two cases would be $1.5 trillion for

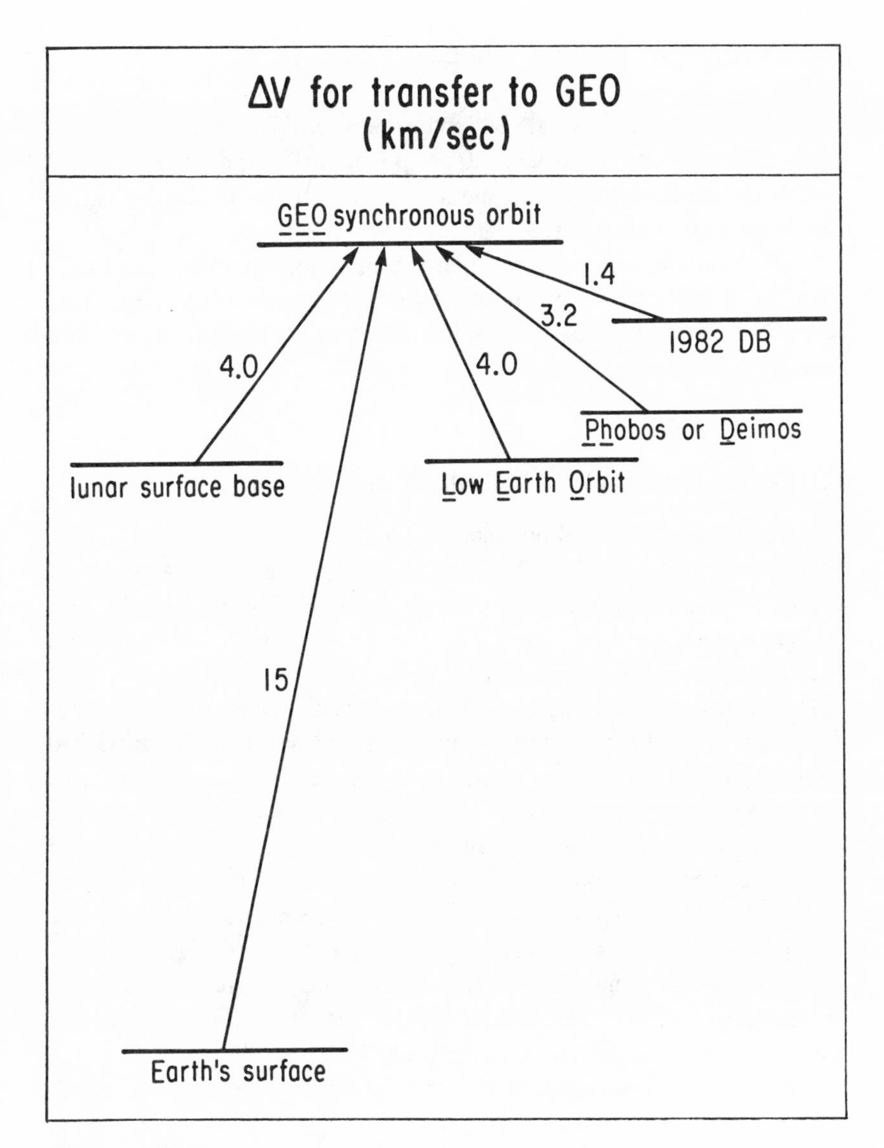

Figure 12.4
delta V for Transfers to Geosynchronous Orbit
If any form of large-scale activity is contemplated in GEO (such as, for example, construction of Solar Power Satellites) the best sources of construction materials and propellants would be the near-Earth asteroids. Phobos and Deimos would also be somewhat better energetically than the Moon. The Moon would be a far more attractive source than the surface of the Earth, *if* rocket propellants are available on the Moon at reasonable cost.

Earth-derived material vs. $1.5 billion for the asteroidal scenario. These estimates both assume the same $250 per pound tariff for the SDLV, which we believe may be two or three times too low to be realistic. But the *relative* cost of the two options is still significant.

Aside from the issue of the environmental impact of 30,000 SDLV launches, I believe the launch cost figures speak for themselves: Solar Power Satellites are economically feasible if they are made from space materials.

ASTEROID-HOPPING TO THE PLANETS

The idea of using hollowed-out asteroids as "traveling hotels" goes back at least as far as J. D. Bernal in 1929. Cox and Cole, writing in 1964, borrowed a number of older illustrations from Roy Scarfo of General Electric. This idea resurfaced later in the book *The High Frontier,* by Gerald K. O'Neill, and can still be encountered from time to time.

The principle is basically to find an asteroid on a convenient orbit for shuttling back and forth between, say, Earth and Mars. The asteroid just passively follows its eccentric orbit, like a great ocean liner too big to stop, and passengers arrive and depart via small, crowded shuttlecraft. The hollowed interior of the asteroid allows a vast amount of internal volume with a high level of protection from radiation and micrometeoroid bombardment. There is food production in solar greenhouses on board the asteroid-ship, and those coming aboard need only bring luggage appropriate for a short outing. Everything is available on the asteroid. A suitably strong, processed asteroid could be spun to provide artificial gravity (up would be down, and down, up . . . more or less).

Some saving on fuel is possible at each terminus because of the very small payload that needs to be transferred, and some is lost because the orbit of an asteroid can never be as energetically efficient as a computer-designed transfer orbit. The big difference is one of the comfort of the traveler. The difference in cost is also significant: imagine what it would cost to hollow out, strengthen, spin up, equip and furnish a multikilometer-sized asteroid! The cost of simply getting the necessary equipment to the asteroid would be enough to bankrupt any average-sized continent.

More plausible in the foreseeable future are asteroids in decent transfer orbits with small permanent installations on them, like service plazas on in-

Figure 12.5
Mom and Pop's Truck Stop

terstate highways. There will be a filling station, a small agricultural module, a restaurant, and maybe even clean rest rooms.

Among the hordes of near-Earth asteroids, few will have both desirable orbits and abundant water. These will be the object of real estate speculation of a new kind.

The availability of filling stations on a number of carefully chosen asteroids should, in short order, open the entire asteroid belt to manned exploration and exploitation. A prospector hired by a major ferrous metals concern would be shipped at great expense to the LEO Space Station.

There he would be transferred to a small "bus" that would run him out a million kilometers or so to the next regularly scheduled outbound asteroid, heading for Mars. Near Mars he will take a transfer to the number 9 bus for Ceres, using a Mars swingby for gravity assist, and a few months later he will be in the heart of the belt. Once he is in space, he represents a considerable investment by his employer—but almost all of the investment was the purchase of the ticket from Cape Canaveral or Tyuratam (Tanegashima, Woomera, Kourou, Sriharikota, Shuang Cheng Tzu . . . .) to LEO. All the rest of the transportation is provided by asteroidal propellants. Without them, none of this makes any economic sense. With them, the Solar System is ours.

BOOTSTRAPPING SPACE INDUSTRIES

So far we have concentrated on the ways that space resources may be used to serve other ends. We have seen that space-fabricated commodities, especially fuels, life-support fluids, and low-tech structural materials such as plates, beams, wires, cables, insulation, etc. would enable large increases in the scale of human activity in space. They do so by replacing simple, massive components that would be extremely costly to transport into space from the surface of the Earth with space materials that require little or no transportation expense.

But now we must allow for a factor that we have not so far considered explicitly: space resources may also be used to make space resources more accessible! This principle is called "bootstrapping."

As the first example of bootstrapping we will consider several different schemes for returning asteroidal material to LEO. These schemes may represent alternative plans for initial development, or, in a more conservative program, they may be viewed as evolutionary steps in a development sequence that takes decades.

Scheme 1. A hydrogen-oxygen rocket stage carrying an asteroid mining machine, a return aerobrake, and a small amount of fuel for the return trip is launched from LEO to the asteroid 1982DB. The ascent engine accelerates the probe to an intercept trajectory to the asteroid, burns out, and is discarded. The vehicle proceeds on to the asteroid, lands, digs enough dirt to fill the cargo allowance of the return aerobrake, and hops off the asteroid on an Earth-return trajectory, leaving the mining machinery behind. The aerobrake is used to decelerate the return vehicle and capture it into a highly eccentric Earth Orbit (HEEO). The heat shield is given a slight kick at apogee and placed in a stable HEEO, and the virtually unshielded payload, after passing through the upper atmosphere many more times, slowly spirals in to LEO. The heat shield will be picked up and reused by the next outbound mission. This mission returns 2.6 pounds of asteroidal material to LEO for every pound originally placed in LEO. (We say that the mass payback ratio, or MPBR, is 2.6.) This scheme involves NO bootstrapping.

Scheme 2. An ion engine with a very high specific impulse is used to lift the payload from LEO and carry it to the asteroid. Everything else is the same as in scheme 1. The mass payback ratio reaches 18. Again there is no bootstrapping.

Scheme 3. The aerobrake from scheme 1 is reused several times. The MPBR reaches about 8.

Scheme 4. A solar thermal engine using hydrogen as the working fluid, with a specific impulse of 1000 seconds, is used for the transfer from LEO to the asteroid, and all else is the same. The hydrogen is made out of asteroidal water returned by previous missions. For the first flight the mass payback ratio is only 4.4 (because there is no hydrogen fuel from previous flights!). After a few flights, the mass payback ratio levels off at about 88. This is an example of fuel bootstrapping.

Scheme 5. The routine in scheme 4 is altered by adding the ability to manufacture crude, inefficient aerocapture heat shields at the asteroid where mining is taking place, and by returning all the heat shields to LEO for processing and consumption there. Since almost all the Earth-derived mass that departs LEO in the previous scheme is the heat shield, this permits a further increase in the mass payback ratio by a factor of about 10 to 20, bringing the mass payback ratio up to somewhere in the range from 900 to 1800. This is an example of bootstrapping fuel and structural (aerobrake plus payload container) material.

But what if we're really ambitious, and we want to make the mass payback ratio infinite? The main reason it isn't infinite is that all the high-tech equipment is made on Earth, and eventually wears out after several years of use in space. But what if the mining vehicle, etc. are designed so that the parts sensitive to wear or degradation are easily removable modules? Then periodic servicing and replacement of modules could be carried out without the presence of men by equipment sent to the asteroid. Thus only the mass of the wear-sensitive parts, a small percentage of the total mass, needs to be replaced. The payload mass returned by a given mining machine before it dies is doubled or quadrupled, while the mass that must be brought up from LEO is increased by only a few percent. The mass payback ratio now is in the range from 2000 to 7000. This is a result of bootstrapping combined with a long-term design strategy that uses self-diagnosis and redundancy, and makes automated repair easy. Considering that the Space Station budget planners already envision spending $800 million on robotics and automation, it is possible that all the diagnosis and repair procedures we need will be fully developed at their expense.

If the transportation cost to lift a pound of Earth material to LEO is brought down to $1000, then the cost of asteroidal materials in LEO can be brought down to less than a dollar a pound. This is comparable to long-distance trucking tariffs.

The moral of this story is clear: returning asteroidal material to LEO for use in other programs is itself very desirable, but investment in improving the mass payback ratio of the mining system provides an absolutely overwhelming leverage. It makes clear economic sense to use a system with a mass payback ratio of 18 to return fuel to LEO for use by a Mars expedition or Solar Power Satellite construction, but it makes even better sense to fold back a percentage of the cost savings into bootstrapping of the space resources program. Such an investment may provide such large improvements in the mass payback ratio that almost anything we can think of doing in space would become economically feasible.

. . . AND ON TO NEW TECHNOLOGIES

The solar thermal engine described in chapter 5 and referred to above is an extraordinary piece of technology: it is very high-tech in a sense, but also disarmingly simple. The three main essentials (the solar collector, rhenium thrust chamber, and liquid hydrogen propellant) would be far more attractive if they could be made in space. We have already lavished considerable attention on the production of hydrogen from asteroidal water. But what about the refractory metal thrust chamber? On Earth, these chambers are made by chemical vapor deposition of rhenium from a gaseous rhenium compound. In space, rhenium and other refractory metals are byproducts of the processing of asteroidal iron-nickel alloy. Thrust chambers could be "grown" in a mold, out of a gas such as rhenium carbonyl chloride, at temperatures of only a few hundred degrees. The only major component we do not yet know how to make is the solar collector mirror.

Another promising technology that we have mentioned for eventual use in space is the use of laser chemical deposition to "write" extremely thin metal films, complete with reinforcing wires, directly onto a substrate such as a plastic film or water-soluble salt surface. A low-power laser, emitting a narrow, intense beam of light, is shot through a vapor such as nickel carbonyl at a target surface. The laser wavelength is chosen to pass through the vapor without absorption, so all of the laser energy is delivered to the target surface, where it heats a tiny spot to temperatures of several hundred degrees. The vapor decomposes on the hot spot, depositing a thin film of nickel metal. As the laser "writes" on the target surface, a trail of thin metal film is left behind. As the metal film builds up it quickly becomes highly reflective, like a metal mirror, and also becomes an excellent conductor of heat. This very precisely limits the thickness of the metal film: beyond a certain thickness, the metal reflects and conducts heat away from the laser spot so fast that the spot is no longer hot enough to decompose the nickel carbonyl gas and deposit more metal. The more intense the laser, the thicker the resulting film.

As an example, let us use a water-soluble substrate of salts extracted by leaching a carboaceous asteroid with liquid water. Let us then imagine "writing" a network of reinforcing wires on a smooth meter-square salt surface with an intense laser beam, then defocusing the beam and "painting" the entire surface with a thin layer of metal. The salt is then gently dissolved away, leaving a super-thin sheet of metal film with wire "handles" at the corners. These sheets could be far lighter than any brought up from the ground because they do not need to be strong enough to suffer the stresses of folding, the crushing acceleration of blastoff from Earth, and the lesser but unavoidable contortions of unfolding in space.

These sheets would be ideal for use in light sails. And here, space-based technology takes a step into a fascinating new realm of possibilities: the performance of a solar sail is not limited by the rocket equation. There is no fuel to carry, no engines, no huge insulated tanks. The solar sail is limited in only three ways. First, it requires sunlight: since the intensity of sunlight drops off with the square of the distance from the Sun, the solar sail is of most use in the inner solar system. (At Saturn, a sail would deliver only 1 percent of the force that it would near Earth's orbit.) Second, the sail's performance increases as the thickness of the sail decreases: there is less mass to move relative to the amount of solar energy being intercepted. For this reason, space fabrication of ultra-thin films is especially promising—in fact, space may be the only place where they can be made. And finally, good (that is, ultra-thin) sails are ultra-fragile and sensitive to any kind of drag forces. They cannot be used casually deep in the gravitational field of any planet that has an atmosphere. They would be stopped dead by even the incredibly low air densities found 1000 km above the Earth's surface. Also, they are vulnerable to being torn and bent by gravitational tidal forces when they are very near a massive body.

Used with proper respect for these limitations, solar sails can deliver very impressive performance. Accelerations can be so high that Hohmann transfers will seem a silly waste of time. Trip times from the Moon to Mars could be cut from eight months to two or three weeks. The lifetimes of the sails might be many years, so lengthy flights could be undertaken without fuel. Micrometeoroid punctures present no problem: a properly designed sail would be segmented and reinforced by occasional wires, so that tears could not propagate. Sails could easily accelerate to speeds well in excess of the escape velocity of the Solar System. And, last but not least, sails could be driven by lasers as well as by the Sun. A century from now, solar sails may be the preferred method of moving cargo between points in the inner Solar System. It is hard to imagine a more ecologically sound method of traveling about.

Oh yes, don't forget solar thermal engines! If very high thrust levels and accelerations are needed, solar sails won't help. But solar thermal propulsion is ideal for such uses: it provides high thrust levels along with specific impulses as high as about 1200 seconds. And now we even know how to make solar collectors for them in space—by using light sail fabrication techniques.

We also should keep in mind that technology for making thin-film solar cells (to convert sunlight into electricity) is advancing rapidly. In space, it may be possible to fabricate solar cell panels that are much lighter than any that could survive the stresses of launch from Earth. If so, Solar Power Satellites may be considerably lighter, and therefore more economical, than everyone has assumed in the past.

13

CURRENT PLANS AND GOALS FOR SPACE DEVELOPMENT

People do not lack strength: they lack will.—Victor Hugo

What do the spacefaring nations of Earth intend to do in space in the next quarter century? It has always been easy to discover American civil plans because of the openness of the US space program, including budgets, congressional hearings, launch schedules, and status reports on programs. American military plans have become much more closely guarded, but enough is leaked to the press (who, after all, are fanatical about publishing news of all kinds of secrets) to keep the interested onlooker informed.

The Soviet space program was, from about 1960, under a rather continuous shroud of secrecy. Their program was also, however, very logically and conservatively structured, so that observers with great patience and keen powers of deduction were frequently successful in deciphering trends and intentions. In recent years, the day-to-day operations of the Soviet civil space program have become progressively more visible. This is probably a result of decades of frank and open American discussion of all of their problems. Such openness involves people in the issues and helps them identify with the participants, who are not (like Korolev) shadowy, remote figures hidden behind a veil of secrecy. Possibly this new candor is also partly due to the devastating experiences of the *Soyuz 1* and *Soyuz 11* disasters, in which four cosmonauts died in flight: revelation of these problems brought forth not ridicule and contempt from the rest of the world, but grief and sympathy for the dead heroes.

Whatever the cause, however, the Soviet Union has become extremely frank about many of its specific future mission plans. While it still helps to have a crystal ball to read their long-range purposes and motivations, it is often sufficient just to have a newspaper to read the list of missions they will attempt in the next decade.

This trend toward open discussion of their future plans in the planetary and manned space programs is accompanied by increasing involvement with Western scientists, especially the very fruitful collaboration they have built up with France. We refer here not to the flight of a French astronaut on *Soyuz T6* (an event of purely ritual political significance), but to French participation in a series of Venera spacecraft missions. This series culminated in the 1985 *Vega 1* and *2* flights, which deployed French balloon-borne payloads to float in the atmosphere of Venus. American instruments have been flown on Soviet Kosmos biosatellites, even during the recent icy phase of Soviet-American relations.

Further, the European Space Agency has been conducting its operations in an accessible and open way, and both Japan and China have chosen to make sweeping disclosures of their programs, warts and all. Readers of certain Western periodicals have recently been startled to see full-color photographs of Chinese satellite launchings, and even of the recovery capsule used by Chinese military reconnaissance satellites. These pictures have been accompanied by detailed accounts of the reasons for certain past launch failures whose occurrence was not even admitted as recently as two years ago. The reason for this frankness is probably simple: all of these agencies either are currently launching payloads for hire, or have announced their intention to do so in the near future. Both China and Japan will offer launch service to GEO, and the Soviet Union has announced plans to market flights of the D1e heavy-lift booster as a source of foreign convertible currency. Everyone wants to appear credible, and frank discussion of problems and their solutions beats crude and sloppy attempts at censorship every time.

In chapter 2 we surveyed the history of the major space races, and in chapter 4 we analyzed the present status quo. Let us now delve into the future, using both the open announcements of the spacefaring nations and the skills developed through decades of second-guessing the launch announcements of Kosmos satellites.

AMERICAN SPACE STATION PLANS

With three Space Shuttle vehicles, *Columbia, Discovery,* and *Atlantis,* now available to fly in regular service, the eyes of NASA are focused on its next big project, the Space Station. The biggest mysteries about the Space Station are, first, why it is being built, and second, what it will do when it is in service.

NASA's official rationale for the Space Station, and the ostensible basis for its eventual approval by Congress, was that the Space Station afforded a new and superior way of doing space science. Given as secondary motives were its use as a laboratory for materials processing experiments and preproduction development work (crystal growing, etc.), and vague references to national security and prestige.

We are inclined to doubt the compelling importance of national prestige, since in its early years the Space Station will be several times smaller than the Skylab orbital station of the mid 1970s. It will have an initial capability (in 1994) not much more impressive than that of the Soviet Salyut-based stations (actually Salyut-Kosmos "Star"-Progress-Soyuz-Mir complexes) that have been succeeding each other in seemingly endless progression since 1971, and probably inferior to what the Soviet program can do in 1986–87 with its new-generation space station. Of course, the Space Station is expected to grow continuously beyond that point, but the funding for such growth is still in the distant future. If we could live with the Salyuts from 1971 to 1996 without maintaining a space station program of our own, and without suffering terminal loss of face, then the prestige argument is plainly silly.

National security is another issue of relevance. Let us save ourselves a lot of tedious guesswork here by taking the word of the Department of Defense and Air Force: they claim they don't need the Space Station, and certainly don't want to be hit for the part of the tab. (This is a hauntingly familiar reprise of the Shuttle story.) Of course, we should not be surprised if a military module should show up on the Space Station some day, just as the military presence on the Shuttle has blossomed now that NASA has safely paid for the "useless" thing. But for now, just ask them: they're not interested in the Space Station.

The fundamental role of the Space Station, then, seems to be to serve the needs of space scientists and crystal growers. By a curious accident of fate, I (JSL) happen to have been a member of the Space Science Board of the National Academy of Sciences at the time that NASA was seeking endorsements for the Space Station, preparatory to selling the concept to Congress. NASA gave the Board a fascinating briefing on the uses of the Space Station, with about 85 percent of the emphasis on its usefulness in space science. The Board sat, with ever-escalating discomfort, through a couple of hours of fairy tales, each more outrageous than the last. Finally the Board offered to write a report on exactly what the Space Station could do for space science, starting with a critical review of all the things that space scientists wanted to do, and keying these needs and desires to what the Station could accomplish.

For starters, the Board found that there are many widely different kinds of scientific areas of interest. Some problems require observations from low-altitude polar orbit, some from equatorial low orbits, some from GEO, some from highly inclined, highly eccentric orbits. Many experiments cannot be done in the presence of disturbances such as vibration, heat, gas emissions, electrical and magnetic transients, etc. When the list of desired near-Earth experiments was completed and screened for compatibility with the Space Station environment, some 80 percent were clearly ruled out. About 10 percent were complex enough, or had so much man-machine interaction, that they would profit from being aboard the Station, and the remaining 10 percent were "don't-cares." However, in order to fly the 20 percent that could stand life on the Space Station, we would need *at least* two Stations, one in the polar orbit and one in a low-inclination orbit! The report concluded that space science had no need for the Space Station, and expressed concern that the cost of the Space Station would, by squeezing the NASA budget hard, kill more space science than it could do. NASA said "Thank you!" and went away. A week later they were testifying before Congress that America needed the Space Station first and foremost because of its ability to serve the needs of the space sciences. (Does this remind anyone of the Space Shuttle story? Or Apollo? This is a blatant repetition of the same lies told to secure funding for both programs!)

Now we know why the Space Station is being built: for crystal growing! We recently heard a talk given by Alan Shepard on the future of the space program. He said that "for the next thirty years our space program will be centered on growing crystals on the Space Station."

The other question we started with is what the Space Station will be able to do when it is placed in orbit. Certainly crystal growing, there's no disputing that, but what else? Of course, all conceivable space science experiments will be placed aboard the Station. Many more, however, will occupy "free flyers" in space well clear of the Station or in radically different orbits. What is a free flyer? It is an unmanned platform on which experiments may be installed. It has most of the advantages of a Space Station environment, but, from the space science point of view, very few of the disadvantages. It is a compromise between the "old way" of doing things, with many small autonomous eggs in their own optimized orbits, and the "Space Station" way of doing things, with a dozen eggs squashed into a box big enough for four.

Now we can ask another question: "Why is NASA *really* building the Space Station? The answer is simple: NASA must at all times have one big project that qualifies as a National Goal. If they didn't, the vultures that patrol Old Foggy Bottom would tear it limb from limb. Weather satellites? Give them to the National Oceanic and Atmospheric Administration (NOAA). Navigation satellites? Let the Navy take care of them. The Earth Resources program? Give it to Interior. Communications satellite research? Let COMSAT pay for it. Basic research? Let those pesky scientists ask the National Science Foundation for money. Space Shuttle operations? Let's set up an operating company to run the show. Planetary exploration? That doesn't seem to belong anywhere—I guess we'll just stop doing it. This is not fiction: the Carter Administration actually produced detailed dismemberment plans for NASA. These plans were dropped in the face of a public furor.

So that's why this great nation is building the Space Station: as an insurance policy against the sudden death of NASA. We are convinced that many members of Congress were knowledgeable enough to see through the sham rationales presented by NASA for the Shuttle and the Space Station. They voted for these programs because, fundamentally, they believed that NASA had a great future. Let us hope that we will be clever enough in using these expensive tools to prove them right.

So what do we think about the Space Station? As a member of the Space Science Board, I (JSL) voted with clear conscience for the statement that the Space Station was of negligible use to space science, and we accept the idea that this large project will compete with space science for funds, and probably win. However, we are not just space scientists. We are residents of Earth, and American citizens. We are convinced that, in the long run, after the silly games of lies and hucksterism have faded from our memory, the Space Station will be doing great things for the people of Earth. We think we know what it will be doing to earn its keep, and that is processing space resources. *That was not even on NASA's list!*

STRATEGIC DEFENSE

The Strategic Defense Initiative (SDI), mocked as "Star Wars" by its enemies, is a research program aimed at changing the balance of terror in the world today into a new balance, based on self-defense rather than on offense.

The present equilibrium between the East and West is the result of vast numbers of nuclear missiles on both sides. Each side has so many missiles that it could ride out a surprise first strike by the other party and still have enough firepower left over to bomb their adversary "back into the Stone Age." This balance is referred to as Mutual Assured Destruction, appropriately abbreviated MAD. The security of each nation rests upon its ability to attack and kill millions of citizens of the other nation.

The reason for our vulnerability to surprise first strikes is simple: our only warning of an attack before the missiles are actually falling comes from a small number of surveillance satellites in space and about five crucial radar installations on the ground. Sabotage against the missile warning radar sites would not be difficult, especially using radar-busting missiles of the sort available for use by several nations. Furthermore, all the satellites in near-Earth space could be killed simultaneously by three multi-megaton explosions set off in low Earth orbit. Almost every satellite ever launched is extremely vulnerable to the Electromagnetic Pulse (EMP) given off by nuclear explosions. The EMP propagates unhindered through space, and travels virtually at the speed of light. For all we know, those three nuclear warheads are already in Earth orbit awaiting detonation. Indeed, there may be two or more sets of them made in different nations! Once these warheads go off, all reconnaissance satellites in space are dead. The best that one can hope is that dedicated nuclear-test monitoring satellites in very high Earth orbit might survive and report that the explosions had taken place. Otherwise, the party under attack doesn't get any news of the loss, but the party that set them off knows it can now launch a surprise first strike undetected. Being blind and knowing it is only marginally better than being blind and not knowing it. A first strike against satellites confers an enormous advantage on the agressor.

But the EMP is not the only hazard faced by satellites. The Soviet Union, as we saw in chapter 2, has a long history of experimentation with killer satellites. These vehicles rendezvous with their targets, inspect them at close range, and then detonate a shaped charge warhead that destroys both vehicles. The American spacecraft most vulnerable to the Soviet antisatellite weapons are ocean surveillance satellites and KH-11 reconnaissance satellites in low altitude, highly inclined orbits.

Also, the Soviet Union has for over ten years maintained an antiballistic missile (ABM) installation on the outskirts of Moscow. This system has recently been undergoing renovation. These ABMs are capable of shooting down American satellites as they pass overhead. While this system is too localized to pose a global threat to American space assets, it could buy precious time by selectively removing one or two critically placed surveillance satellites. In the nuclear age, one hour of freedom from observation is enough time to stage World War III. This perilous scenario is a greater threat to America than to the Soviet Union: first, the United States has no operational antisatellite capability, while the Soviets do. Second, in the summer of 1985 the United States had only a single KH-11 in orbit and one ready for launch. The program had been due for one more KH-11 launch in August 1985, to be followed by a gap of about a year before the larger, heavier KH-12 satellite would have been ready to be launched by the Space Shuttle. But the August 1985 launch of the last flight-ready KH-11 was terminated by the explosion of its Titan 34D booster. In a frantic hurry, a partial flight-spare spacecraft was rebuilt and upgraded for launch in April of 1986—but its Titan 34D booster also exploded, when its solid rocket boosters failed. The Titan 34D was then grounded for at least a year, just as the Space Shuttle had been grounded a couple of months earlier. This left the United States with only a single photographic reconnaissance satellite in orbit, no flight-ready spacecraft, and no booster larger than the Atlas Centaur (an ancient workhorse dating from 1963). This situation could not possibly be alleviated in less than a year. During that time, loss of a single satellite, for whatever reason, could leave the United States blind to Soviet missile and aircraft deployments.

Satellites are extremely vulnerable to collisions with "slow-moving" objects. The orbital speed of a satellite in LEO is about 8 kilometers per second, so a cloud of buckshot lofted into its path by a small, vertically launched sounding rocket would destroy it utterly. The American antisatellite (ASAT) missile under testing for use by specially modified F-15 fighters carries no warhead: it simply pops up to orbital altitude, gets in front of a satellite, and lets the enormous kinetic energy of the collision blow the target to bits. This ASAT is useful only against small numbers of selected low-altitude Soviet reconnaissance satellites, such as the Soviet nuclear-powered ocean surveillance radar platforms.

Finally, satellites are vulnerable to energy-beam weapons. Historically, satellites have been delicate, of light construction. Lasers or neutral particle beams capable of penetrating their shells can be made and fired on the ground. Deploying such heavy, power-hungry systems on satellites is another story.

Shielding satellites against EMP, kinetic energy, and beam weapons is an expensive proposition: a reasonable degree of protection would require a meter or so of shielding. Shielding the basic communications, command, control, and intelligence (CCCI) satellites of a nation would use hundreds of thousands of tons of material, an amount impossible to lift into orbit.

In early 1983, President Reagan announced the beginning of SDI. He did so in the full expectation of acquiescence from the USSR, which had long proclaimed that defensive measures were not negotiable: "Offense is immoral; defense is moral." He appointed a blue-ribbon panel, headed by former NASA Administrator James Fletcher, to conduct a preliminary study of potential defensive technologies that would enable the United States to assure its own security by deploying defensive systems to destroy ballistic missiles, while at the same time decreasing our arsenal of offensive weapons. In the fall of 1984, during the campaign, Reagan emphasized the willingness of the United States to share defensive technology with the Soviet Union.

SDI, since it is a research program, is actively considering a wide range of potential future systems for deployment at least ten years in the future. It is a certainty that research will uncover serious shortcomings of some of these systems and unexpected advantages of others.

In the summer of 1983, Carolyn Meinel, a long-time space activist, and I (JSL) met with Dr. Fletcher in his office in Virginia. We presented to him a quite specific plan for the return of large amounts of asteroidal material to near-Earth space. This material would be used jointly by the US and the USSR to armor all their essential CCCI satellites against attack. The level of protection that could be afforded by, say, a million tons of shielding is quite impressive. Any attempt to remove the other side's space assets would itself look like the beginning of World War III. Thus a surprise first strike would no longer be possible for either side: everybody would know what was going on at all times.

With the possibility of a massive surprise attack removed, the sizes of the nuclear arsenal of both nations could at once be reduced by about a factor of five: that is the factor by which the arsenals were inflated to ensure that sufficient retaliatory weapons survive a surprise attack.

The Fletcher Committee, in response to this idea, funded a brief, two-week study of the defense uses of space resources, held in La Jolla, California, in August of 1983. A panel of about 40 astronomers, lunar and asteroid experts, chemical engineers, defense analysts, and aerospace and systems engineers heard testimony on this subject, debated the evidence, and then wrote a report strongly supportive of the technical feasibility and economic potential of using nonterrestrial resources in near-Earth space. The study group, of which I (JSL) was a member, produced a report urging the Fletcher Committee to endorse basic research in utilization of space resources. The conclusions of the La Jolla study were adopted virtually verbatim by the Fletcher Committee.

Fletcher talked to Hans Mark, Deputy Administrator of NASA, a former Secretary of the Air Force during the Carter administration, and a one-time director of NASA Ames Research Center in California. Mark, whom we had already informed of the idea, concurred, and a major 10-week study was set up for the following summer under NASA direction.

Thus, in the summer of 1984, the California Space Institute in La Jolla was again the host of a workshop on space resource utilization, this time for NASA's Johnson Space Center in Houston. NASA Headquarters had funded JSC, which was NASA's lead center for Lunar Base studies, to run the workshop. The workshop was structured about a "resident faculty" of about 20, who were assisted in their deliberations by well over 100 imported experts who were brought in for brief visits while the areas of their expertise were under study. The faculty was characterized by a very broad mix of management experts, an architect, a metallurgist, a professor of space law, an asteroid fanatic (guess who), an astronomer who did compositional studies on asteroids and the moon, and several lunar scientists. The study was directed by David McKay, a lunar scientist on the staff at JSC. James R. Arnold, the director of CalSpace, placed Stuart Nozette in charge of

CalSpace's part of the study. Nozette had recently received his PhD from the Department of Earth and Planetary Sciences at MIT, where his dissertation work had been supported by me (JSL).

The study was marked by a remarkable consensus on the potential of space resources, including a general acceptance of the evidence that the Moon's main significance for resources was for local use. The study agreed that the lunar ilmenite process for making oxygen for export, while far from proven in detail, looked like an excellent target for development. There was also general agreement that, over the long term, asteroids were exceedingly attractive targets for exploitation because of their energetic accessibility and richly diverse compositions. Finally, the study voted, on three separate occasions, to give equal billing to asteroidal and lunar resources (because of their distinct uses), and to make no endorsement of the Lunar Base program. Conclusions about the Moon were to be phrased, "If NASA embarks on a program to build a large lunar base, then . . . ."

Former astronaut Buzz Aldrin, who with Neil Armstrong made the first manned landing on the Moon, came to lobby for a lunar base and seven or eight other, unrelated things. He gave a talk in which he made a great point of how much easier it was to get to the Moon than to any asteroid, because the Moon was at the bottom of its own gravitational well, and therefore closer to Earth's level than the asteroids. Evidently he forgot the lengthy descent stage engine burn that enabled him to land on the Moon alive. It sure beat arriving in free fall at three kilometers per second! Descent to the Moon doesn't save a blessed thing—it *costs* a large amount of energy. And of course, if you want to return to Earth, you get to pay that cost all over again when you take off from the Moon!

An interesting highlight of the summer was a debate, staged before television cameras, between three major schools of thought on NASA's future. The position of advocacy for the Lunar Base was taken by Jack Schmitt, who, prior to his election to the Senate, had been the last astronaut (and first geologist) to set foot on the Moon. Bert King, a professor of geology at the University of Houston, and a well-known expert on the Moon and meteorites, took the part of a program of manned exploration of Mars. I (JSL) advocated development of the technology for using space resources, focused on near-Earth asteroids and Phobos, with the end in view of lowering the cost of ambitious missions to Mars and beyond. Schmitt was gracious enough to agree with us at the end: he is by no means a narrow partisan.

When we all departed from La Jolla to return to our usual duties the final report, which was pictured as a document of about 300 pages, had been only partly written. A draft of the executive summary report had been written weeks earlier by Jim Burke of JPL, an ardent but objective advocate of a polar lunar base. Burke's draft was of course written before we reached many of our conclusions, so its tone and contents required constant revision. A draft incorporating all of our comments was supposed to be circulated in September, and the final draft was to receive approval by the entire faculty in time for presentation in late October.

The schedule was conditioned by the timing of a large Lunar Base Conference, held in the hall of the National Academy of Sciences in Washington. The conference, reasonably enough, was organized by the Lunar Base study team from JSC. September came and went with no draft report. October waxed and waned. Correspondence addressed to JSC went unanswered.

Finally it was time for the Lunar Base Conference. We all showed up, and nothing happened. Quite a number of speakers gave presentations, ranging from excellent to silly, on future lunar activities. Edward Teller gave an astonishing and amusing paper on launching cubic kilometers of dirt from the Moon to the lunar Lagrange points for use as shielding. I (JSL) gave a paper with Carolyn Meinel on processing of native metals from asteroids and the Moon. Just as we were about to leave at the end of the conference, a reporter mentioned in passing that he had seen our report in the conference *press kit!* We were shocked. He showed us the text, and it was missing most of the recent changes that had been submitted. More irritating, it closed with the slogan, "On to the Moon!" I caught a few of the summer study participants before they left, and they were equally taken aback. Carolyn Dry, the architect of our group, said "What do you expect?" Bert King, a summer convert from the lunar base to asteroid resources and Mars expeditions, was badly bent out of shape. While the things he said were probably accurate, we hesitate to quote them here, gentle reader, because the paper of this book has not been fireproofed.

The ensuing exchange of letters was intense and brief. JSC absorbed all insults like a black hole. Nary a photon escaped to inform the rest of the universe what was going on. In fact, more than two years have now passed since that fateful week, and not a single word from our summer study has yet seen the light of day—except in that infamous press kit!

The question of the usefulness of space resources in defense has been answered twice by impartial panels of experts. The answer both times has been "yes," but the SDI program has become obsessed with giant space-based beam weapons, X-ray lasers, super-super-computers, and hyper-high-tech hardware. The climate is not right for cheap, low-tech endeavors without weapons. General Abrahamson, the director of SDI, has decreed (against the opinions of the large majority of his staff) that space resources must, for now, be ignored.

THE SOLAR SYSTEM EXPLORATION COMMITTEE

In 1983, NASA received a report from its blue-ribbon Solar System Exploration Committee. SSEC, chaired by Dr. David Morrison of the University of Hawaii, was charged with the responsibility of reassessing the goals and means of the Solar System exploration program and recommending a prioritized schedule of future planetary missions. SSEC began from a broad strategy plan formulated by the Space Science Board's Committee on Planetary and Lunar Exploration (COMPLEX) and fleshed it out with a detailed consideration of many flight opportunities.

Before the SSEC report, the sole purpose of the Solar System exploration program was considered to be the scientific study of the planetary system. SSEC made a profound change in that philosophy by adding a second purpose to the basic one of pure science: "the provision of a scientific basis for the future utilization of resources available in near-Earth space."

In support of the general strategy formulated by COMPLEX, SSEC proposed a specific set of "core missions." These missions emphasize the use of two major families of spacecraft, within each of which there is a high degree of commonality of design. Assuming the already funded Galileo Jupiter orbiter and probe mission in 1986, the core program begins with four initial missions:

1. Venus Radar Mapper (Magellan) 1989
2. Mars Geoscience/Climatology Orbiter (MO) 1989
3. Comet Rendezvous/Asteroid Flyby (CRAF) 1990
4. Titan Probe/Radar Mapper 1992

Subsequent core missions are grouped according to their destinations. Missions to the outer Solar System would be a Saturn Orbiter, a Saturn flyby with an entry probe, and a similar Uranus flyby with a probe. Missions to the terrestrial planets would include a Mars aeronomy orbiter, an advanced Venus atmospheric entry probe, a Lunar Geosciences Orbiter, and a network of Mars surface probes. Finally, small Solar System bodies would be the targets of a comet atomized-sample return (preferably linked to the 1990 CRAF mission), one or more multiple main-belt asteroid orbiter-flyby missions, and a rendezvous with an Earth-approaching asteroid. Of these, the Mars Orbiter, The Comet/Asteroid mission, the Lunar Orbiter, the Mars surface probes, and the entire set of small-body missions would be of great value for resource evaluation.

The entire SSEC program, including supporting research and analysis, mission operations and flight data analysis, and the cost of all of the core missions, is constrained to fit within a budget of $300 million per year in 1984 dollars. It is interesting that, for the entire period from 1962 to 1977 the planetary program averaged about $600 million per year, and never dropped below $300 million. From 1979 through 1985, because of severe financial stress imposed by the development costs of the Space Shuttle, not a single planetary mission was launched.

The SSEC strategy, aimed at inexpensive missions, high degrees of commonality between missions, and level program funding, is in fact a conservative program, shaped in consultation with NASA management, Congress, and the Office of Management and the Budget (OMB) to be acceptable to all parties.

A GREAT LEAP BACKWARD

The early days of the planetary exploration program were conducted under a similar philosophy: most missions belonged to one of two "stables" of spacecraft, the three-axis stabilized Mariners and the spinning Pioneers. A high enough level of launch activity was maintained so that the unit costs of spacecraft were not excessive. But, as budgets waned in the post-Apollo years, the system broke down.

It has become the rule over the past decade to build single spacecraft of unique design, each of which must be certain of success, since no backup spacecraft will be available in the event of a failure. This has led to the crowding of every possible instrument onto each spacecraft, since a missed flight opportunity may mean a decade's delay. It also has fed the growth of a "gold-plated" design philosophy to assure near-perfect reliability. As a result of these pressures, the unit cost of planetary missions has escalated out of sight. Pioneer Venus, originally conceived as a low-budget mission to be lifted by a Delta booster, increased in cost to about four times the original estimate as the designs were modified to squeeze in more instruments and guarantee success. This necessitated more weight, volume, and power. The payload, consisting ultimately of six independent spacecraft, soon grew to the point where two large, expensive Atlas Centaur boosters were required for launch.

The Voyager spacecraft, launched in 1977, were originally planned to be the first of a continuing family of outer-planet spacecraft that would realize lower unit costs because of the efficiencies of "production-line" assembly. *Voyager 1* and *2* were originally conceived as Mariner-type Jupiter and Saturn flybys, with instruments optimized for observing these planets. A second pair of almost identical Voyagers was to be built with two instruments slightly modified to optimize them for the study of colder planets. These "Mariner Uranus-Neptune flyby" spacecraft were to be launched two years after *Voyagers 1* and *2*. Farther in the future, modified Voyager spacecraft were to be used for a Jupiter orbiter, a Saturn orbiter, and possibly as probe carriers to Jupiter, Saturn, and Saturn's huge moon, Titan. But by 1975 the NASA budget had ebbed so severely that only the first two Voyagers were left. It was decided to target *Voyager 2* so that, as it flew by Saturn, that planet's gravity would accelerate and deflect *Voyager 2* toward the next planet, Uranus. Upon arrival there, in January 1986, it would again be "kicked" by Uranus' gravity toward a 1989 flyby of Neptune. This decision was advertised as a great cost-saver. In fact, because the instruments were designed to look at the much warmer Jupiter and Saturn, crucial areas of science were lost. While Voyager surely accomplished much on its flyby of Uranus and will probably do so at Neptune, the fact that we have carried out these second-rate flybys will probably make it politically impossible to get an adequate mission to either planet in the foreseeable future.

A similar "cost-saving" move placed the United States in a truly humiliating position in the study of Halley's comet. The Soviet Union dedicated two multiton spacecraft, *Vega 1* and *2*, to close flybys of Halley's comet. The European Space Agency sent its first deep-space mission, Giotto, on a similar mission. Japan dispatched two spacecraft to fly by the comet, also their first ventures into deep space. The United States, by comparison, could not afford to participate. Instead, to save money, an old spacecraft called ISEE (International Sun-Earth Explorer) in a very high

Earth orbit was diverted to carry out an encounter with the small periodic comet Giacobini-Zinner in September of 1985, six months before the arrival of the armada at Halley. The sincerity of this effort, ostensibly demonstrated by renaming the spacecraft ICE (International Cometary Explorer; pronounced "icy," the same as ISEE), may rationally be questioned, since the spacecraft was designed and launched for a radically different type of mission, and was quite incapable of carrying out most of the science requirements for an early comet flyby as spelled out years ago by the Space Science Board and by NASA's own advisory groups. The diversion of ICE was, quite simply, a face-saving venture.

The second function of this leap backward was to create the popular impression that we had done a comet mission, and therefore do not need to try again to do it right. It seemed clear to us, when the ICE diversion plan was announced, that all foreseeable cometary missions had received the kiss of death. As if in depressing confirmation of this prophecy, *Aviation Week and Space Technology* leaked a story less than one week after the ICE flyby that James Beggs, then Administrator of NASA, had removed the next possible comet mission, CRAF, from the budget to be submitted to Congress. CRAF (Comet Rendezvous, Asteroid Flyby) was a scientifically exciting but relatively inexpensive mission, scheduled for launch in 1992, which would not simply have made a very short, incredibly fast flyby of a comet like the 1986 Halley missions, but would, after flying by a belt asteroid, go on to rendezvous with a comet and follow it around most of an orbit. The spacecraft would have carried a small, javelin-like instrumented "penetrator," which would have pierced the surface of the comet nucleus and planted scientific instruments inside it to report on its composition and thermal state. We maintain that ICE was the lame excuse for the attempted murder of the far more competent CRAF mission, just as the "extended mission" of *Voyager 2* to Uranus and Neptune was an excuse for killing the much more valuable Voyager Uranus-Neptune spacecraft. The next time we are offered such a second-rate sop, we are not likely to be enthusiastic.

If you don't invest in the future, you can hardly expect to own it. The Soviet Union and the European Space Agency have addressed space exploration with new vigor, and Japan has begun to invest in space, at a time when the American presence is waning through lack of vision and resolve.

If the ICE-CRAF incident is indicative of how seriously the Solar System

Exploration Committee's plan is taken by NASA management and by Congress, and if it truly reflects the nation's commitment to excellence in space science, then we are in very serious trouble.

PLANS WE'VE GOT, BUT GOALS . . . ?

At NASA headquarters the one clear theme is the desire to keep NASA alive and intact. Conventional wisdom holds that this is best done by having one giant program of political interest, and letting the great diversity of valuable small programs die away. It is this philosophy which has left NASA in its present state of wandering in the desert; it has caused NASA to change from a wellspring of brilliant, innovative, and daring ideas in the 1960s to the dry hole of the 1980s. SSEC dares to dream cheap dreams that would keep our activity in space alive at very modest cost, but even that prospect finds no support.

It is not correct to say that the Shuttle was a goal: it is merely a tool made for a very poorly defined purpose. The Space Station is not a goal: it is another very expensive, and potentially very powerful, tool, made for ostensible purposes that range from silly tales to outright lies. *We have no goal in space, and we have had none since the Apollo program.*

SALYUT AND SONS

The Soviet space program has undergone several trials as traumatic as any suffered by the United States since the flights of *Sputnik 1* and Yuri Gagarin. The complete loss of its manned lunar program, the dismal failure of the G1 Saturn 5 class booster, the expropriation of the juiciest Mars science by Viking after the *Mars 4, 5, 6* and *7* debacle of 1973, and the *Soyuz 1* and *Soyuz 11* disasters all spring to mind. Yet the Soviet Union has swallowed these difficulties, planned far ahead, and built upon its strengths.

The early days of the Salyut space station program were dogged with serious problems. The only two visits to *Salyut 1* in 1971 were failures: *Soyuz 10* failed to dock with Salyut, and *Soyuz 11* killed three cosmonauts on the way home. *Salyut 2* tumbled wildly and could not be manned, and *Kosmos 557* may have been another of the same. *Salyut 3* was visited briefly by *Soyuz 14,* but *Soyuz 15,* a month later, was unable to dock and returned after two days. *Salyut 4* in 1974 was visited and manned for one month by

Soyuz 17, and later (after a Soyuz launch abort eavesdropped on by American reconnaissance satellites) for two months by *Soyuz 18,* but by no other missions. Only three Soyuz missions visited *Salyut 5,* and one of these, *Salyut 23,* had to abort its docking attempt. The first mission to *Salyut 6* in 1977, *Soyuz 25,* failed to dock also.

Yet the Soviet Union persevered and realized the tremendous achievements of *Salyut 6* and *7,* which captured the world duration record from *Skylab 4* and then broke and rebroke their own records. The development of the automatic docking technique used by the Progress resupply vehicles and the Kosmos "Star" modules, and the improved *Soyuz T* spacecraft, greatly multiplied Soviet manned capability.

Now, as *Salyut 7* passes its fifth birthday in April 1987, and as the new Mir space station has begun its life, we have a good basis for discussion of the future of the Soviet space station program. The six docking ports on Mir conjure up visions of a very large orbital complex (see fig. 13.1). Star modules and even *Salyut 7* itself may be docked with it, as well as Soyuz crew transport vehicles. The likely crew size on the mature Mir station is probably 12 to 15, and at that level of activity it becomes inefficient to use little two- and three-man Soyuz craft for crew rotation. A larger booster and a shuttle craft would make more sense.

The refurbishment of the two huge G1 launch pads at Tyuratam, begun in 1983, and the construction of a very long, wide runway nearby, are both of interest in this connection. There are persistent reports from the intelligence community of a massive "tank farm" to accommodate very large volumes of cryogenic propellants at Tyuratam. One theory is that they are developing the capability for "ripple" launching, with many vehicles sent aloft in a short time. Another is that they are preparing for simultaneous launches of giant boosters with immense thirst for liquid oxygen.

Our favored interpretation, however, is that the great tank volume is due to the impending introduction of liquid hydrogen propellant. The density of liquid hydrogen is very low, and tanks to hold it are correspondingly huge. The introduction of hydrogen-oxygen propulsion, which greatly increases the payload capacity of existing boosters, should more than double the capacity of the D1 (Proton) booster to about 40–50 tons. This breakthrough in propulsion technology was made by the United States with the Centaur vehicle in the 1960s with the first successful flight of the Centaur on November 27, 1963. There is some evidence that the Soviet Union attempted to fly hydrogen-oxygen upper stages in the G1 booster test series in

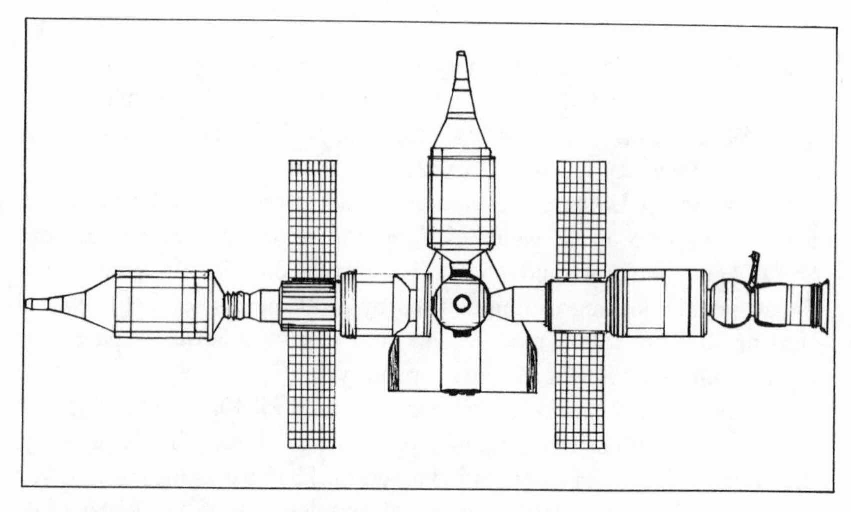

Figure 13.1
The Mir Space Station Complex
The Mir Space Station, with its six docking ports, may easily be extended to configurations containing eight or more Mir, Salyut, Soyuz T, Kosmos-Star and Progress modules, and can accommodate either the Soviet Space Shuttle or the Hermes minishuttle. This is a schematic rendering of one such configuration that could hold a permanent crew of six and as many as six to nine short-term visitors. This particular configuration has a Star module on the far left, docked to the front port of the Mir space station, the Mir block (with solar panels extended), carrying the spherical docking adapter on its aft end. To the right are a Salyut space station module with a Soyuz T crew transfer vehicle docked at its rear port. A second Star module is attached to one of the ports on the central docking adapter (top center), and a French Hermes mini-shuttle is docked at the port on the far side of the adapter. Two ports (facing us and "below" the adapter) remain vacant to accommodate other modules. Except for the Hermes spaceplane, which will first be launched by an ESA Ariane 5 vehicle after 1995, this configuration could be operational by 1989.

1969 to 1972. Reports leaked from the intelligence community describe three failures, with indefinite postponement of the program occurring in 1972. To date, no Soviet high-energy stage is confidently known to have been flown successfully in space. There have recently been rumors from the intelligence community that a hydrogen-oxygen stage (the J-1) with about

1.3 million pounds of thrust has been tested several times in suborbital flights since early 1985 and that *Kosmos 1767,* launched in July 1986, may have been launched by that booster.

Another interesting development is that two Myasischev Mya-4 heavy strategic bombers have been photographed with conversions that would allow them to carry external stages, similar to the modifications of the Boeing 747 to make it into a carrier for the Space Shuttle (see fig. 13.2). The new, very long and wide runway at the Tyuratam launch complex near the two recently restored G1 superbooster pads could be used both for delivery

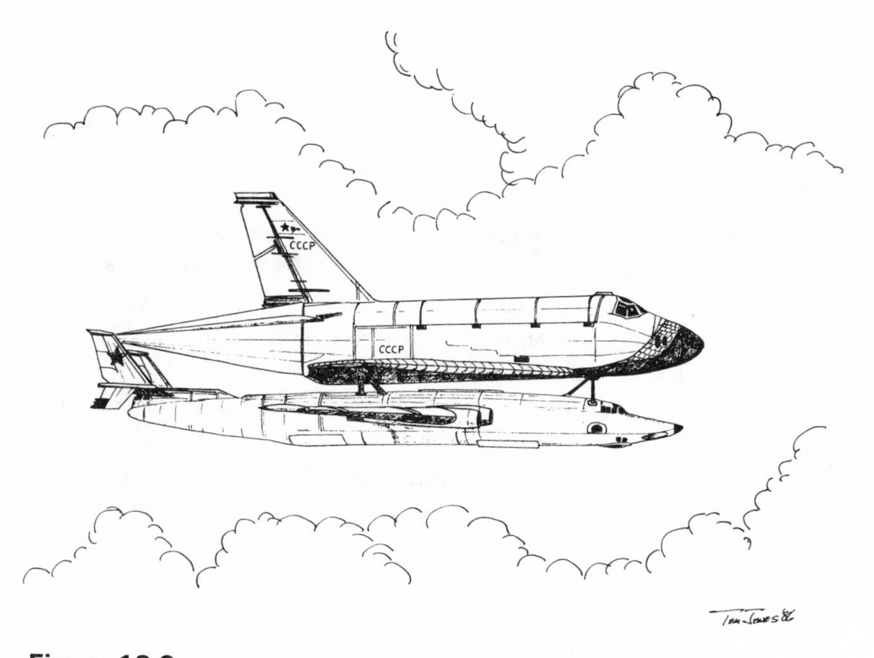

Figure 13.2
The Soviet Space Shuttle
This drawing, by Thomas D. Jones, is based on intelligence reports that the Soviet Union has been conducting drop tests of a Space Shuttle orbiter from atop a Myasischev-4 strategic bomber. The apparent size disparity should not be a cause for alarm—the Soviet shuttle does not carry the main engine cluster that the American Shuttle has, and hence has an empty weight under 50 tons, within the payload capacity of the Mya-4. The first orbital tests of Soviet Shuttle launch vehicles should come in 1987.

of large shuttlecraft by their Mya-4 carriers and for direct landings of the Soviet shuttle during return from orbit. Indeed, American photographic reconnaissance satellites have already photographed two full-sized Soviet shuttles, almost exactly the same size as the American Space Shuttle orbiters, at Ramenskoe near Moscow. One of these shuttles was left in plain sight for several days when its Mya-4 carrier plane ran off a taxiway at Ramenskoe and became stuck in the mud! The first Soviet space shuttle orbiter made its debut at Tyuratam in late 1986, where it underwent compatibility test with the super booster on its launch pad.

At least one Soviet shuttlecraft has made its appearance in space. Anatal Skripko, speaking at the American Astronautical Society meeting in February 1982, said that Soviet shuttle development activity would be "more evident" in two or three years, and that full-scale flights would be underway in five years. The program has indeed begun. The *Kosmos 1374* mission of June 1982 was a one-orbit test of a spaceplane model. Excellent photographs of the recovery operation in the Indian Ocean were secured by a Royal Australian Navy Orion patrol plane (see fig. 13.3). A second such flight, *Kosmos 1445,* in March 1983, also landed in the Indian Ocean. A third spaceplane, flown as *Kosmos 1517* in December 1983, landed in the Black Sea, as did *Kosmos 1614* in December 1984. All four tests appear to have been successful. These tests involved small models of about 1000 kilograms maximum weight, launched by the C1 booster from Kapustin Yar. It is possible that this is a military reentry vehicle, unrelated to the manned shuttle.

In addition, there have been several truly astonishing launches on D-class heavy-lift boosters. Speculation has been rampant, and just about the only thing everyone agrees on is that the flights were related to the manned program. The *Kosmos 146* and *154* flights of 1967, the *Kosmos 300* and *305* flights of 1969, and *Kosmos 379* and *382* in 1970 appear to have been associated with the Soviet manned lunar program and the use of leftover hardware developed for it. Then came, after a gap of several years in which every D1 booster launch is well understood, the double launch of *Kosmos 881* and *882* aboard a single booster, on December 15, 1976. These payloads, in the 7- to 10-ton weight range, were placed in a Salyut-type orbit for a single pass around the Earth and then recalled. No explanation of the significance of this mission was offered.

Next, in 1978, the *Kosmos 997–998* launch, again on a D1 heavy-lift booster, retraced the steps of the earlier mission. In 1979 *Kosmos 1100* and *1101,* yet another pair of payloads in the 10-ton class, were launched: one

Figure 13.3
The Soviet Spaceplane

This aerodynamic reentry vehicle, which has been tested four times in orbital flight, may not be intended for manned use, but rather, according to a provocative argument by Art Bozlee, as a fractional-orbit bombardment system (FOBS) to target American carrier task forces. This picture was taken by Australian Orion ocean patrol aircraft during recovery operations near the Comoro Islands northwest of Australia.

was recovered after a single orbit, and the other after two orbits. The only later flights on the D1 that may be related are *Kosmos 1267* (in 1981), *1443* (in 1983), and *1669* and *1686* (in 1985). These are the "Star" expansion modules that rendezvous and dock with Salyut space stations. The most recent of these has a large reentry vehicle attached, similar in size and shape to an American Gemini capsule. Some observers have linked the *Kosmos 881-2, 997-8,* and *1100-1* missions to the development of a Soviet manned shuttlecraft. This is without substantiating evidence, and the link to the "Star" modules, which appeared in use shortly after those tests, seems more plausible.

Several other recent D1e launches are of great interest. These began with *Kosmos 1603* in September 1984 and *Kosmos 1656* in May of 1985. Both payloads were placed in low orbits with inclinations of 51.6 degrees, then immediately (within the first orbit) diverted to higher orbits with inclinations of 71.1 degrees. This requires a huge velocity change, and is a thoroughly ridiculous way of reaching such an inclination: why not launch directly into it? These flights may possibly have been flight tests of a new high-energy escape stage for the D1e. The velocity changes required for these orbit changes would be nearly sufficient to escape from Earth! These flights could enable a return to the Zond-type lunar flyby or lunar-orbiting missions planned originally for 1968.

The Star modules have about the same mass as the Salyut station itself. A complete "stack" of Salyut, Star, Soyuz, and a Progress vehicle place the total mass of the assembly in the 55-ton range. We may soon expect much more elaborate combinations of modules, using the Mir station as the core, to be assembled in orbit. Such construction schemes need not be limited by the sizes of the present component vehicles: with hydrogen-oxygen upper stages, the D1 booster could put individual unmanned payloads of 40 to 50 tons in a Salyut-type orbit.

In early 1979, the Soviet Novosti press agency described plans for the development in the 1980s of a bigger space station with a lifetime of up to five years. This larger station would have a capacity of 12 to 24 cosmonauts, and would be serviced by visits from Soyuz T and Progress spacecraft, and also by the Soviet Space Shuttle. In mid-1980, *Aviation Week and Space Technology* (AW&ST) summarized and interpreted information released by Soviet sources concerning this Space Station. They estimated a weight of about 110 tons for the 12-man version of the station, and predicted that it would be launched by a cluster of several J-class boosters. AW&ST also predicted an attempt to launch the J1 booster as early as 1983. (In retrospect, they do seem to have done everything necessary to prepare the two launchpads by early 1983, but no launch materialized until 1985.) With the space station dependent upon the success of the J1, this would seem to indicate a two-year setback in the large space station. AW&ST also predicted developments leading to a manned shuttlecraft in the mid 1980s. Note that they were writing two years before the first orbital test of the Soviet model shuttlecraft in June 1982. It now seems clear that the first of these large space stations, Mir, relies on the D1 booster and is to be launched and assembled piecemeal, not all at once. We do not yet know for certain whether the D booster was upgraded to launch more than the 20-25 tons carried by the D1, but the evidence weighs against any substantially higher performance.

If the new L-class super-booster, with its first-stage thrust of 3.5 million kilograms (7.7 million pounds) and payload capacity of 100 metric tons (110 tons), makes its flight debut soon, then the capabilities conferred by it would be very similar to those of the Saturn 5 booster used in the Apollo lunar landing program with its 3.4 million kg of thrust. The Proton D1 booster, as it demonstrated during the *Zond 4–8* flights, can send a manned Soyuz around the Moon and back on a flyby mission. With new high-energy upper stages (which would make it the D2), the lunar payload of the D booster would be roughly doubled to 15 tons. This is close to the mass of a Salyut space station. Two D2 boosters could rendezvous in LEO to send a payload complex weighing 30 tons to fly by or orbit the Moon. This is more than the mass of a Salyut plus a lunar Soyuz-Zond capsule.

The suspected hydrogen-oxygen J1 booster that has recently been test-flown is probably the unit that will be clustered to make the first stage of this new Soviet superbooster.

FUTURE UNMANNED EXPLORATORY MISSIONS

For nearly 15 years the only future planetary missions openly discussed by the Soviet Union were spacecraft in the Venera and Vega series. In the fall of 1983, stories began to circulate among planetary scientists that the USSR would launch a mission to the Martian moon Phobos in the 1986 or 1988 launch windows.

These stories were not substantiated in detail until March 1985, when a team of Soviet scientists gave a detailed and astonishingly frank presentation on their future plans at the Lunar and Planetary Science Conference in

Houston. The program that they announced was remarkable in a number of ways.

First, the next planned planetary mission was indeed to be sent to Phobos. The mission involved the launch of two sets of spacecraft by D1e boosters. Each set would contain a Mars orbiter and a probe to be dropped onto a Martian satellite. Both sets would be placed in orbit about Mars. One would be maneuvered into an orbit that permits repeated close encounters with Phobos, and then would rendezvous with Phobos in a station-keeping orbit. It would approach within tens of meters of the surface and illuminate the surface with bursts of energy from lasers and particle beams. Ions emitted from the surface would be analyzed. Landing probes or instrumented penetrators could also be deployed. If the first spacecraft was successful with its study of Phobos, the second would be targeted on Deimos, where it would carry out the same program. This mission would be the first Soviet return to Mars in 15 years.

The second new mission would be a 1991 lunar polar orbiter. The purpose of the spacecraft would be to make a detailed compositional, structural, and magnetic map of the Moon. This spacecraft would be very similar to the long-awaited Lunar Polar Orbiter spacecraft sought by American lunar scientists since the early 1970s. Past Soviet performance suggests that two or more spacecraft will be built for this mission. As with the Mars mission in 1988, this would mark the first Soviet mission to the Moon in 15 years.

The third new mission would be a joint Soviet-French exploration of a main-belt asteroid. This ambitious plan includes global mapping from orbit, followed by a soft landing on its surface. Again, two or more spacecraft may be expected. This would be the first asteroid rendezvous, orbiting, or landing mission by any nation. The target discussed in presenting preliminary mission plans was Vesta, but the Soviet presenters made it clear that the actual target for the mission has not yet been selected.

No explicit mention of sample return from Phobos, Deimos, or the asteroid has been made. Such returns are actually easier than the automated lunar sample return missions carried out by the Soviet *Luna 16, 20,* and *24* missions. We could expect such attempts after the Soviet Union has accumulated some flight experience with the new spacecraft.

Curiously, these missions are precisely those that would be demanded by a space program that has as its main goal the assessment and exploitation of space resources. The only factor missing from the list of new missions is the near-Earth asteroid family. Their absence may be caused by the fact that the first of the highly attractive asteroids were not discovered until 1982. It also at least partly reflects the fact that all known research on these bodies has been conducted in the United States. Because of that absence, it is amusing to note that, shortly after the *Vega 2* spacecraft emerged alive from its encounter with Halley's Comet, the Soviets announced that they had retargeted it to carry out a flyby of the near-Earth asteroid Adonis.

In the summer of 1986 two new and extremely ambitious Soviet planetary missions were unveiled. The first is a proposed 1994 Mars surface rover, based loosely on Soviet experience with the Lunokhod lunar roving vehicles. The other is a 1998 automated Mars surface sample return mission. Both of these lie well beyond the scope of present American plans. Unlike the American Mars Observer, which represents a retreat to smaller, less expensive spacecraft, the Soviet program clearly is moving toward much more complex and ambitious undertakings. These two new missions must be regarded as essential precursors to manned exploration of Mars.

MEN TO THE MOON AND BEYOND

The Soviet popular press has always emphasized two main elements of its long-range program, manned flight and planetary exploration. Countless articles have appeared in the domestic press featuring manned Mars missions. It is now certain that the Soviet Union had a vigorous and well-funded manned lunar program in the late 1960s, aimed at flying by or orbiting the Moon before Apollo. A manned lunar landing program, dependent upon the G1 booster, was also in development in the event that Apollo was cancelled or seriously delayed. It is instructive to review exactly what capabilities the Soviet Union may exercise within the next few years, and to see how these developments may contribute to a long-term goal of sending men to Mars.

The following goals, and the accompanying timetables, appear feasible with the qualifications given:

Mission 1. A one- or two-man lunar flyby and return

Equipment. D1e booster and "heavy Zond" Soyuz reentry capsule.

Earliest Date. Capability has existed since early 1969.

Probable Fate. This mission is regarded as second-best and too late, clearly inferior to Apollo. It will not be flown except as a first step to large-scale manned lunar and planetary operations.

Mission 2. A 2-3 man lunar-orbiting Salyut space station

Equipment. D2 boosters with high-energy upper stages; modified Salyut space station and lunar Soyuz manned ferry.

Earliest Date. About two years after the first D2 flight: the booster need not be man-rated because the men could come up on a Soyuz T after the Salyut is in LEO.

Probable Fate. This mission could spend months in orbit about the Moon, mapping the surface. If this were intended, the unmanned lunar polar orbiter mission planned for 1991 would make little sense except as a "pathfinder." Unlikely before 1995.

Mission 3. 12- to 15-man permanently occupied LEO space station

Equipment. D1 booster, 22-ton Mir with multiple docking ports, and added Star modules, Soyuz T and Progress ferries, and K 1686-type return capsules.

Earliest Date. 1987 for 3–6 men; later stepwise growth; permanent occupancy to be achieved within 2 to 4 years; new duration records of up to 12 months to be set; routine crew rotation.

Probable Fate. The path of least resistance, independent of the development of new launchers—an excellent bet.

Mission 4. 24-man permanently occupied space station

Equipment. L1 superbooster, 110-ton main module, with support from Soyuz T and Progress and J1- or L1-launched modules.

Earliest Date. Two years after successful flight test of the L1; 1989 at earliest, but more likely 1990.

Probable Fate. A reasonable long-term development, especially if hydrogen-oxygen D2 upper stages prove too hard or impossible, or if superbooster development is pressed instead.

Mission 5. Permanently manned space station with Shuttle

Equipment. Soviet Space Shuttle with 50-ton payload to LEO; 100-ton space station modules; lifted by L-class booster. Resupply by Star modules more likely than by Progress series.

Earliest Date. Three years after flight debut of heavy Soviet shuttlecraft; 1990 at earliest.

Probable Fate. Seems to have better basis in recent Soviet activity than Mission 4.

Mission 6. Manned lunar landing "to prepare for Mars"

Equipment. Components launched by J1 or Spaceki Shuttleski-L1; 200–400 ton assembly in LEO including lunar ascent and descent stages. High-energy propellant version would be 220 tons lifted by dual L1 launches.

Earliest Date. Early 1990s

Probable Fate. Insufficient hardware commonality with a Mars lander to make this a compelling choice. Too much like a reprise of Apollo a quarter-century later. Unlikely unless a lunar base is the ultimate objective.

Mission 7. Manned lunar base

Equipment. Multiple Shuttle or J1 launches of components into LEO.

Earliest Date. Early 1990s

Probable Fate. Little basis in Soviet press or public opinion. Could be done, but probably won't be.

Mission 8. Manned Mars flyby

Equipment. 25–50 ton vehicle including Salyut and a Soyuz/Zond Earth-reentry capsule. LEO departure weight about 110–220 tons. Mission duration under 12 months for plausible flight profiles using outbound lunar swingby, Mars swingby at perihelion, and ballistic return to Earth.

Earliest Date. 1998?

Probable Fate. Extremely plausible; could be assembled in the near future using D1 launches, or put up in one or two installments with the L1. Provides the most impressive use of present capabilities with minimal new technology.

Mission 9. Manned Mars Landing and Return

Equipment. Many versions possible—the nature and mass of equipment is highly sensitive to whether space production of propellants is used. Probable minimum mass is 200 tons.

Earliest Date. Late 1990s

Probable Fate. USSR may view this as too expensive to attempt alone. Fate probably hinges on the use of Phobos- and Mars-derived propellants. This makes the mission easier but delays it to 2000+.

Our clear preference for the time interval 1986–2001 is a program building on Salyut technology, with early establishment of a permanent manned presence in space. The choice of boosters is, as we see it, secondary. We fully expect to see a Soviet space shuttle flight within two years, and would not be surprised by a first full-scale booster test flight in 1987. Expect up to 13 cosmonauts and astronauts in space simultaneously by early 1988, breaking the present record of 11 (7 on the Space Shuttle and four on Salyut 7 at the same time). Thereafter, On to Mars!

It is entirely possible that the United States will be encouraged to plan a Space Shuttle mission to Mir. Do not, however, expect reciprocity. Mir and Salyut orbits cover the entire continental United States (remember the early heavy emphasis on military functions on Salyut), and are accessible from Cape Canaveral (fig. 13.4). The planned American Space Station orbit is, however, at so low an inclination (28.6 degrees) that it never overflies Soviet territory and is virtually inaccessible to launches from the Soviet Union.

A powerful measure of the ambition and competence of the Soviet space program will be provided by seeing how much of their resources they devote to Mars. A return to unmanned operations in the Mars system, complete with Phobos, Deimos, and Mars surface sample returns, is an essential prerequisite for manned operations on these bodies. The manned Mars flyby is in a separate category: it is a very daring use of available resources,

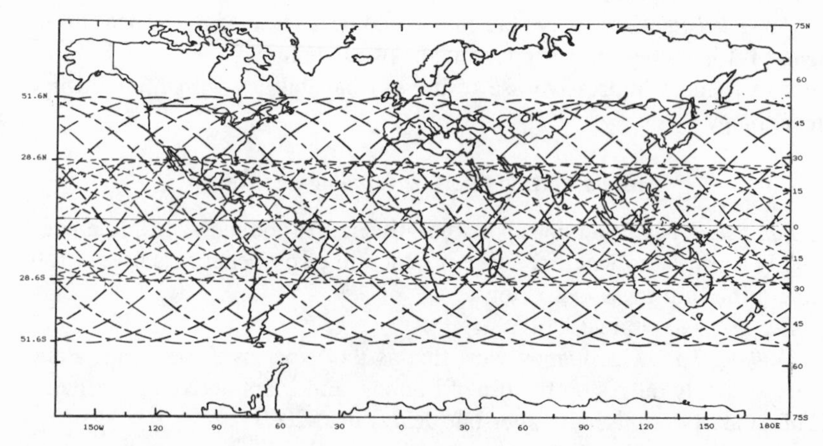

Figure 13.4
Space Station Geographic Coverage
Soviet Salyut and Mir space stations are launched from the Tyuratam cosmodrome in central Asia (marked with an x on the map). The D1 Salyut/

and may be preceded by distinctive unmanned operations to test spacecraft and procedures before committing men.

SOLAR POWER SATELLITES

The Soviet Union has a research program on Solar Power Satellites, but few details have been made public. The most interesting revelation was a

---

Mir booster is launched on an azimuth slightly north of due east to prevent upper stages from the launch vehicle from falling in China. The inclinations of the orbits of the Soviet space stations are all 51.6 degrees, which means that each spacecraft gets as far north as 51.6 N and as far south as 51.6 S on each orbit, and that its orbit always cuts the equator at an angle of 51.6 degrees relative to true east. The plane of the orbit is fixed in space, and the Earth rotates beneath the orbiting spacecraft. Salyut stations have orbital periods of a little over 90 minutes. Thus, after one orbit, Earth has rotated almost exactly 1/16 revolution (22.5 degrees) to the east. This causes the ground track of Salyut to shift about 22.5 degrees west on each orbit. After 16 orbits the launch site has moved back into the orbital plane, and the satellite passes almost exactly over the launch site. During each 24-hour period, the Soviet space station executes a ground track that covers the globe rather thoroughly with a network that extends from 51.6 S to 51.6 N. Such an orbit covers all of the continental United States, almost all of western Europe, plus Japan, Africa, China, South America, Australia, New Zealand, and almost all of the inhabited portion of Canada. This makes the use of Salyut space stations for military reconnaissance a serious temptation. Indeed, at least two Salyuts were exclusively devoted to military activities.

The American Space Station, like almost all Shuttle flights, will be launched due east from the Kennedy Space Center in Florida to take maximum advantage of the rotation speed of the Earth. It will be injected into an orbit with an inclination of 28.6 degrees (the latitude of the launch site). In this orbit, the Space Station would provide about 1 percent coverage of the United States and would never overfly the USSR, Poland, Hungary, Czechoslovakia, the DDR, Bulgaria, Romania, Mongolia, Albania, or North Korea. It would overfly Cuba, Vietnam, Kampuchea, and the southernmost 10 percent of the People's Republic of China. The only military temptation would be to monitor Soviet naval activites in the tropical and subtropical oceans, a function that would be better filled by existing satellites. The best orbit for a reconnaissance satellite is at an inclination of about 90 degrees (polar orbit).

January 1985 press release describing, in fractured English, Soviet plans to have a working SPS "the size of a small town" by 1999. It is hard to guess what constitutes a small town, but let us take an area one square kilometer as being one standard town. The one million square meters of this town would intercept just over one billion watts of solar power. After conversion to electricity, this would provide about 200 million watts of electrical power.

Soviet sources are vague on how this power would be apportioned between use in space and transmission to the ground. However, power usages in excess of 20 kilowatts in space would be impressive. What could they do with 10,000 times as much power? Weapons such as lasers and particle beams spring to mind at once, but do not stand up under analysis. Large solar arrays are very fragile and vulnerable even to simple countermeasures. One doesn't normally make battle stations out of tissue paper. Maybe the energy is all coming down to the ground for eventual commercial use. If there is a genuine Soviet interest in establishing independence from foreign supplies of energy, then an SPS system makes sense.

But SPS does not make any sense at all without two new developments that should be in place before the end of the century. One of these is a heavy-lift launch vehicle, such as the L1 or the Soviet Space Shuttle (quite possibly with the latter sitting on top of the former). The other is a careful assessment of the technical and economic feasibility of using space resources. Possibly, from the SPS perspective, the overall unity of the future Soviet program is more evident than it seemed before.

OTHER NATIONS, OTHER PLANS

Japan dispatched two probes toward Halley's comet that arrived in early 1986, the first launches beyond GEO for that nation. The larger H class booster will be operational in a few years, and larger and more ambitious planetary missions will be possible. At the top of their study list is a lunar polar-orbiting resource survey mission for the 1990s (see figs. 13.5 and 13.6).

The European Space Agency also sent its Giotto spacecraft to view Halley's comet. ESA has built a satellite to be launched on an ambitious highly inclined solar orbit mission. The launch, originally scheduled to take place from *Challenger* in May of 1986, has been postponed at least until 1989 as a result of the *Challenger* explosion. Plans to modify the *Discovery* orbiter

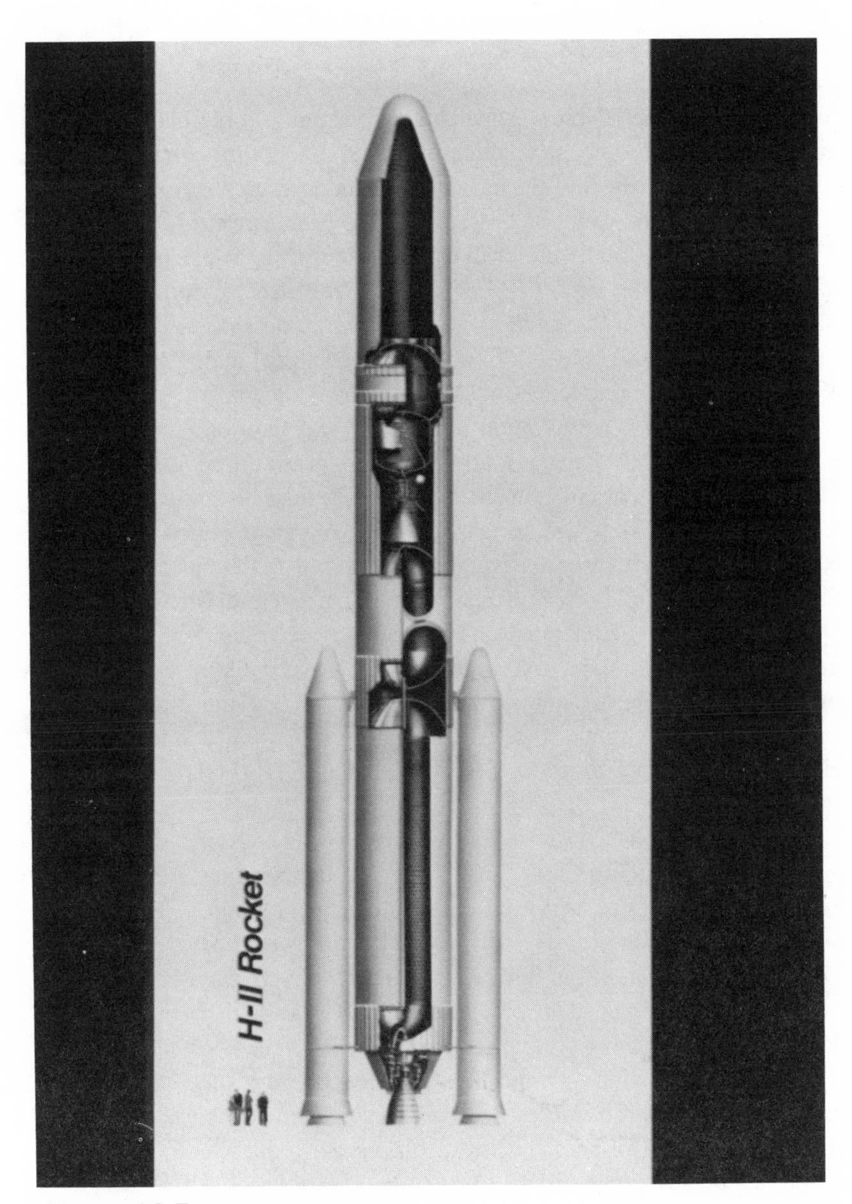

Figure 13.5
The Japanese HII Booster
This huge booster, larger than the American Titan 34D, is powered by very advanced hydrogen-oxygen main engines. When the HII becomes operational in the mid-1990s it will provide Japan with the ability to launch almost any conceivable civil, scientific, or commercial payload. Present constraints on the use of the current Japanese N2 booster, due to its derivation from the American Delta booster, would not apply to this all-Japanese product.

to acommodate the Ulysses payload and its Centaur G′ high-energy stage, an adaptation of the ancient Centaur of the 1960s to fit into (and nearly fill) the payload bay of the Shuttle, have been abandoned because of safety considerations, and the Centaur G′ program has been cancelled.

Ulysses, when if finally flies, will fly by Jupiter, using the planet's gravity to flip it into an orbit almost perpendicular to the plane of the orbits of the planets, passing over the poles of the Sun where no spacecraft has gone before. ESA presently has under study a polar orbiting lunar mission for the 1990s, with the working name POLO (invoking the memory of Marco Polo as well as serving as an acronym for Polar Lunar Orbiter).

ESA also has under serious study a mission, based on Giotto hardware, to return a sample of cometary material to Earth. This mission, dubbed Caesar, would be launched as early as 1989 to a short-period comet. Interestingly, all such comet-sampling missions are more energetically demanding than sample return from a near-Earth asteroid.

ESA has recently elaborated on its plans for the Ariane 5 booster, their next-generation workhorse. One version will be capable of carrying the French-designed Hermes minishuttle. Meanwhile, England is pursuing its design efforts on a single-stage-to-orbit unmanned winged vehicle called HOTOL. They feel that the first test flight of HOTOL could be as early as 1996.

China has established its ability to recover sizable payloads from LEO, and to orbit satellites in GEO. The Chinese government has recently been extremely open about its space program. China has announced its intention to market launch services to GEO in competition with ESA's Ariane (fig. 13.7), NASA's Shuttle, the Soviet D1e, and Japan's H-class booster. The first commercial payload to be launched by China will be the Swedish Mailstar communications satellite. China has not yet indicated any plans for deep-space activities beyond GEO, and we do not expect any until at least the late 1990s. The largest present Chinese launch vehicle, the CSS-4 ICBM, is adequate for launching small deep-space probes, but not for more ambitious undertakings (fig. 13.8).

At present, the only other nation capable of launching its own Earth satellites is India, but that nation has as yet shown no interest in operations beyond GEO. Indeed, India's only spacecraft in GEO, in the INSAT series, were launched by other nations.

Australia, which launched a satellite of its own as early as 1967 using modified American rocket hardware, is now seriously contemplating a

Figure 13.6
Proposed Japanese Space Station Module
This module has been proposed for Japanese participation in the American Space Station program, along with a module from the Eurpoean Space Agency and an orbital assembly and repair facility to be built by Canada. All modules would be launched by the Space Shuttle. Both Japan and China have astronauts in training for future flights on the Shuttle.

return to space. The British and American activities at the Woomera rocket range in the 1960s, and the struggles of the European Launcher Development Organization (ELDO) in the 1970s, made Australia an early and vigorous participant in space research, but the field has languished now that the big boys have gone home. Australian payloads will again be flying in the near future (one is tentatively scheduled for a flight on the Space Shuttle in 1989), and earth-resources payloads are under active discussion. Australians have identified space as a means of strengthening the national technological base and a natural contributor to the development of their vast, poorly explored land. But, at least for the near future, Australia will not be able to launch its own payloads.

Table 13.1 presents a list of satellites launched by nations other than the United States and the Soviet Union.

AMERICAN GOALS IN SPACE: THE PRESIDENTIAL COMMISSION

In early 1985 President Reagan, with the encouragement of Congress, appointed a commission for the express purpose of identifying and advocating future goals for the American space program. The full Commission report was released in May 1986, having been delayed by the Challenger tragedy.

In general, the Commission favors a broad program of development of basic technology, closely similar in its overall nature to that advocated in our closing chapter. The Commission is clearly cognizant of the the damage done to NASA in the past by monomaniacal giant programs, such as Apollo, that spent vast amounts of precious resources without contributing

Figure 13.7
ESA's *Ariane* Blasts Off
This photo shows an *Ariane 3* taking off from the Kourou launch site, on its way to GEO with a commercial cargo. The main present competitor of the Space Shuttle, Ariane has a full flight manifest through early 1988. The loss of the Space Shuttle Challenger and the two-year delay in the shuttle program have provided ESA with the incentive to expand its launch capability even further. A much larger booster, the *Ariane 5,* is planned for the mid-1990s. It will serve as the booster of the French Hermes mini-shuttle, which is being designed to dock with the Soviet Mir Space Station.

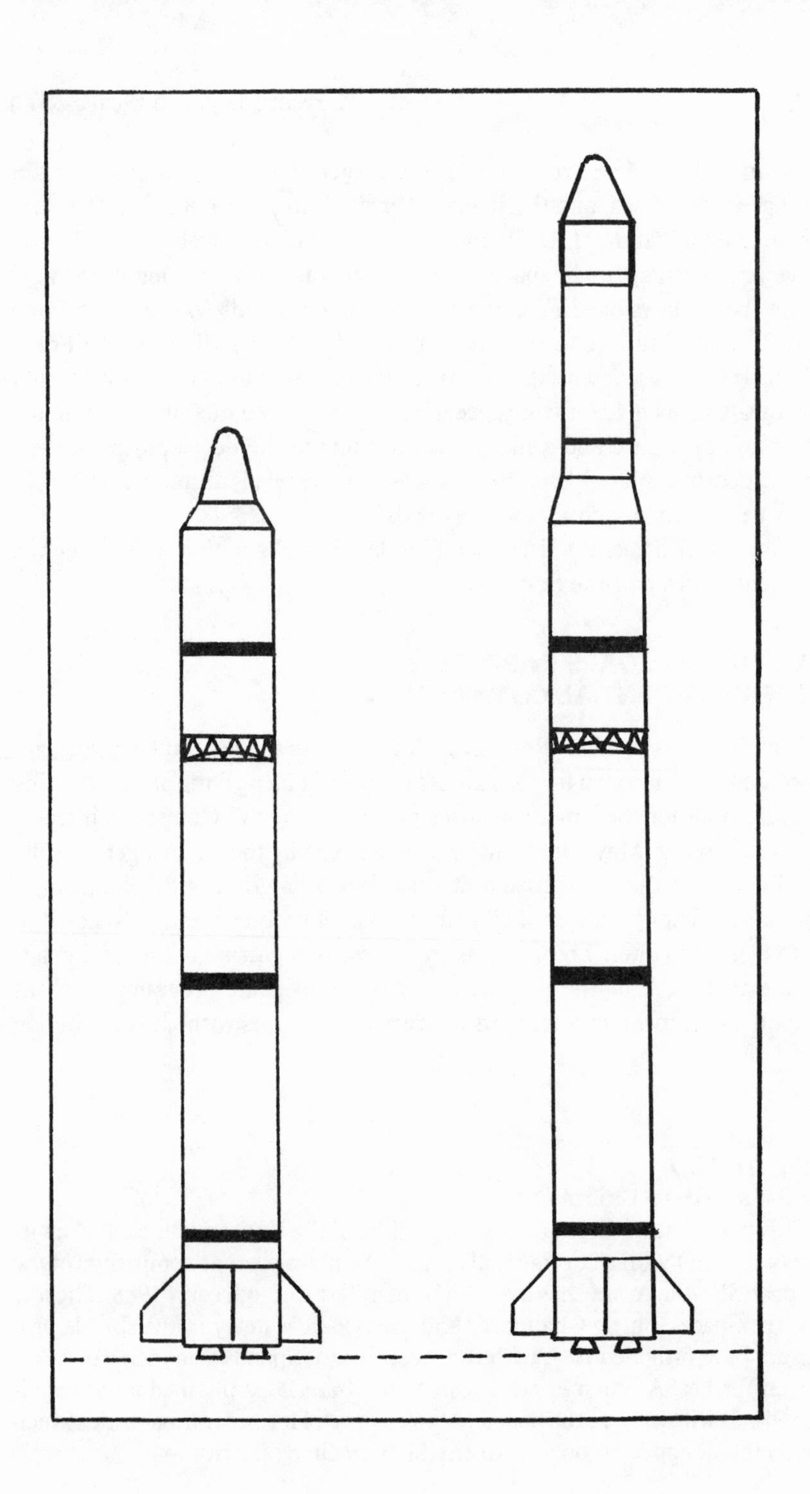

Table 13.1
Satellites Launched by Other Nations (Non-USA; Non-USSR)

Nation	*Spacecraft*	*Launch*	*Booster*	*Site*	*Remarks*
FR	A-1	26 Nov 65	Diamant	Ham	Test
FR	D1A	17 Feb 66	Diamant	Ham	Geodetic research
FR	D1C	8 Feb 67	Diamant	Ham	Geodetic research
FR	D1D	15 Feb 67	Diamant	Ham	Geodetic research
IT	San Marco 2	26 Apr 67	Scout	SM	Research
AU	WRESAT1	29 Nov 67	Sparta	Woo	Solar radiation research
JA	Osumi	11 Feb 70	Lambda 4S	Uch	Engineering test
GE	DIAL 1	10 Mar 70	Diamant B	Kou	Research
FR	MIKA	"	"	"	Engineering test
CH	China 1	24 Apr 70	LM 1	SCT	Test
US	Explorer 42	12 Dec 70	Scout	SM	Small Astronomy Satellite (Uhuru)
FR	PEOLE 1	12 Dec 70	Diamant B	Kou	Meteorological data relay
JA	Tansei 1	16 Feb 71	Mu 4S	Uch	
CH	China 2	3 Mar 71	LM 1	SCT	Test
FR	Tournesol 1	15 Apr 71	Diamant B	Kou	
IT	San Marco 3	24 Apr 71	Scout	SM	Atmospheric research
JA	Shinsei	28 Sep 71	Mu 4S	Uch	Scientific research
UK	Prospero	28 Oct 71	Bk Arrow	Woo	
JA	Denpa	19 Aug 72	N 4S	Kag	Ionospheric research
JA	Tansei 2	16 Feb 74	Mu 3C	Uch	
IT	San Marco 4	18 Feb 74	Scout	SM	Research
UK	Ariel 5	15 Oct 74	Scout	SM	X-ray observatory
FR	Starlette	6 Feb 75	Diam. BP4	Kou	Geodetic research
JA	Taiyo	24 Feb 75	Mu 3C2	Uch	Research
US	Explorer 53	7 May 75	Scout	SM	Small astronomy satellite
FR	Pollux	12 May 75	Diam. BP4	Kou	Technology development
FR	Castor	"	"	"	Scientific research
CH	China 3	26 Jul 75	LM 2	SCT	Military recon.
JA	Kiku	9 Sep 75	N	Tan	Applications technology

(table continued)

Figure 13.8
Long March (Chang Zheng)
These two vehicles, based on the People's Republic of China CSS-4 ICBM, are the current Chinese contenders to offer commercial launch services into LEO and GEO, respectively. The first-stage thrust of both vehicles is about 660,000 pounds.

Nation	Spacecraft	Launch	Booster	Site	Remarks
FR	D2B	27 Sep 75	Diam. BP4	Kou	Solar UV research
CH	China 4	26 Nov 75	LM 2	SCT	Military recon.; 6 d recovery
CH	China 5	16 Dec 75	LM 2	SCT	Military recon.
JA	Ume	29 Feb 76	N	Tan	Ionospheric sounder
CH	China 6	30 Aug 76	LM 2	SCT	Electronic ferret
CH	China 7	7 Dec 76	LM 2	SCT	Military recon.; 2 d recovery
JA	Tansei 3	19 Feb 77	Mu 3H	Uch	
JA	Kiku 2	23 Feb 77	N	Tan	Technology
CH	China 8	26 Jan 78	LM 2	SCT	Military recon.
JA	Kyokko	4 Feb 78	Mu 3H	Kag	Auroral research
JA	Ume 2	16 Feb 78	N 1	Tan	Ionospheric sounder
JA	Jiki'ken	16 Sep 78	Mu 3H	Kag	Research
JA	Ayame	6 Feb 79	N 1	Tan	Experimental comsat
JA	Hakucho	21 Feb 79	Mu 3C	Kag	X-ray research
ES	CAT	24 Dec 79	Ariane	Kou	Vehicle flight test
JA	Tansei 4	17 Feb 80	Mu 35	Kag	Research
JA	Ayame 2	22 Feb 80	N 1	Tan	Communications
IN	Rohini 1	18 Jul 80	SLV-3	Sri	Test
JA	Kiku 3	11 Feb 81	N 2	Tan	Technology
JA	Astro	21 Feb 81	Mu 3	Uch	Research
IN	Rohini 2	31 May 81	SLV-3	Sri	38 kg earth resources payload
IN	Meteosat 2	19 Jun 81	Ariane	Kou	Indian comsat
ES	Apple	"	"	"	
ES	CAT 3	"	"	"	Flight test instrumentation
JA	Himawari 2	10 Aug 81			Weather satellite in GEO
CH	China 9	19 Sep 81	LM 2	SCT	Scientific research
	China 10	"	"	"	
	China 11	"	"	"	
ES	Marecs 1	20 Dec 81	Ariane	Kou	Maritime comsat in GEO
ES	CAT 4	"	"	"	Test instrumentation
JA	Kiku 4	3 Sep 82	N	Tan	Systems technology
CH	China 12	9 Sep 82	LM 2	SCT	4 day military recon.; recovered
JA	Sakura 2A	4 Feb 83	N 2	Tan	Comsat in GEO 130E
JA	Tenuna	20 Feb 83	Mu 3S	Kag	X-ray observatory
IN	Rohini 3	17 Apr 83	SLV-3	Sri	Earth resources
ES	ECS1	16 Jun 83	Ariane	Kou	Comsat in GEO 10E
	Oscar 10	"	"	"	Amateur radio relay
JA	Sakura 2B	5 Aug 83	N 2	Tan	Comsat in GEO 135E
CH	China 13	19 Aug 83	LM 2	SCT	Satellite recovery experiment
ES	Intelsat 5(F7)	19 Oct 83	Ariane	Kou	Commercial comsat in GEO
CH	China 14	29 Jan 84	LM 2	SCT	

Nation	Spacecraft	Launch	Booster	Site	Remarks
JA	Ohzora	14 Feb 84	Mu 3S	Tan	Aeronomy research
ES	Intelsat 5(F8)	5 Mar 84	Ariane	Kou	Commercial comsat in GEO
CH	China 15 (STW1)	8 Apr 84	LM 3	SCT	Comsat test in GEO 125E
ES	Spacenet 1	23 May 84	Ariane	Kou	Comsat in GEO 120W
JA	Himawari 3	2 Aug 84	N 2	Tan	Weather satellite in GEO 140E
FR	Telecom 1A	4 Aug 84	Ariane 3	Kou	Comsat in GEO 8W
ES	ECS 2	"	"	"	TV relay in GEO 7E
CH	China 16	12 Sep 84	LM 2	SCT	Military recon.
FR	Marecs B2	10 Nov 84	Ariane 3	Kou	Maritime comsat in GEO 177.5E
ES	Spacenet 2	"	"	"	Commercial comsat in GEO 69W
JA	Sakigake	7 Jan 85	Mu 3S	Kag	Comet Halley distant flyby
BR	Brasilsat	8 Feb 85	Ariane 3	Kou	Comsat in GEO 65W
AR	Arabsat	"	"	"	Comsat in GEO 19E
ES	GSTAR 1	8 May 85	Ariane 3	Kou	
FR	Telecom 1B	"	"	"	Comsat
ES	Giotto	2 Jul 85	Ariane 3	Kou	Comet Halley flyby, Jan 86
JA	Suisei	18 Aug 85	Mu 3S	Kag	Planet A; Halley close flyby
CH	China 17	21 Oct 85	LM 2	SCT	17-day photo recon.
CH	China 18	1 Feb 86	LM 3	Xi	Comsat in GEO
JA	BS-2B	12 Feb 86	N 2	Tan	TV relay satellite in GEO; 770 lb
FR	SPOT 1	22 Feb 86	Ariane 3	Kou	French Earth resources satellite
SW	Viking	"	"	"	Swedish research satellite
US	GSTAR 2	28 Mar 86	Ariane 3	Kou	American private comsat in GEO
BR	Brasilsat 2	"	"	"	Comsat in GEO 70W
JA	EGP(Ajisai)	12 Aug 86	H-1	Tan	Geodetic research
	JAS-1(Fuji)	"	"	"	Amateur radio
	MABES(Jindai)	"	"	"	Flywheel experiment
CH	China 19	6 Oct 86	LM 2	Jiu	5-day Earth resources

Abbreviations

NATION: AR = Arab Satellite Communications Organization; AU = Australia; BR = Brasil; CH = People's Republic of China; ES = European Space Agency; FR = France; GE = German Federal Republic; IN = India; IT = Italy; JA = Japan; SW = Sweden; UK = United Kingdom; US = United States

LAUNCH SITE: Ham = Hammaguir, Algeria; Jiu = Jiuquan, China; Kag = Kagoshima, Japan; Kou = Kourou, Guiana; SM = San Marcos (Italian launch platform off coast of Kenya); SCT = Shuang Cheng Tzu, China; Sri = Sriharikota, India; Tan = Tanegashima, Japan; Uch = Uchinoura, Japan; Woo = Woomera, Australia; Xi = Xichang, China

Remarks: GEO = geosynchronous orbit (24-hour period); longitudes of satellite rest points are given as 19E, 95W, etc. Dittoes indicate multiple spacecraft launched by the same booster.

to any long-range goals. The Commission does not, unfortunately, come to grips with the fiscal and managerial stranglehold that the Space Shuttle has on NASA today. They make no suggestions about how to restore NASA to its original purpose as expressed in its charter: basic research and development.

The Commission tacitly accepts the general timing and nature of the Space Station program as now conceived by NASA, although it is not too late for informed, pointed criticism to improve their designs significantly. Perhaps they were told, "We don't want your advice on the Space Shuttle *or* the Space Station." Or perhaps they realized that such advice would, in any case, be ignored.

The Commission's position on the Lunar Base program seems to be somewhat ambivalent. The appropriate cautions are made about starting with a modest, research-oriented facility to study the origin and evolution of the Moon, map resources, and conduct experiments on the use of lunar resources. But elsewhere in their document the Commission seems to take for granted that the "resource wealth" of the Moon will necessarily drive rapid growth of the lunar base to very large size.

If there is any fundamental problem faced by the Commission, it is the difficulty of communicating and selling its overall program to a public that often seems to want only bread and circuses. One has difficulty imagining mobs surging down the Mall toward the Capitol, chanting "Give us a broadly based technology development program!" The phrase fits poorly on crudely-lettered placards, and stumbles awkwardly off the tongue. Yet it would be foolish to rush to the opposite error and propose a sharply focused, monomaniacal dead-end program simply because it would be more easily grasped by the "public imagination." We have made this mistake before, and cannot afford to do it again.

Finally, the Commission emphasizes the disastrous effects of the one-year American budget cycle on space missions and programs that take six to twenty years to execute. Some form of long-range budgetary planning must be instituted in order to permit NASA to work effectively and efficiently, free of the enormous waste caused by abrupt, whimsical, and violent upward and downward excursions of the budget.

14

A PROPOSAL FOR A RENEWED SPACE PROGRAM

Those who cannot remember the past are condemned to repeat it.—George Santayana

When a fool avoids one error, he rushes to its opposite.—Horace

We must take stock of the lessons of history before attempting to plan the future. We must be aware of the mistakes made in the past, especially the passionate claims of exclusivity made by advocates of one extreme plan or another, and we must avoid all manner of destructive excesses. The history recounted in this volume provides a rich arena for thought and analysis: We will summarize what we see as the principal lessons to be learned from it.

PRINCIPLES

Among the first lessons to be found in the history of space is that political support is the key to rapid advancement. Whenever political support has been weak, as in Germany after Hitler's bad dream of 1939, the USSR after the fall of Khrushchev, or in the USA after Apollo, progress has slowed to a crawl. Whenever political support has been strong, as it was after Hitler's complete reversal in 1943, during Khrushchev's bid for world recognition, in Kennedy's advancement of Apollo for reasons of "national prestige," progress has occurred with dazzling speed.

The second lesson, dramatically driven home by these examples, is the arbitrariness and irrationality of political support. Space projects are advanced for the most remote and irrelevant reasons. Almost never is space exploration given a higher priority because of the benefits of exploring space. These benefits are rarely known or understood by political decision-makers. Instead, the purposes of politicians in advancing space exploration and exploitations have usually been primarily political, and secondarily military.

The third lesson is that institutions that conduct space operations are shockingly willing to sell their souls for money, almost irrespective of what they are asked to do. In this respect, they show a similarity to most other human institutions. In the 1930s, Soviet rocket expert Fridrikh Tsander, upon arriving at work each morning, would throw open the door and shout, "On to Mars!" Stalin bought Tsander's products, but they were ICBMs, not Mars missions. Wernher von Braun and his team at Peenemuende dreamed of Mars at night, but spent their days bombarding London. NASA, chartered as a research and development agency without any operational roles, finds itself driving a delivery truck for commercial interests and the Department of Defense. In fact, it is actually paying for the truck (the Space Shuttle) by decimating its research and development programs.

The fourth lesson is that political domination of a space agency can lead to a series of gigantic decade-long programs that utterly overwhelm and destroy the adaptability, diversity, and overall viability of that agency. This is the phenomenon, described earlier, of the "slaughter of the innocents." The most immense programs, most monomaniacally pursued, are the most lethal to the innocents. They also tend to freeze and fossilize technology at one particular level, which leads to the withering of innovation and the exodus of the most creative, up-to-date scientific and technical personnel to more attractive positions elsewhere. In the United States, the sharp, young, innovative aerospace engineer naturally goes where he's most wanted—the military space business.

A corollary of this lesson is that broad-gauge technical progress across the widest possible spectrum of fields is the natural victim of voracious, obsessive, giant programs. The rich diversity of space science before Apollo suffered terribly at the hands of that giant competitor.

Another corollary, and one less obvious to outsiders, is that a high budget does not necessarily mean a healthy space program. Boom-and-bust cycles kill at both extremes: when there is little money, nothing flies; and when there is lots of money, only the giant project flies.

A fifth observation, and the one hardest to bear, is that the United States does not have any goals in space. We wander from one giant project to another without any clear notion why we are doing it, and with absolutely no idea of where we are headed. The corresponding lesson is that goals are absolutely essential for a healthy and vigorous program. Further, *if we have no goals, then we also have no standards against which to judge our progress.* It is no surprise, then, that we should wander.

The sixth lesson is that the evolution of technology is like biological evolution: the arbitrary extinction of a major area of competence leads to the loss of many abilities that will be badly missed in years to come, and which may not be readily restored. Avoid evolutionary *cul-de-sacs:* do not kill vigorous, already-paid-for, working technologies when you have nothing in hand to replace them. We made a devastating error when we abandoned Skylab, scrapped the Saturn 1 and Saturn 5, and threw away all the technology developed in the Apollo program. If we wanted Apollo capabilities again, we would have to pay for them again. The idea of "picking up where Apollo left off" in lunar exploration is a chimera. There is nothing to pick up: when we dropped it, it broke.

The seventh lesson is that not all new technology is equally valuable. Sometimes, as with the Space Shuttle, we spend billions to develop a capability that is then grossly misused. We paid dearly for the shuttle orbiter, for those fantastic but delicate heat-shield tiles, for the manned crew compartment and its life-support equipment, all in the name of making spaceflight cheaper. The concept of reusability brought potential shuttle operating costs down by a factor of about four relative to disposable launchers. But because the 70-ton orbiter and its crew compartment absolutely had to be lifted into space and recovered on each flight, and because the decision was made not to carry the 35-ton External Tank and its fuel reserves into orbit on each flight, the *useful* payload of the shuttle turned out to be only 25 to 30 tons, less than a quarter of the *total* weight that could be placed in orbit. This completely removed the hoped-for cost advantage of the Shuttle. We invested $10 billion of precious resources in a system that used a lot of expensive new technology to accomplish *no reduction in launch costs.*

This incredible blunder was made for several reasons: first, there was no competition between schemes to lower launch costs. Only one concept was allowed to race, and it couldn't lose (the only standard of comparison was the "bad old way" of using disposable booster rockets: Shuttle costs rose freely until they reached the "old" levels—which is where they ended up). Second, we had no long-range plans that required economy: the real purpose of the Shuttle program was not to *save* money; it was to *spend* enough money to keep NASA alive. Third, in order to make it seem that we needed the shuttle, NASA dishonestly assigned to it a wide range of payloads that did not in any way profit from being launched by a manned vehicle, while it failed to provide for the kind of mission that the Shuttle was ideally suited to perform: servicing a manned space station. Skylab was allowed to die without any replacement in sight because all the Apollo-era technology had been abandoned—but the Apollo and Saturn hardware was abandoned to *force* Congress to fund the Shuttle!

An eighth lesson, based on this experience, must be that competition works and monopolistic systems stagger along without recourse in the event of a failure. The Soviets are acutely (if not vocally) aware of this principle. They have a great fear of American competition, which they rightly believe to be capable of working wonders in open races. Evidence of this subconscious awareness surfaces daily in *Pravda,* where cartoons always depict Americans as filthy rich, and where they are always labeled as "monopolists." This view of American capitalism is of course a good half century out of date. It belongs in the trash barrel with the Stalinesque representation of Soviet life seen in American political cartoons. But let's face it: the Soviet state is a monopoly of the first magnitude. One could make a case that the extraordinary achievements of the Soviet military and space establishments were made possible only by inventing several different design bureaus that, in imitation of the Western model, compete against each other [the autonomous Tupolev, Ilyushin, Antonov, Mikoyan-Gurevich (MiG), Beriev, Kamov, Mil, Myasischev and Sukhoi bureaus in aircraft, and the Korolyev, Yangel, and Chalomei semi-autonomous bureaus in rocketry]. Believers in free competition, whether Soviet or American, should beware of subsidies and other gross manipulations of the market. This applies to launch tariffs on the Shuttle, the D1e, and Ariane as much as to tobacco subsidies at home, or steel and integrated circuit chip prices in the world market.

A ninth lesson, also exposed by the great Shuttle debate, is the issue of the relationship between manned and unmanned space missions. Some of the most vocal critics of the Space Shuttle and the Space Station, notably James Van Allen, favor the elimination of manned activities and the return to exclusively unmanned activities. This extreme position is based upon an analysis of Shuttle economics that we agree with almost completely; however, we do not share Van Allen's conclusions. The polar nature of his stance is understandable if one reflects that the real state of affairs, paid for at enormous expense, is one in which virtually every satellite launching must drag along seven largely irrelevant astronauts.

But we believe that a program in which *no* missions are manned in as senseless as one in which *all* missions are manned. Manned interplanetary travel is around the corner. While it is true that early scientific reconnaissance of the Solar System can be done enormously less expensively by small unmanned spacecraft, we are clearly at a stage in many of our exploratory programs where we need much more depth. Automated unmanned sample returns from, for example, Mars, are under study, but the costs of such missions are close to the costs of enormously more effective manned missions. The time for men and women has arrived—but the era of unmanned exploratory probes will never expire. We need to achieve an appropriate balance between manned and unmanned endeavors. If we disagree with Van Allen's conclusions, we should heed the warning of Horace and not rush mindlessly to the opposite conclusion.

Now let us examine very specifically the principles by which the Soviet and American space programs operate. These principles have been exposed in earlier chapters, and need only be quickly summarized here.

Principle number ten is that the Soviet space program has, since Khrushchev's downfall, adopted a low profile. This low profile is in part a legacy of Soviet xenophobia and paranoia. Long-range goals and purposes are only hinted at, and short-range plans are being discussed more frankly than ever before. The wall of secrecy about Soviet space activities, while still tall and thick, has been breached in some respects in order to create an impression of openness. The Soviet program has long-range goals, and makes steady, stepwise, conservative progress toward them. Almost certainly, the construction of a large permanently manned space station will occupy the next few years, leading to a manned expedition to Mars.

The eleventh principle, and the guiding principle of the American space program, is that Americans respond to Pearl Harbors with incredible energy and originality. A corollary may be that Americans don't respond to anything *but* Pearl Harbors. Combined with the second corollary of principle number four (boom-and-bust cycles kill at both extremes), this seems to suggest that the American space program could never be healthy. Actually, in the American system, the wild extremes are avoided mainly in the context of continuing meaningful competition or continuing commercial contractual obligations. Open competition and lasting covenants are good for business—and they also lead to the best climate for innovation.

This leads us to a twelfth principle: competition and cooperation are both good. But when and how can one establish a cooperative relationship?

Khrushchev supplied us with principle number thirteen: cooperation is an activity that can be engaged in only by near equals. American-Soviet cooperation must be founded upon equality or, at the very least, complementary strengths that, combined, enable missions that would not be possible to either party alone. Cooperation means helping toward a common goal: it doesn't mean giving away the store or begging on doorsteps. Cooperating with the Soviet Union means acting with generosity and sincerity, with compassion for the terrible trials that their land has suffered, and establishing the broadest possible human contact between individuals on both sides. Cooperating with the Americans means demanding firm, contractual commitments, not depending on the goodwill of American scientists—the scientists love to cooperate, but they aren't the ones who pay the bills! The sad European experience with the Solar Polar Mission (now Ulysses) should serve as a cautionary note: once an equal partnership between ESA and NASA, with each contributing one spacecraft, budgetary strictures forced the precipitous cancellation of the American half. Our European colleagues report that they learned of the cancellation of the mission by reading about it in the newspapers.

Our fourteenth principle is that, after two decades of loud professions of belief in lowering the cost of launching payloads from Earth into space, the costs remain essentially unchanged. We regard this not as proof of the futility of attempts to lower costs, but of the present system's toleration for high costs. We see no reason why, using the competitive principle espoused above, costs could not be lowered by a factor of four or even ten. But this will not happen until cost reduction becomes a top priority.

Fifteenth, we have seen several clear examples how the cost of materials in near-Earth space may be cut not only by a factor of ten, but by factors of 100 or 1000, by the use of extraterrestrial resources. The costs of bringing these resources down to LEO from nearby asteroids are much lower than the costs of lifting them up out of the deep gravitational well of Earth.

Several additional principles have emerged from our discussions of extraterrestrial resources. These include the following:

Seek extraterrestrial resources that are rich in desirable materials. We may call this the sixteenth or Willy Sutton principle: when asked by a reporter why he robbed banks, Sutton replied, "Because that's where the money is."

The seventeenth principle is that the ideal sites for resource mining operations for near-Earth use must be easily accessible from LEO. Expressed in terms of energy, the near-Earth asteroids are best, Phobos and Deimos are next, and the Moon is last. Expressed in terms of trip time, the Moon is closest, and the near-Earth asteroids and the Martian moons are comparable to each other.

The eighteenth principle is that resources should come from sources where delivery to their place of use would be least expensive. This means that materials needed on the Moon should be derived from the Moon whenever possible, and that materials needed for activities in (and returning from) the Mars system should be derived from Phobos, Deimos, or Mars itself. Materials required in Low Earth Orbit should be derived mainly from the near-Earth asteroids, with the Martian moons a poor second and the Moon an even poorer third.

The nineteenth principle is that the initial desirability of space resources derives from their use *in space.* The crucial issue is to reduce the proportion of the mass of future space constructions and activities that must be launched from Earth.

The twentieth principle is that the most important space resources in the near future are those materials which will be needed in great masses, but which require only low-technology farbrication. The principal classes of use are propellant, metal structures, shielding, and life support fluids. Almost all of the needed materials can be categorized as volatiles or metals. The Martian moons and the C-type asteroids are excellent sources of volatiles, but the Moon may be unable to supply any volatile except oxygen. Free metals are about 100 times more abundant in chondritic asteroids than they are in the lunar regolith. For this reason, the principal use of lunar resources would seem to be support of operations on the Moon: the Moon's resource poverty with respect to free metals and volatiles may make any large operation on the Moon dependent upon nonlunar sources of these materials.

Principle number twenty-one states that extraterrestrial mining operations, once begun for other reasons, may supply Earth with small tonnages of extremely valuable critical and strategic metals. But we see no clear rationale for initiating the exploitation of extraterrestrial resources solely for the purpose of providing such materials to Earth.

Finally, we conclude as our 22nd principle that new initiatives are most attractive when they can be begun with familiar technology, develop in an open-ended matter with specific milestones to verify progress, and when by their nature, they provide rich rewards for technological innovation. It is this principle that connects our past and present to the future.

PROPOSALS

We propose that the central focus of human progress in space be to achieve the greatest possible independence of space activities from Earth by minimizing the cost of providing raw materials in near-Earth space. The present high cost of space activities is due in part to serious flaws in the present role of NASA, in part to the dearth of technological progress in NASA since 1970, and in part to the lack of long-range goals for the Agency. We propose the following essential steps to correct many of these deficiencies and to contribute to the goal of lowering dramatically the cost of space operations, beginning with programmatic considerations related to the lowering of launch costs from Earth:

1. Congress, the Executive Department, and NASA must confirm the commitment of NASA to abide by its charter. The NASA charter defines it as a research and development agency, not responsible for operational programs such as fully developed weather, navigation, communication, or transportation systems.
2. In compliance with NASA's charter, NASA should develop immediate plans for full divestiture of the Space Shuttle transportation system to an operating consortium. The negative impact of the resulting substantial decrease in the NASA budget will be more than offset by the reawakening of NASA to its proper and legally mandated purpose, fundamental research and development. It is absolutely essential that the new, smaller, tighter NASA be protected during its transition by a firm Congressional and Executive commitment to keep it safe from the wolves who would dismember it. The purpose of divestiture is to allow NASA to do what it was designed to do, and what it does best, not to provide an excuse for its destruction by hostile forces.
3. Until such time as the divestiture of the Shuttle may be achieved, NASA

should adopt a policy of full-cost pricing for the Space Shuttle. This will surely cause the loss of many Shuttle payloads to foreign competitors, but there is a serious shortage of global launch capacity over the next two or three years because of the loss of *Challenger,* the grounding of the Shuttle fleet, the grounding of the Titan 34D, and the slowing of both Shuttle and Ariane operations due to flight problems. Many payloads are now queued up awaiting launchers: the Shuttle should still be able to fill its flight manifest even while withdrawing from the commercial launch business.

4. NASA's Office of Space Flight, charged with responsibility for manned space activities, should be thoroughly redirected toward exploratory and developmental missions such as development of the Space Station and manned lunar and planetary flights, rather than toward routine operational control of the Space Shuttle.
5. NASA must not under any circumstances be permitted to fund another Shuttle orbiter (although DoD or an operating consortium may choose to do so). This would place severe strain on the NASA budget, and would further entrench it in doing the wrong things. The sole allowable exception would the prompt placement of orders for critical long-lead-time components of a new Shuttle orbiter, in anticipation of the early divestment of all Shuttle-related activities to a new operating consortium.
6. The additional costs incurred by NASA in its attempts to recover from the loss of Challenger, including expenditures for a new Shuttle orbiter, reopening of assembly lines for conventional boosters, and development of new low-cost launch vehicles, must not be allowed to cause another "slaughter of the innocents" in NASA basic research and development programs. This is a real and frightening possibility, so important that we insert here a quotation from a March 1986 staff working paper of the Congressional Budget Office:

   > One possible area in which to reduce NASA spending over the next two or three years is the research and development (R&D) function, since the loss of shuttle capacity will lead to a dramatic reduction in shuttle flights available to launch R&D payloads during this period. Once the system returns to regular operations, pressing national security needs will fill most of the available flights for several years, thus delaying the launch of major scientific payloads—which require large R&D expenditures—by as much as three years beyond their original launch dates.

7. NASA must immediately reopen assembly lines for a variety of small boosters, drawing from candidates such as the Scout, Delta, Atlas F, Centaur, and Titan 34D, to restore the launch capability lost by the ex-

plosion of *Challenger,* by lifting those payloads currently on the Shuttle launch manifest that do not require or profit from launch by the Shuttle. Launch of these payloads on disposable boosters, although as expensive as launch on the Shuttle, has two major advantages: first, no single-point failure, such as the grounding of the Shuttle, could damage more than a small fraction of the overall national space effort, and second, there would be no need to risk human lives on routine launches in which human presence is unnecessary.

8. The bulk of the funds that would otherwise be spent on the construction of a fourth Shuttle orbiter to replace *Challenger* (commonly quoted as $1.5 to 2 billion, but actually closer to $2.8 billion) should be devoted to accelerating competitive design efforts for the Heavy Lift Launch Vehicle (HLLV), an unmanned Shuttle-Derived Launch Vehicle (SDLV), and the Aerospace Plane. The winner of this competition would be chosen on the basis of its ability to minimize the cost of launching payloads of up to 100 or 200 tons into LEO.

Several of our proposals concern enhancement and redirection of manned activities in near-Earth space, especially the Space Station and Solar Power Satellites:

9. Since each External Tank (ET) carried into orbit increases the effective payload capacity of the Space Shuttle by about 35 to 40 tons at no additional cost, NASA must establish a vigorous program for conserving Space Shuttle ETs in orbit, with the following provisos:
   a. NASA, or a Shuttle operating company, should decree that space Shuttle flights must routinely carry their External Tanks into orbit, where they should be treated as a valuable national resource.
   b. The ETs accumulated in orbit should be "mined" for their residual propellant content, with the propellant to be used for orbit maintenance of the Space Station and the ET cluster and for fueling missions to GEO and to escape.
   c. A propellant-handling facility should be constructed at the Space Station to refrigerate, store, and dispense liquid hydrogen and liquid oxygen scavenged from the ETs and, later, manufactured from carbonaceous asteroidal water and lunar dirt.
   d. ETs should be used in the construction and expansion of the Space Station, in building hydroponic "greenhouses," as modules to house processing facilities, for lease to industrial and academic institutions for research, educational, development, and commercial purposes, and as raw material for refabrication into other useful items.

10. NASA must redirect the design of the Space Station to permit maximum use of ETs and to achieve reduced costs of construction and operation.
11. NASA must redirect Space Station design activities to provide the capability to process large masses of extraterrestrial raw materials, especially the production of propellants and structures for use in expanding and enhancing the Space Station and other operations in near-Earth space.
12. NASA and other American and foreign governmental agenices, in cooperation with the aerospace industry, must immediately establish an international program for preliminary design and economic assessment of Solar Power Satellites. The design must be predicated upon the extensive use of lunar and/or asteroidal resources, with minimal reliance upon materials and components launched from Earth.

Several additional recommendations focus on the development of a wide range of basic enabling technology, especially in areas that will be essential for lowering the cost of large-scale manned space endeavors. These points highlight a number of capabilities that are essential ingredients of any large-scale and long-term use of space resources.

13. NASA must immediately greatly accelerate the development of advanced, high-performance propulsion systems for interplanetary travel, taking full advantage of the availability of solar energy and space sources of propellant. A minimal response to this imperative would be a substantial increase in development work on solar thermal, solar electric, nuclear electric and light-sail technology, with early in-space feasibility demonstrations of each promising technology.
14. NASA must immediately begin design and testing of techniques for food production in space. The Space Shuttle, with its practical limit of about one week in space, is virtually useless as a platform for such studies. Waiting for the Space Station to become operational before beginning such research would guarantee the Soviet Union nearly a 20-year lead in this crucial technology.
15. NASA must undertake rapid development of short-tether technology to make possible the production of "artificial gravity" by spinning space structures, and of long-tether techniques for effecting orbital transfers without use of rocket propellants. The only experiment in this area since the Gemini era, the Italian-American tethered satellite experiment, has been delayed by the grounding of the Space Shuttle.

The next group of proposed actions focuses upon the identification, characterization, and economic assessment of space resources:

16. NASA must undertake a greatly accelerated program for the discovery, characterization, and economic assessment of resources on asteroids and on the satellites of Mars and Earth. This program consists of several parts:

    a. NASA must accelerate Earth-based telescopic searches for near-Earth asteroids. This search would be in part motivated by the positive desire to discover economically attractive bodies for resource exploitation, and in part by the concern that a multi-megaton Earth impact by an undetected near-Earth asteroid might trigger World War III.
    b. Earth-based spectroscopic and radar studies of near-Earth asteroids must be accelerated.
    c. At least one Lunar Polar Orbiter spacecraft must be launched to carry out compositional and geophysical mapping of the entire lunar surface.
    d. Phobos and Deimos missions capable of landing on their surfaces and determining their subsurface composition and physical properties are required. We await the results of the Soviet Phobos mission with great interest.
    e. At least one multiple-asteroid rendezvous mission must be launched to visit at least four selected nearby asteroids of diverse compositions.
    f. The ability to land automated sample return vehicles on the Moon must be revived, and
    g. the ability to land instrumented, remotely controlled roving vehicles on the Moon must be revived in order to provide scientific data and prospectors' samples of the lunar surface in regions too remote, hazardous, or numerous for manned visits.
    h. The abilities described in items f and g must be coupled together, so that lunar samples collected by roving vehicles may be returned to Earth.
    i. Automated sample return from Phobos, Deimos, and selected near-Earth asteroids is an urgent adjunct to rendezvous and landing missions. Since return from several near-Earth asteroids is so easy, they may logically be the first targets.
    j. Design efforts for future manned lunar activities should be focused on a small, episodically manned lunar base with ability to carry out

a wide range of scientific and materials-processing experiments, and on a long-range manned roving vehicle.

k. Three basic types of standard unmanned spacecraft, designed respectively to carry out orbital mapping, sample return, and penetrator missions to small bodies, must be built in continuing series to minimize mission costs and maximize scientific and economic return per dollar.

17. NASA must immediately begin the study of techniques for processing and extraction of useful materials, especially water, hydrogen, oxygen, and native metals, from space resources, both on Earth and on the Space Shuttle. These efforts include several distinct activities:

    a. Laboratory experiments must be done to explore methods of extraction of useful materials from a wide range of lunar and asteroidal materials and their simulants.
    b. Experiments must be done both on Earth and in LEO to explore the use of solar furnaces to extract volatiles from carbonaceous materials and to melt meteoritic and lunar samples.
    c. Experiments must be done aboard the Space Shuttle to develop experience in the handling and processing of space resources in a microgravity environment.
    d. Experiments on simple schemes for the in-space fabrication of ferrous metal structural components, including beams, plates, wires, cables, thin films, insulation, and shielding, should be initiated at once in laboratories on Earth and in space.
    e. Experiments on the electrolysis of water and liquefaction of hydrogen and oxygen in zero gravity must be begun at the earliest possible date. Ideally, the Space Station should be spun to produce a small artificial gravity; however, the Space Station design group already seems to have ruled out this possibility.

18. In order to avoid perpetuating the kinds of programmatic blunders made by NASA in recent years, any plans for a lunar base must be rigorously reviewed in competition with a wide variety of other space endeavors before being considered seriously for funding. The lunar base program must at all times be viewed with informed skepticism by an Administration and Congress who are unwilling to see a promising scientific and technological experiment become bloated by political interference into a $100 billion monster that destroys the rest of NASA. Likewise, we must be vigilant lest the naïve idea that "we've done the

Moon already" should deter us from giving it the attention it deserves.

19. NASA should issue a standing order to buy a stated number of tons per year of water, liquid hydrogen, liquid oxygen, or ferrous metals, at the Space Station in LEO, from any supplier. This would provide a crucial incentive to potential space industries to invest in the technology of extracting and processing of space resources. The guaranteed market would create an atmosphere in which venture capital could be raised at moderate interest rates.
20. NASA must immediately implement the balanced SSEC plan for exploration of the Solar System, with bipartisan political support for long-term, stable funding, *but not necessarily as an exclusively American venture.*

This point raises the issue of international collaboration as a means of achieving a greater global understanding of space at lower cost to each nation. Several proposals come to mind:

21. NASA should take the lead in immediately establishing a permanent international planning group for the coordination of both unmanned and manned deep-space missions. In the face of rapidly deteriorating USA-ESA relations (the Solar Polar/Ulysses fiasco, especially) and rapidly improving Soviet relations with European national space agencies (as on many Venera spacecraft, the Vega missions to Halley's comet, and the planned Phobos-Deimos landing mission), this measure is an absolute necessity to prevent the permanent loss of European collaboration.
22. As a means of restoring mutual cooperative space activities, NASA should make an early proposal to the USSR for a visit by the Space Shuttle to the Mir space station complex. This could easily be done as a minor modification of any of the planned Spacelab missions, which are to fly in orbits with inclinations similar to that of Mir. This should have been done two or three years ago, but the chill in American-Soviet relations made it impossible. Now, with the Shuttle unlikely to return to full operational status before late 1988, there is plenty of time to plan such a visit.
23. NASA should demonstrate its sincere interest in collaborative space exploration by tendering an immediate offer to the USSR for the United States to provide hardware for use in extending Soviet peaceful exploratory capabilities. An example would be the offer of American instrumented penetrators, survivable landers, or sample return vehicles

to be carried by future Soviet planetary and asteroidal missions, with full data access from all mission instruments for both nations.

24. Several of the functions and missions listed under proposal 15 above could be carried out by other nations or international agencies in collaboration with the United States. Among the possibilities that come to mind are the revival of Soviet automated lunar return vehicles in conjunction with American automated lunar rovers and Japanese lunar-orbiting mapping missions. Each of these vehicles might carry already-developed instruments from other nations to minimize development costs, and Japanese payloads too heavy to be launched by existing Japanese boosters could be carried by American Titan Centaur or Soviet D1e boosters. ESA might concentrate on electric propulsion and instrument packages for long-duration multiple-body flyby, rendezvous, and sample-return missions to comets and asteroids.
25. As a specific focus for international cooperation in space, we propose a series of joint manned expeditions to Mars' system in the 1990s. These flyby, orbiting, and landing missions would be built mainly upon existing Soviet and American capabilities. They may feature a large Soviet mission module, based on Salyut and Mir experience, for the lengthy trip from Earth to Mars and back, American landing vehicles, Japanese processing modules for manufacture of the propellant for the return trip on Phobos, Deimos, or even Mars itself, and ESA manned rovers and unmanned sample-retrieval vehicles. An early emphasis on Phobos and Deimos exploration and propellant production provides a bridge not only to the Martian surface, but to the entire solar system.

In view of Soviet concerns regarding the Strategic Defense Initiative, and because of the extremely unstable nature of the Mutual Assured Destruction deterrence scheme, every attempt should be made to divert Soviet and American military energies toward establishment of a stable defense-oriented balance, and to give the Soviets clear assurance that the US is not attempting to develop a first-strike capability. This may permit the use of some substantial amount of military space funding in both nations to increase the stability of our military balance without developing new weapons systems, and to do so while opening up space resources that will lessen the incentive to dominate mineral supplies in nations on Earth. There are several steps that could be taken to begin:

26. The USA should immediately commit to sharing all technology developed for the Strategic Defense Initiative with the USSR.

27. The USA should immediately commit to linking the phase-out of American offensive strategic weapons to the deployment of defensive space systems.
28. A joint Soviet-American panel charged with proposing jointly acceptable guidelines for shifting from an offensively oriented deterrent posture (MAD) to a defensively oriented goal of Mutual Assured Survival (MAS) should be established by appointment at the highest political levels.
29. Both nations should make a vigorous study of defensive technologies that do not require the deployment of any weapons in space.
30. A conscious effort should be made by the President and Congress to divert a significant fraction of the advanced-technology research funding of DoD away from weapons and toward peaceful space endeavors that increase national security. An example would be the armoring of essential Soviet and American military communications, command, control, and intelligence (CCCI) satellites with asteroidal metals.

BEYOND 2010: BUILDING ON NEW CAPABILITIES

The program of restructuring NASA's philosophy and operations described above would have several salutary effects by the end of the quarter century we are considering.

First of all, NASA will have realized an enormous reduction in the cost of space operations. The worst case would see a reduction in real launch costs by a factor of five due to the introduction of the SDLV, HLLV, or Aerospace Plane. In this worst-case scenario, the only operational space-resources retrieval scheme would be the simplest version of retrieval of carbonaceous dirt or metal dust from near-Earth asteroids, with a mass payback ratio of about five to one. Overall, this combination of factors would lower the cost of large-scale activities in near-Earth space by a factor of 25.

At this level, Solar Power Satellites would probably be irresistible attractions. A program to generate Earth's electric power in space would almost certainly be under way. The cost of energy for the industrialization of the most remote and impoverished parts of Earth would be on its way down—forever.

In this worst-case scenario, the cost of a manned planetary expedition would be reduced to well below the cost of each moon landing in the Apollo era. The cost would be low enough so that a single major nation could afford to carry out such expeditions.

However, even in the worst case, there would be strong incentives for building planetary expedition plans about international consortiums. The first manned expeditions to the Mars system, whether American, Soviet, or multinational, would have already been done before 2010.

Human activities on the Moon, at first centered about one or two small "camps," would have given us mastery of the art of extracting oxygen and iron from the lunar surface. We would be on the verge of producing glass products, sintered refractory heat shields, titanium, manganese, and chromium. The principal export commodity, liquid oxygen, may by that time be profitable enough to pay for the operating costs of the lunar surface activites—unless asteroidal water is by then available in LEO in quantities sufficient to meet the demand.

In this worst case, only the most profitable space ventures could pay their own way. Space colonies would not be economically feasible. The only manned expeditions likely to have occurred would have been to the Moon, Mars, Phobos and Deimos, and possibly a few of the choicest asteroids. Manned flybys of Venus, en route to other targets on gravity-assisted trajectories, would also be possible. Solar Power Satellites and processing industries in LEO would be the largest space operations. *All this could be done for a real cost less than the present NASA budget.*

But this is the most conservative consequence of a vigorous technology- and resource-oriented program: this is the worst case. What would the *best* reasonable case provide in 2010?

The best reasonable reduction in launch costs achievable by 2010 is about a factor of 20. We could then lift hardware from Earth into LEO for a real total cost of about $200 per pound. Lunar resources produced for local use would include all those listed above plus others not yet imagined. Export of lunar materials would be made much less expensive by construction of a long mass driver on the lunar surface. The mass driver would be powered by electricity from large solar-cell arrays on the lunar surface.

Asteroidal water and metal would be recovered by mining vehicles dispatched from LEO with electrical propulsion, or from a higher Earth orbit with light sails. Heat shields for aerocapture would be manufactured by microwave sintering on the asteroid. Each ton of equipment dispatched to an asteroid would return about 1000 tons to LEO. The combination of lower real transportation costs from Earth to LEO and the 1000:1 mass payback ratio for asteroidal volatiles and metals would permit lowering the cost of large-scale space activities in near-Earth space by a factor of 20,000.

A million tons of material for a factory or habitat in LEO that would cost $8 trillion to lift into place with current technology (33,000 Shuttle flights!) would cost about $0.4 billion in this best-care scenario. Allowing about five tons per occupant, this much mass could build a habitat to accommodate a population of 200,000 people. The cost of the raw materials would be $2000 per person. This is far less than the cost of materials to build housing on Earth, and is less than the handling, processing, and construction costs. In effect, the raw materials for space industrialization and colonization will have become free.

In this best-case scenario, the cost of operating interplanetary missions will be so low that any attractive goal will be visited. The main barrier in the way of exploration and exploitation of the Solar System will be its sheer size: it will not be possible before 2010 to build and dispatch enough missions to do more than lightly scratch the riches of the asteroid belt. There are tens of thousands of asteroids larger than a few kilometers in diameter. Each one-kilometer-sized metallic asteroid will provide a billion tons of iron, 200 million tons of nickel, 10 million tons of cobalt, and 20,000 tons of platinum metals: net market value, about $1 trillion.

The transportation systems available in this best case include very light, high-performance light sails fabricated in space out of asteroidal materials. These sails will be so thin that high accelerations will be achievable: trips from Earth to Mars or Venus will be cut to a few weeks from many months. Colonies will be established on Mars, and filling stations on the Martian moons and a few selected carbonaceous asteroids.

By 2010, manned expeditions to the satellite and ring systems of the outer planets will be possible. The number of Solar system bodies that will have been visited by humanity will be increasing exponentially.

The cost of this staggering enterprise may by then be a hundred billion dollars. Over a period of 25 years, it will have cost Earth a few billion dollars a year. But the benefits will have been many trillions of dollars! The *net* cost will be negative: the job will pay for itself many times over. We need only be careful that we choose our investments wisely, so that we do not have to spend too much before we reach the break-even period. We need to identify early profit-makers. The key products, as we have seen, are propellants, metals, life-support fluids, and electrical energy and strategic metals for export to Earth. The rest can wait. If we develop these resources intelligently and swiftly, the wait will be very short indeed. But, if we squander our precious resources on pointless posturings and politically motivated dead-end spectaculars, we will exhaust the patience of humanity before we realize any of these vast benefits.

By 2010, if we have done what we can do, we will be planning the urbanization of the Solar System—and it will be within our grasp.

APPENDIX: MAN-RELATED LAUNCHINGS

Spacecraft	*Launch*	*Mission*	*BR*	*Results*
Sputnik Korabl 1	15 May 60	Recovery	A1	Orbital unmanned test of 5-t. Vostok capsule: Recovery attempt failed
Discoverer 13	10 Aug 60	Recovery	TAA	First successful recovery from orbit-small instrument capsule
Discoverer 14	18 Aug 60	Recovery	TAA	Midair recovery of small capsule
Sputnik Korabl 2	19 Aug 60	Recovery	A1	Capsule, 2 dogs recovered after 1 d.
Sputnik Korabl 3	1 Dec 60	Recovery	A1	Capsule and dog lost during reentry
Sputnik Korabl 4	9 Mar 61	Recovery	A1	Capsule and dog recovered
Sputnik Korabl 5	25 Mar 61	Recovery	A1	Recovered after 1 orbit
Vostok 1	12 Apr 61	Manned	A1	First manned space flight: Gagarin in space for 1 orbit, 1.5 hrs
Mercury Atlas 3	25 Apr 61	Unmanned	AD	Launch failure with 1-t. Mercury
Merc. Redstone 3	5 May 61	Ballistic	R	Suborbital flight test: Shepard
Merc. Redstone 4	21 Jul 61	Ballistic	R	Suborbital flight: Grissom
Vostok 2	6 Aug 61	Manned	A1	Recovered after 1 day: Titov
Mercury Atlas 4	13 Sep 61	Recovery	AD	Unmanned: recovered after 1 orbit
Mercury Atlas 5	29 Nov 61	Recovery	AD	Capsule and monkey recovered: 3 hrs
Mercury Atlas 6	20 Feb 62	Manned	AD	Recovered after 3 orbits: Glenn
**Kosmos 4*	26 Apr 62	Test	A1	Recovered after 3 d: unmanned
Mercury Atlas 7	24 May 62	Manned	AD	Recovered after 3 orbits: Carpenter
**Kosmos 7*	28 Jul 62	Test	A1	Recovered after 4 d: Vostok test
Vostok 3	11 Aug 62	Manned	A1	Recovered after 4 d: Nikolayev
Vostok 4	12 Aug 62	Manned	A1	Recovered after 3 d: Popovich—dual flight with *Vostok 3*
**Kosmos 9*	27 Sep 62	Unmanned	A1	Recovered after 4 d: probable Vostok-derived military satellite
Mercury Atlas 8	3 Oct 62	Manned	AD	Recovered after 6 orbits: Schirra
**Kosmos 10*	17 Oct 62	Unmanned	A1	Recovered after 4 d
**Kosmos 12*	22 Dec 62	Unmanned	A1	Recovered after 8 d: extension of Vostok beyond 4 day limit?
**Kosmos 13*	21 Mar 63	Unmanned	A1	Recovered after 8 d: military?
**Kosmos 15*	22 Apr 63	Unmanned	A1	Recovered after 5 d: Vostok test
**Kosmos 16*	28 Apr 63	Unmanned	A1	Recovered after 10 d: military Vostok-derived recon. satellite
Mercury Atlas 9	15 May 63	Manned	AD	1.5 day flight: Cooper

Spacecraft	*Launch*	*Mission*	*BR*	*Results*
**Kosmos 18*	24 May 63	Unmanned	A1	Recovered after 9 d: military?
Vostok 5	14 Jun 63	Manned	A1	Recovered after 5 d: Bykovskii
Vostok 6	16 Jun 63	Manned	A1	3 d: 1st woman in space: Tereshkova: Dual flight within 5 km of *Vostok 5*
Saturn SA5	29 Jan 64	Test	S1	Booster test: orbited 19 t
Gemini Titan 1	8 Apr 64	Test	T2	Orbital test of Gemini Titan launcher
Saturn SA6	28 May 64	Test	S1	Booster test: orbited 18.6 t
Saturn SA7	18 Sep 64	Test	S1	Orbited 18.3 t
Kosmos 47	6 Oct 64	Unmanned	A2	1-d flight: probable Voskhod test
Voskhod 1	12 Oct 64	Manned	A2	6 tons, 1 d: 1st multimanned flight: Komarov, Feoktistov, Yegorov
Pegasus 1	16 Feb 65	Test	S1	11.5 ton satellite: Saturn test
Kosmos 57	22 Feb 65	Unmanned	A2	Exploded after 1 d?
Voskhod 2	18 Mar 65	Manned	A2	6.3 t; 1 d: Belyayev and Leonov: First spacewalk
Gemini 3	23 Mar 65	Manned	T2	3.3 t; 3 orbits: Grissom and Young
Pegasus 2	25 May 65	Test	S1	11.5 t: Saturn test flight
Gemini 4	3 Jun 65	Manned	T2	3.5 t: 4 d: White and McDivitt
Proton 1	16 Jul 65	Test	D1	13.5 t: test of huge new D booster
Gemini 5	21 Aug 65	Manned	T2	8-day duration record: Cooper, Conrad
Gemini 6 Target	25 Oct 65	Unmanned	AAD	Launch failure: Docking target for G6
**Proton 2*	2 Nov 65	Test	D1	13.4 t: orbital test of D booster
Gemini 7	4 Dec 65	Manned	T2	4 t, 14 d duration record: served as docking target for *Gemini 6:* Borman and Lovell
Gemini 6	15 Dec 65	Manned	T2	1 d; came within 1 foot of *Gemini 7:* Schirra and Stafford
**Kosmos 110*	22 Feb 66	Unmanned	A2	2 dogs recovered after 22 days
Gemini 8 Target	16 Mar 66	Unmanned	AAD	Docking target for G8: orbited
Gemini 8	16 Mar 66	Manned	T2	Electrical problems, early return after 11 hrs: no docking attempt: Armstrong and Scott
Gemini 9 Target A	17 May 66	Unmanned	AAD	Booster malfunction—no orbit
Gemini 9 Target B	1 Jun 66	Unmanned	AD	Fairing failed to separate: in orbit
Gemini 9	3 Jun 66	Manned	T2	3 d, rendezvous and EVA test: Stafford and Cernan
Apollo 2	6 Jul 66	Test	S1	29.7 t: booster test
**Proton 3*	6 Jul 66	Test	D1	14 t: test of D booster
Gemini 10 Target	18 Jul 66	Unmanned	T2	3.5 t: Docking target and engine
Gemini 10	18 Jul 66	Manned	T2	4.1 t; 3 d: rendezvoused with GI0T and G8T; used G10 engine: Young and Collins
Gemini 11 Target	12 Sep 66	Unmanned	AAD	3.5 t: target and engine for G11

Spacecraft	*Launch*	*Mission*	*BR*	*Results*
Gemini 11	12 Sep 66	Manned	T2	4.2 t; 3 d: Conrad and Gordon
Gemini 12 Target	11 Nov 66	Unmanned	AAD	Target and engine for G12
Gemini 12	11 Nov 66	Manned	T2	4 d: EVA, docking: Lovell, Aldrin
**Kosmos 133*	28 Nov 66	Unmanned	A2	2 d: Soyuz precursor
**Kosmos 140*	7 Feb 67	Unmanned	A2	2 d; Soyuz precursor
**Kosmos 146*	10 Mar 67	Unmanned	D1	8 d: possible manned precursor
**Kosmos 154*	8 Apr 67	Unmanned	D1	2 d; Sozond precursor
Soyuz 1	23 Apr 67	Manned	A2	Recovery failed; Komarov killed
**Kosmos 186*	27 Oct 67	Docking	A2	Automated docking with K188
**Kosmos 188*	30 Oct 67	Docking	A2	K186 docking target
Apollo 4	9 Nov 67	Test	S5	139.3 t orbited in Saturn 5 test
Apollo 5	22 Jan 68	Test	S1b	15.8 t including Lunar Module
**Zond 4*	2 Mar 68	Test	D1e	Launch into secret high orbit
Apollo 6	4 Apr 68	Test	S5	42.5 t; Saturn 5 3-stage test
**Kosmos 212*	14 Apr 68	Docking	A2	Automated docking with K213
**Kosmos 213*	15 Apr 68	Docking	A2	7 t; K212 docking target
**Kosmos 238*	28 Aug 68	Unmanned	A2	4 d; Probable Soyuz precursor
**Zond 5*	15 Sep 68	Lunar	D1e	6 d; lunar flyby and recovery
Apollo 7	11 Oct 68	Manned	S1b	22.5 t, 11 d; first manned Apollo: Schirra, Cunningham, Eisele
Soyuz 2	25 Oct 68	Target	A2	Unmanned rendezvous target for S3
Soyuz 3	26 Oct 68	Docking	A2	4 d; docking failure: Beregovoi
Zond 6	10 Nov 68	Flyby	D1e	7 d; lunar flyby and recovery
Apollo 8	21 Dec 68	Orbit	S5	6 d; first manned lunar orbit: Borman, Lovell and Anders
Soyuz 4	14 Jan 69	Docking	A2	7 t, 3 d: Shatalov docked with S5, returned with Khrunov, Yeliseyev
Soyuz 5	15 Jan 69	Docking	A2	Volynov (3 d); Khrunov, Yeliseyev
Apollo 9	3 Mar 69	LM test	S5	40 t, 10 d: Earth orbit rehearsal of lunar landing mission
Apollo 10	18 May 69	LM test	S5	47.2 t, 8 d; rehearsal of lunar landing mission in lunar orbit Stafford, Young, Cernan
Apollo 11	16 Jul 69	Landing	S5	48.3 t, 8 d; first manned lunar landing and return; Armstrong and Aldrin: Collins in Command Module
**Zond 7*	7 Aug 69	Flyby	D1e	6.6 t, 6 d.; lunar flyby and recovery
Soyuz 6	11 Oct 69	Triple	A2	6.6 t, 5 d; triple flight with *Soyuz 7* and *8:* Shonin and Kubasov
Soyuz 7	12 Oct 69	Docking	A2	5 d; docking target for *Soyuz 8;* Filipchenko, Volkov, Gorbatko
Soyuz 8	13 Oct 69	Docking	A2	5 d; failed to dock with *Soyuz 7;* Shatalov, Yeliseyev
Apollo 12	14 Nov 69	Landing	S5	48.4 t, 10 d; lunar landing; Conrad, Bean, Gordon

Spacecraft	*Launch*	*Mission*	*BR*	*Results*
Apollo 13	11 Apr 70	Landing	S5	55.1 t, 6 d; landing aborted; Lovell, Swigert and Haise
Soyuz 9	2 Jun 70	Duration	A2	6.6 t, 17.7 d; duration record Nikolayev and Sevastyanov
**Zond 8*	20 Oct 70	Flyby	D1e	lunar flyby; recovered after 7 days
**Kosmos 379*	24 Nov 70	Test	D1	booster test for Salyut
**Kosmos 382*	2 Dec 70	Test	D1	Proton launch; possibly man-related
Apollo 14	31 Jan 71	Landing	S5	55.6 t, 9 d; lunar landing; Shepard, Mitchell, Roosa
**Kosmos 398*	26 Feb 71	Test	D1	Proton launch into eccentric orbit
Salyut 1	19 Apr 71	Station	D1	20.4 t; first space station
Soyuz 10	23 Apr 71	to *Sal. 1*	A2	Docking with Salyut 1 fails; 1 d Shatalov, Yeliseyev, Rukavishnikov
Soyuz 11	6 Jun 71	to Sal 1	A2	24 d visit to *Salyut 1;* Patsayev, Dobrovolskii, Volkov die on entry
Apollo 15	26 Jul 71	Landing	S5	53.6 t, 12 d; lunar landing Scott, Irwin, Worden
Apollo 16	16 Apr 72	Landing	S5	11 d; lunar landing and return; Young, Duke, Mattingly
**Kosmos 496*	26 Jun 72	Test	A2	6 d; Unmanned Soyuz test
Apollo 17	7 Dec 72	Landing	S5	12 d; lunar landing and return; Cernan, Schmitt, Evans
Salyut 2	3 Apr 73	Station	D1	Disintegrated after 11 d
**Kosmos 557*	11 May 73	Station	D1	20 t; probable Salyut failure
Skylab 1	14 May 73	Station	S5	82.2 t; huge laboratory
Skylab 2	25 May 73	to Sklb	S1b	22 t, 28 d; Conrad, Kerwin, and Weitz
**Kosmos 573*	15 Jun 73	Test	A2	2 d; unmanned Soyuz test
Skylab 3	28 Jul 73	to Sklb	S1b	22.1 t, 2 mo.; Bean, Garriott and Lousma
Soyuz 12	27 Sep 73	Test	A2	2 d; Lazarev and Makarov
Skylab 4	16 Nov 73	to Sklb	S1b	22.8 t, 84 d; Carr, Gibson, and Pogue
**Kosmos 613*	30 Nov 73	Test	A2	2 month unmanned Soyuz (see *Soyuz 18)*
Soyuz 13	18 Dec 73	Orbit	A2	8 days; Klimuk and Lebedev
**Kosmos 638*	3 Apr 74	Test	A2	10 d unmanned Soyuz test
**Kosmos 656*	27 May 74	Test	A2	7.1 t, 2 d unmanned Soyuz
Salyut 3	24 Jun 74	Station	D1	20.3 t; military space station
Soyuz 14	3 Jul 74	to *Sal 3*	A2	16 d mission to *Salyut 3;* Popovich and Artyukhin
**Kosmos 672*	12 Aug 74	Test	A2	ASTP unmanned precursor; 6d
Soyuz 15	26 Aug 74	to *Sal 3*	A2	Salyut docking failed; 2 d; Sarafanov and Dyomin
Soyuz 16	2 Dec 74	Test	A2	6 d; test for ASTP mission; Filipchenko and Rukavishnikov
Salyut 4	26 Dec 74	Station	D1	20.3 t
Soyuz 17	10 Jan 75	to *Sal 4*	A2	7.1 t, 1 mo; docked with Salyut 4; Gubarev, Grechko

Spacecraft	*Launch*	*Mission*	*BR*	*Results*
Soyuz 18	24 May 75	to *Sal 4*	A2	2 mo. visit to *Salyut 4;* Klimuk and Sevastyanov
Soyuz 19	15 Jul 75	ASTP	A2	6 d linkup with *Apollo 18;* Leonov and Kubasov
Apollo 18	15 Jul 75	ASTP	S1b	9 d, 16.2 t; linkup with *Soyuz 19;* Stafford, Slayton, Brand
**Kosmos 772*	29 Sep 75	Test	A2	Unmanned 3-d Soyuz test
Soyuz 20	17 Nov 75	Biosat	A2	3 mo. unmanned biosatellite; automatic docking with Salyut 4
Salyut 5	22 Jun 76	Station	D1	20.9 tons; military space station
Soyuz 21	6 Jul 76	to *Sal 5*	A2	49 day visit to *Salyut 5;* Volynov and Zholobov
Soyuz 22	15 Sep 76	Novel	A2	8 day mission in 65 degree orbit; Bykovskii and Aksionov
Soyuz 23	14 Oct 76	to *Sal 5*	A2	2 d aborted linkup with *Salyut 5;* Zudov and Rozhdestvenskii
**Kosmos 869*	29 Nov 76	Test	A2	18 d unmanned Soyuz-T test
**Kosmos 881*	15 Dec 76	Unknown	D1	One orbit: Kosmolyot test?
**Kosmos 882*	"	"	"	Double flight with *K881;* see *K997*
Soyuz 24	7 Feb 77	to *Sal 5*	A2	18 d mission; shut down *Salyut 5*
**Kosmos 929*	17 Jul 77	Station	D1	Probable Salyut/space tug engine test
Salyut 6	29 Sep 77	Station	D1	20.9 t space station
Soyuz 25	9 Oct 77	to *Sal 6*	A2	2 d; failed to link with *Salyut 6* Kovalyonok, Ryumin
Soyuz 26	10 Dec 77	to *Sal 6*	A2	Romanenko and Grechko spend 96 d aboard *Salyut 6*
Soyuz 27	10 Jan 78	to *Sal 6*	A2	Double docking with *Salyut 6/Soyuz 26* Dzhanibekov, Makarov
Progress 1	20 Jan 78	Resupply	A2	Automatic docking with *Salyut 6*
Soyuz 28	2 Mar 78	to *Sal 6*	A2	8d visit; Gubarev and Remek
**Kosmos 997*	30 Mar 78	Unknown	D1	Possible Kosmolyot test; 1 d; K 1100
**Kosmos 998*	"	"	"	Double flight with K 997; see *K 881*
**Kosmos 1001*	4 Apr 78	test ?	A2	11 day unmanned *(Soyuz T?)* test
Soyuz 29	15 Jun 78	to *Sal 6*	A2	Kovalyonok and Ivanchenkov reoccupy *Salyut 6* for 140 days
Soyuz 30	27 Jun 78	to *Sal 6*	A2	Klimuk and Giermaszewski; 7 d visit
Progress 2	7 Jul 78	Resupply	A2	Automatic docking with *Salyut 6*
Progress 3	7 Aug 78	Resupply	A2	Automatic docking with *Salyut 6*
Soyuz 31	26 Aug 78	to *Sal 6*	A2	Bykovskii and Jaehn; 8 d visit
Progress 4	3 Oct 78	Resupply	A2	Automatic docking with *Salyut 6*
**Kosmos 1074*	31 Jan 79	Test	A2	60 day apparent unmanned Soyuz
Soyuz 32	25 Feb 79	to *Sal 6*	A2	Lyakhov and Ryumin reoccupy *Salyut 6* for 175 d
Progress 5	12 Mar 79	Resupply	A2	Automatic docking with *Salyut 6*
Soyuz 33	10 Apr 79	to *Sal 6*	A2	Rukavishnikov and Ivanov; 2 d; *Salyut 6* docking and s/c exchange fail

Spacecraft	*Launch*	*Mission*	*BR*	*Results*
Progress 6	13 May 79	Resupply	A2	Automatic docking with *Salyut 6*
**Kosmos 1100*	22 May 79	Unknown	D1	Possible Kosmolyot test; 1 day only
**Kosmos 1102*	"	"	"	Double flight with *K 1100;* see *K 998*
Soyuz 34	6 Jun 79	to *Sal 6*	A2	Unmanned launch; *Soyuz 32* crew return
Progress 7	28 Jun 79	Resupply	A2	Automatic docking with *Salyut 6*
Soyuz T1	16 Dec 79	to *Sal 6*	A2	7.7 t; unmanned docking
Progress 8	27 Mar 80	Resupply	A2	Automatic docking with *Salyut 6*
Soyuz 35	9 Apr 80	to *Sal 6*	A2	Popov and Ryumin reopen for 185 d
Progress 9	27 Apr 80	Resupply	A2	Automatic docking with *Salyut 6*
Soyuz 36	26 May 80	to *Sal 6*	A2	Kubasov and Farkash to *Salyut 6;* 8 d
Soyuz T2	5 Jun 80	to *Sal 6*	A2	Malyshev and Aksionov to *Sal 6;* 4 d
Progress 10	29 Jun 80	Resupply	A2	Automatic docking with *Salyut 6*
Soyuz 37	23 Jul 80	to *Sal 6*	A2	Gorbatko and Pham to *Sal 6;* 8 d
Soyuz 38	18 Sep 80	to *Sal 6*	A2	Romanenko and Tamayo to *Sal 6;* 8 d
Progress 11	28 Sep 80	Resupply	A2	Automatic docking with *Salyut 6*
Soyuz T3	27 Nov 80	to *Sal 6*	A2	Kizin, Makarov, Strekalov; 13 d
Progress 12	24 Jan 81	Resupply	A2	Automatic docking with *Salyut 6*
Soyuz T4	12 Mar 81	to *Sal 6*	A2	14 d; Kovalyenok and Savinykh
Soyuz 39	22 Mar 81	to *Sal 6*	A2	8 d; Dzhanibekov and Gurragcha
Columbia STS-1	12 Apr 81	Test	SS	2 d, 93.2 t; Young and Crippen
**Kosmos 1267*	25 Apr 81	to *Sal 6*	D1	16.6 t; "Star" module docked w/S6
Soyuz 40	14 May 81	to *Sal 6*	A2	8 day visit; Popov and Pruhariu
Columbia STS-2	12 Nov 81	Test	SS	2 d; Engle and Truly
Columbia STS-3	22 Mar 82	Test	SS	7 d; Lousma and Fullerton
Salyut 7	19 Apr 82	Station	D1	21 t space station
Soyuz T5	13 May 82	to *Sal 7*	A2	211 d; Berezovoi, Lebedev open *Sal. 7*
Progress 13	23 May 82	Resupply	A2	Automatic docking with *Salyut 7*
**Kosmos 1374*	3 Jun 82	Test	C1	1 orbit test of spaceplane
Soyuz T6	24 Jun 82	to *Sal 7*	A2	8 d visit to *Salyut 7;* Dzhanibekov, Ivanchenkov, Chretien
Columbia STS-4	27 Jun 82	Test	SS	7 d; Mattingly and Hartsfield
Progress 14	10 Jul 82	Resupply	A2	Automatic docking with *Salyut 7*
Soyuz T7	9 Aug 82	to *Sal 7*	A2	8 d; Popov, Serebrov, Savitskaya
Progress 15	18 Sep 82	Resupply	A2	Automatic docking with *Salyut 7*
Progress 16	31 Oct 82	Resupply	A2	Automatic docking with *Salyut 7*
Columbia STS-5	11 Nov 82	Shuttle	SS	5 d; Brand, Overmyer, Allen, Lenoir
**Kosmos 1443*	2 Mar 83	to *Sal 7*	D1	20 t "Star" module docked to Salyut
**Kosmos 1445*	15 Mar 83	Test	C1	1.5 orbit test of 1-ton spaceplane
Challenger STS-6	4 Apr 83	Shuttle	SS	5 days; Weitz, Bobko, Musgrave, Peterson; TDRS launch
Soyuz T8	20 Apr 83	to *Sal 7*	A2	2 days; Salyut docking failed; Titov, Strekalov, Serebrov
Challenger STS-7	18 Jun 83	Shuttle	SS	6 days; Crippen, Hauck, Ride, Fabian and Thagard; Anik C, Palapa B, SPAS

Spacecraft	Launch	Mission	BR	Results
Soyuz T9	27 Jun 83	to *Sal 7*	A2	Lyakov, Aleksandrov open *Sal 7;* 150 d
Progress 17	17 Aug 83	Resupply	A2	Automatic docking with *Salyut 7*
Challenger STS-8	30 Aug 83	Shuttle	SS	5 d; Truly, Brandenstein, Gardner, Bluford, Thornton
**Soyuz*	26 Sep 83	to *Sal 7*	A2	Titov and Strekalov; launch abort
Progress 18	20 Oct 83	Resupply	A2	Automatic docking with *Salyut 7*
Columbia STS-9	28 Nov 83	Spacelab	SS	10 d; Young, Shaw, Garriott, Parker, Lichtenberg, Merbold; spacelab
**Kosmos 1517*	27 Dec 83	Test	C1	1 orbit test of spaceplane
Soyuz T10	8 Feb 84	to *Sal 7*	A2	Kizim, Solovyov and Atkov reopen Salyut 7; stay 237 d
Progress 19	21 Feb 84	Resupply	A2	Automatic docking with *Salyut 7*
Soyuz T11	3 Apr 84	to *Sal 7*	A2	8 d; Malyshev, Strekalov, Sharma
Progress 20	15 Apr 84	Resupply	A2	Automatic docking with *Salyut 7*
Progress 21	7 May 84	Resupply	A2	Automatic docking with *Salyut 7*
Progress 22	23 May 84	Resupply	A2	Automatic docking with *Salyut 7*
Soyuz T12	17 Jul 84	to Sal 7	A2	12 d; Dzhanibekov, Volk, Savitskaya
Progress 23	14 Aug 84	Resupply	A2	Automatic docking with *Salyut 7*
Discovery STS41D	30 Aug 84	Shuttle	SS	6 d; Hartsfield, Coats, Hawley, Mullane, Resnick, Walker
Challenger STS41G	5 Oct 84	Shuttle	SS	8 d; Crippen, McBride, Leetsma, Ride, Sullivan, Scully-Power, Garneau
Discovery STS51A	8 Nov 84	Shuttle	SS	8 d; Hauck, Walker, Fisher, Allen, Gardner
**Kosmos 1614*	19 Dec 84	Test	C1	1 d test of 1000 kg spaceplane
Challenger STS51C	24 Jan 85	Shuttle	SS	3 d; Mattingly, Shriver, Onizuka, Buchli, Payton
Discovery STS51D	12 Apr 85	Shuttle	SS	Bobko, Garn, Seddon, Hoffman, Griggs, Williams, Walker; 7 d
Challenger STS51B	29 Apr 85	Shuttle	SS	Overmeyer, Gregory, Lind, Thagard, Thornton, Vandenberg, Wang; 7 d; spacelab 3
Soyuz T13	6 Jun 85	to Sal 7	A2	Dzhanibekov and Savinykh; Reopening of *Salyut 7*
Discovery STS51G	17 Jun 85	Shuttle	SS	6 d; Brandenstein, Creighton, Fabian, Lucid, Nagel, Baudry, El Saud
Progress 24	21 Jun 85	Resupply	A2	Automatic docking with *Salyut 7*
**Kosmos 1669*	19 Jul 85	to *Sal 7*	D1	20 t "Star" module docked to *Salyut 7*
Challenger STS51F	29 Jul 85	Shuttle	SS	Fullerton, Bridges, Musgrave, Acton, England, Henize, Bartoe, 9 d; spacelab 2
Discovery STS51I	27 Aug 85	Shuttle	SS	Engle, Covey, van Hoften, Fisher, Lounge; 7d
Soyuz T14	17 Sep 85	to Sal 7	A2	Vasyutin, Grechko, Volkov, crew and Soyuz rotation. 65 d.
Atlantis STS51J	3 Oct 85	Shuttle	SS	Bobko, Grabe, Hilmers, Stewart Pailes; military crew; 4 d

Spacecraft	*Launch*	*Mission*	*BR*	*Results*
Challenger STS61A	30 Oct 85	Shuttle	SS	Harsfield, Nagel, Buchli, Dunbar, Bluford, Messerschmid, Furrer, Ockels; 6 d flight
Atlantis STS61B	27 Nov 85	Shuttle	SS	Shaw, Ross, Spring, Neri, O'Connor, Cleave, Walker; 7 d flight
Columbia STS61C	12 Jan 86	Shuttle	SS	Gibson, Bolden, Hawley, G. Nelson, W. Nelson, Cenker, Chang-Diaz; 6 d
Challenger STS51L	28 Jan 86	Shuttle	SS	Scobee, Smith, McNair, Resnick, Onizuka, McAuliffe, Jarvis; exploded 1 min into flight; all killed
Mir	20 Feb 86	Station	D1	22 t station with 6 docking ports
Soyuz T15	13 Mar 86	to *Mir*	A2	Kizim and Solovyev; to open *Mir* and visit Salyut 7; 125 d stay
Progress 25	21 Mar 86	resupply	A2	Automatic docking with *Mir*
Progress 26	23 Apr 86	resupply	A2	Automatic docking with *Mir*
Soyuz TM1	21 May 86	to *Mir*	A2	Unmanned test; docked with *Mir*
Progress 27	16 Jan 87	resupply	A2	Automatic docking with *Mir*
Soyuz TM2	6 Feb 87	to *Mir*	A2	Romanenko and Lavyeikin to reopen *Mir*
Progress 28	3 Mar 87	resupply	A2	Automatic docking with *Mir*

Abbreviations

BR (Booster Rocket): AAD = Atlas Agena D (Atlas ICBM first stage, Agena D second stage); AD = Atlas D ICBM; A1 = Soviet SS6 ICBM; A2 = SS6 ICBM with large second stage; C1 = small Soviet launch vehicle based on the SS5 LRBM; D1 = "Proton" heavy-lift booster; D1e = Proton booster with added kick stage; ICBM = Intercontinental Ballistic Missile (6000 mile range); IRBM = Intermediate Range Ballistic Missile (1500 mile range); LRBM = Long Range Ballistic Missile (2500 mile range); R = Redstone SRBM; SRBM = Short Range Ballistic Missile (600 mile range); SS = Space Shuttle; S1 = Saturn 1 booster with 1.5 million pounds of thrust; S1b = Saturn 1b, an improved version of the Saturn 1; S5 = Saturn 5 superbooster with 7.5 million pounds of first-stage thrust and high energy hydrogen-oxygen upper stages; TAA = Thor Agena A (Air Force Thor IRBM first sstage plus an Agena A second stage derived from the B-58 Hustler bomb pod); T2 = Titan 2 ICBM. Dittoes indicate spacecraft launched together on the same booster.

*Unannounced missions

INDEX